D1929169

FH-Bibliothek Schweinfurt
083000507700

Springer Complexity

Springer Complexity is an interdisciplinary program publishing the best research and academic-level teaching on both fundamental and applied aspects of complex systems – cutting across all traditional disciplines of the natural and life sciences, engineering, economics, medicine, neuroscience, social and computer science.

Complex Systems are systems that comprise many interacting parts with the ability to generate a new quality of macroscopic collective behavior the manifestations of which are the spontaneous formation of distinctive temporal, spatial or functional structures. Models of such systems can be successfully mapped onto quite diverse "real-life" situations like the climate, the coherent emission of light from lasers, chemical reaction-diffusion systems, biological cellular networks, the dynamics of stock markets and of the internet, earthquake statistics and prediction, freeway traffic, the human brain, or the formation of opinions in social systems, to name just some of the popular applications.

Although their scope and methodologies overlap somewhat, one can distinguish the following main concepts and tools: self-organization, nonlinear dynamics, synergetics, turbulence, dynamical systems, catastrophes, instabilities, stochastic processes, chaos, graphs and networks, cellular automata, adaptive systems, genetic algorithms and computational intelligence.

The two major book publication platforms of the Springer Complexity program are the monograph series "Understanding Complex Systems" focusing on the various applications of complexity, and the "Springer Series in Synergetics", which is devoted to the quantitative theoretical and methodological foundations. In addition to the books in these two core series, the program also incorporates individual titles ranging from textbooks to major reference works.

Editorial and Programme Advisory Board

Péter Érdi
Center for Complex Systems Studies, Kalamazoo College, USA
and Hungarian Academy of Sciences, Budapest, Hungary

Karl Friston
National Hospital, Institute for Neurology, Wellcome Dept. Cogn. Neurology, London, UK

Hermann Haken
Center of Synergetics, University of Stuttgart, Stuttgart, Germany

Janusz Kacprzyk
System Research, Polish Academy of Sciences, Warsaw, Poland

Scott Kelso
Center for Complex Systems and Brain Sciences, Florida Atlantic University, Boca Raton, USA

Jürgen Kurths
Nonlinear Dynamics Group, University of Potsdam, Potsdam, Germany

Linda Reichl
Department of Physics, Prigogine Center for Statistical Mechanics, University of Texas, Austin, USA

Peter Schuster
Theoretical Chemistry and Structural Biology, University of Vienna, Vienna, Austria

Frank Schweitzer
System Design, ETH Zürich, Zürich, Switzerland

Didier Sornette
Entrepreneurial Risk, ETH Zürich, Zürich, Switzerland

Understanding Complex Systems

Founding Editor: J.A. Scott Kelso

Future scientific and technological developments in many fields will necessarily depend upon coming to grips with complex systems. Such systems are complex in both their composition – typically many different kinds of components interacting simultaneously and nonlinearly with each other and their environments on multiple levels – and in the rich diversity of behavior of which they are capable.

The Springer Series in Understanding Complex Systems series (UCS) promotes new strategies and paradigms for understanding and realizing applications of complex systems research in a wide variety of fields and endeavors. UCS is explicitly transdisciplinary. It has three main goals: First, to elaborate the concepts, methods and tools of complex systems at all levels of description and in all scientific fields, especially newly emerging areas within the life, social, behavioral, economic, neuro- and cognitive sciences (and derivatives thereof); second, to encourage novel applications of these ideas in various fields of engineering and computation such as robotics, nano-technology and informatics; third, to provide a single forum within which commonalities and differences in the workings of complex systems may be discerned, hence leading to deeper insight and understanding.

UCS will publish monographs, lecture notes and selected edited contributions aimed at communicating new findings to a large multidisciplinary audience.

QH 700 H474+1

D. Helbing (Ed.)

Managing Complexity: Insights, Concepts, Applications

With 145 Figures and 13 Tables

Ausgesondert

S13 69

Springer

Dirk Helbing

TU Dresden
Institut für Wirtschaft und Verkehr
Andreas-Schubert-Str. 23
01062 Dresden
Germany
helbing1@vwi.tu-dresden.d

ETH Zurich
Chair of Sociology, in particular of Modelling
and Simulation
UNO D11, Universitätstr. 41
8092 Zurich
Switzerland
dhelbing@ethz.ch

Library of Congress Control Number: 2007937091

ISBN: 978-3-540-75260-8 e-ISBN: 978-3-540-75261-5
©2008 Springer-Verlag Berlin Heidelberg

This work is subject to copyright. All rights are reserved, whether the whole or part of the material is concerned, specifically the rights of translation, reprinting, reuse of illustrations, recitation, broadcasting, reproduction on microfilm or in any other way, and storage in data banks. Duplication of this publication or parts thereof is permitted only under the provisions of the German Copyright Law of September 9, 1965, in its current version, and permission for use must always be obtained from Springer. Violations are liable to prosecution under the German Copyright Law.

The use of general descriptive names, registered names, trademarks, etc. in this publication does not imply, even in the absence of a specific statement, that such names are exempt from the relevant protective laws and regulations and therefore free for general use.

Cover Design: WMXDesign GmbH, Heidelberg

Printed on acid-free paper

9 8 7 6 5 4 3 2 1

springer.com

Editorial

Dirk Helbing[1,2]

[1] Chair of Sociology, in particular of Modeling & Simulation, ETH Zurich UNO D11, Universitätstrasse 41, 8092 Zurich, Switzerland dhelbing@ethz.ch
[2] Collegium Budapest – Institute for Advanced Study Szentháromság utca 2, H-1014 Budapest, Hungary

As promising as the recent discoveries regarding the functioning, adaptiveness and robustness of complex physical, biological, and socio-economic systems may be, their potential has not yet been broadly recognized in business, industry, public organizations, administrations, and governments. This is mostly because these findings have not yet entered curricula of standard study directions and due to the scarcity of real-world applications, at least published ones.

Therefore, this book aims at increasing the awareness for the relevance of the science of complex systems for

- collective decision making and cooperation in social networks,
- disaster, crisis and risk management,
- the operation of information and traffic systems,
- the resilience of supply systems,
- counter-intuitive effects of management strategies,
- the prediction and control of systems with non-linearities, delays, and imperfect information,
- the interaction of boundedly rational agents, and many more problems.

This book presents the most interesting contributions to the international workshop on "Potentials of Complexity Science for Business, Governments, and the Media" held at the Collegium Budapest—Institute for Advanced Study from August 3–5, 2006. It was further enriched by contributions reflecting other important activities in the field that could not be covered in Budapest. Altogether, this volume intends to tranfer the recent knowledge of the behavior of complex systems to various fields of practical applications. At the same time, it intends to stimulate more intensive research in this promising and important area of science in the future. The presentation is oriented at a general, scientifically interested audience in order fascinate and inspire both, newcomers and experts by the large variety of concepts and ideas presented.

Budapest, July 19, 2006 *Dirk Helbing*

Acknowledgments: D.H. would like to thank the Coordination Action GIACS for financial support within the EU NEST program, Imre Kondor and the Collegium Budapest for their great hospitality, Peter Felten, Tünde Szabolcs, Dietmar Huber, Fred Girod, Agnes Forgo, Vera Kempa, Martina Seifert, Anne Höner and the GIACS team at ISI Torino for their phantastic organizational support, which made this workshop possible.

Contents

Part III Traffic

Part IV Critical Infrastructures and Systemic Risks

Part V Information Systems

Part VI Conflict and Consensus

Part VII Network Design

Managing Complexity: An Introduction

Dirk Helbing[1,2], Stefan Lämmer[3]

[1] Chair of Sociology, in particular of Modeling & Simulation, ETH Zurich, UNO D11, Universitätstrasse 41, 8092 Zurich, Switzerland dhelbing@ethz.ch
[2] Collegium Budapest – Institute for Advanced Study Szentháromság utca 2, H-1014 Budapest, Hungary
[3] Institute for Transport & Economics, TU Dresden, Andreas-Schubert-Str. 23, 01062 Dresden, Germany laemmer@vwi.tu-dresden.de

1 What Is Special About Complex Systems?

Many of us have been raised with the idea of cause and effect, i.e. some stimulus-response theory of the world. Particularly, small causes would have small effects and large causes would have large effects. This is, in fact, true for "linear systems", where cause and effect are proportional to each other. Such behavior is often found close to the equilibrium state of a system. However, when complex systems are driven far from equilibrium, non-linearities dominate, which can cause many kinds of "strange" and counter-intuitive behaviors. In the following, we will mention a few. We all have been surprised by these behaviors many times.

While linear system have no more than *one* stationary state (equilibrium) or *one* optimal solution, the situation for non-linear systems is different. They can have *multiple* stationary solutions or optima (see Fig. 1), which has several important implications:

- The resulting state is history-dependent. Different initial conditions will not automatically end up in the same state [1]. This is sometimes called "hysteresis".
- It may be hard to find the best, i.e. the "global" optimum in the potentially very large set of local optima. Many non-linear optimization problems are "NP hard", i.e. the computational time needed to determine the best state tends to explode with the size of the system [2]. In fact, many optimization problems are "combinatorially complex".

1.1 Chaotic Dynamics and Butterfly Effect

It may also happen that the stationary solutions are unstable, i.e. any small perturbation will drive the system away from the stationary state until it

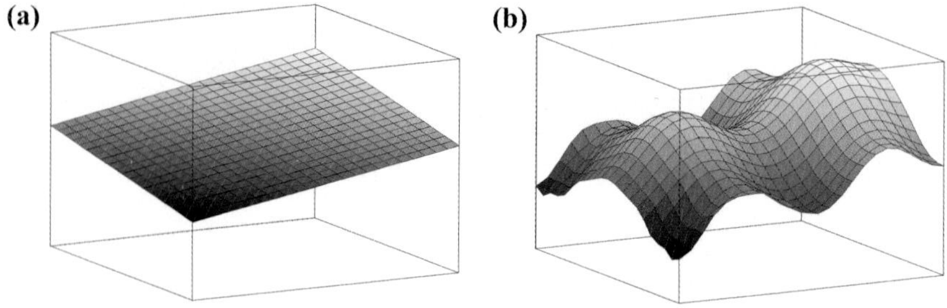

Fig. 1. Illustration of linear and non-linear functions. While linear functions have one maximum in a limited area (**left**), non-linear functions may have many (local) maxima (**right**)

is attracted by another state (a so-called "attractor"). Such attractors may be other stationary solutions, but in many cases, they can be of oscillatory nature (e.g. "limit cycles"). Chaotically behaving systems [3] are characterized by "strange attractors", which are non-periodic (see Fig. 2). Furthermore, the slightest change in the trajectory of a chaotic system ("the beat of a butterfly's wing") will eventually lead to a completely different dynamics. This is often called the "butterfly effect" and makes the behavior of chaotic systems unpredictable (beyond a certain time horizon), see Fig. 3.

1.2 Self-Organization, Competition, and Cooperation

Systems with non-linear interactions do not *necessarily* behave chaotically. Often, they are characterized by "emergent", i.e. spontaneous coordination or synchronization [4, 5, 6]. Even coordinated states, however, may sometimes be undesired. A typical example for this are stop-and-go waves in freeway traffic [13], which are a results of an instability of the traffic flow due to the delayed velocity adjustments of vehicles.

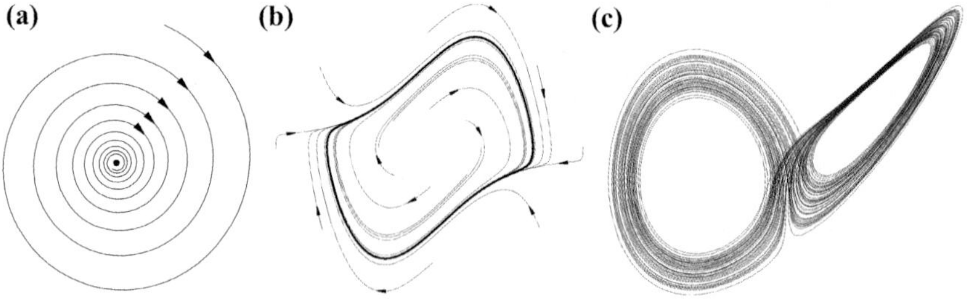

Fig. 2. Illustration of trajectories that converge towards (**a**) a stable stationary point, (**b**) a limit cycle, and (**c**) a strange attractor

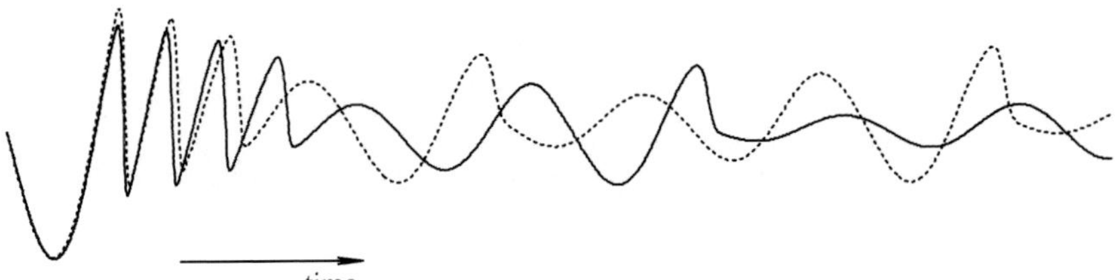

Fig. 3. Illustration of the "butterfly effect", i.e. the separation of neighboring trajectories in the course of time

Self-organization is typical in driven many-component systems [13] such as traffic, crowds, organizations, companies, or production plants. Such systems have been successfully modeled as many-particle or multi-agent systems. Depending on the respective system, the components are vehicles, individuals, workers, or products (or their parts). In these systems, the energy input is absorbed by frictional effects. However, the frictional effect is not homogeneous, i.e. it is not the same everywhere. It rather depends on the local interactions among the different components of the system, which leads to spatio-temporal pattern formation.

The example of social insects like ants, bees, or termites shows that simple interactions can lead to complex structures and impressive functions. This is often called "swarm intelligence" [8]. Swarm intelligence is based on local (i.e. decentralized) interactions and can be used for the self-organization and self-steering of complex systems. Some recent examples are traffic assistance [9] systems or self-organized traffic light control [9, 19]. However, if the interactions are not appropriate, the system may be characterized by unstable dynamics, breakdowns and jamming, or it may be trapped in a local optimum (a "frustrated state").

Many systems are characterized by a competition for scarce resources. Then, the question whether and how a system optimum is reached is often studied with methods from "game theory" [11, 12, 13]. Instead of reaching the state that maximizes the overall success, the system may instead converge to a user equilibrium, where the success ("payoff") of every system component is the *same*, but lower than it *could* be. This happens, for example, in traffic systems with the consequence of excess travel times [14]. In conclusion, if everybody tries to reach the best outcome for him- or herself, this may lead to overall bad results and social dilemmas [15] (the "tragedy of the commons" [16]). Sometimes, however, the system optimum can only be reached by complicated coordination in space and/or time, e.g. by suitable turn-taking behavior (see Fig. 4). We will return to this issue in Sec. 2.4, when we discuss the "faster-is-slower" effect.

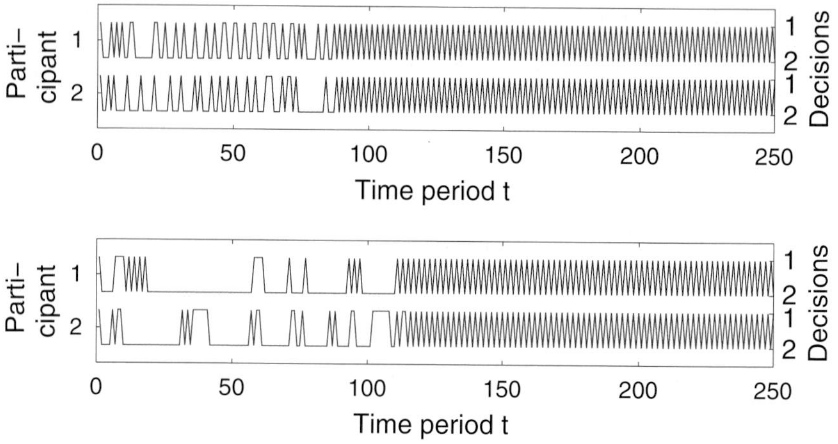

Fig. 4. Emergence of turn-taking behavior: After some time, individuals may learn to improve their average success by choosing both possible options in an alternating and coordinated way (see Ref. [14] for details)

1.3 Phase Transitions and Catastrophe Theory

One typical feature of complex systems is their robustness with respect to perturbations, because the system tends to get back to its "natural state", the attractor. However, as mentioned above, many complex system can assume different states. For this reason, we may have transitions from one system state ("phase" or attractor) to another one. These phase transitions occur at so-called "critical points" that are reached by changes of the system parameters (which are often slowly changing variables of the system). When system parameters come close to critical points, small fluctuations may become a dominating influence and determine the future fate of the system. Therefore, one speaks of "critical fluctuations" [1].

In other words, large fluctuations are a sign of a system entering an unstable regime, indicating its potential transition to another system state, which may be hard to anticipate. Another indicator of potential instability is "critical slowing down". However, once the critical point is passed, the system state may change quite rapidly. The relatively abrupt change from one system state to an often completely different one is studied by "catastrophe theory" [17]. One can distinguish a variety of different types of catastrophes, but we cannot go into all these details, here.

1.4 Self-Organized Criticality, Power Laws, and Cascading Effects

At the critical point itself, fluctuations are not only dominating, they may even become *arbitrarily* large. Therefore, one often speaks of "scale-free" behavior, which is typically characterized by power laws [21, 19]. Note that, for power laws, the variance and the expected value (the average) of a variable may be undefined!

One possible implication of power laws are cascading effects. The classical example is a sand pile, where more and more grains are added on top [21]. Eventually, when the critical "angle of repose" is reached, one observes avalanches of sand grains of all possible sizes, and the avalanche size distribution is given by a power law. The angle of repose, by the way, even determines the stability of the famous pyramids in Egypt.

Cascading effects are the underlying reason for many disasters, where the failure of one element of a system causes the failure of another one (see Fig. 5). Typical examples for this dynamics are blackouts of electrical power grids and the spreading of epidemics, rumors, bankruptcies or congestion patterns. This spreading is often along the links of the underlying causality or interaction networks [20].

"Self-organized criticality" [21, 22] is a particularly interesting phenomenon, where a system is driven towards a critical point. This is not uncommon for economic systems or critical infrastructures: Due to the need to minimize costs, safety margins will not be chosen higher than necessary. For example, they will be adjusted to the largest system perturbation that has occurred in the last so-and-so many years. As a consequence, there will be no failures in a long time. But then, controllers start to argue that one could safe money by reducing the standards. Eventually, the safety margins will be low enough to be exceeded by *some* perturbation, which may finally trigger a disaster.

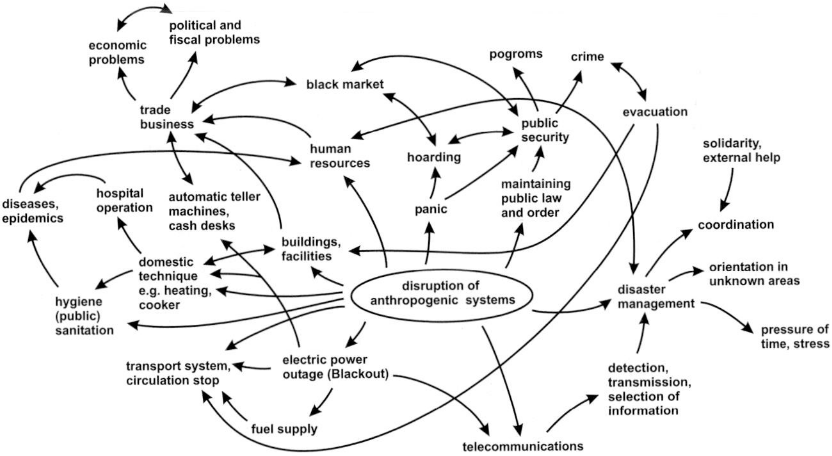

Fig. 5. Illustration of the interaction network in athropogenic systems. When the system is seriously challenged, this is likely to cause cascading failures along the arrows of this network (after [20])

Waves of bankruptcies [23, 24] are not much different. The competition for customers forces companies to make better and better offers, until the profits have reached a critical value and some companies will die. This will reduce the competitive pressure among the remaining companies and increase the profits again. As a consequence, new competitors will enter the market, which eventually drives the system back to the critical point.

2 Some Common Mistakes in the Mangement of Complex Systems

The particular features of complex systems have important implications for organizations, companies, and societies, which are complex multi-component systems themselves. Their counter-intuitive behaviors result from often very complicated feedback loops in the system, which cause many management mistakes and undesired side effects. Such effects are particularly well-known from failing political attempts to improve the social or economic conditions.

2.1 The System Does Not Do What You Want It to Do

One of the consequences of the non-linear interactions between the components of a complex system is that the internal interactions often dominate the external control attempts (or boundary conditions). This is particularly obvious for group dynamics [25, 26].

It is quite typical for complex systems that, many times, large efforts have no significant effect, while sometimes, the slightest change (even a "wrong word") has a "revolutionary" impact. This all depends on whether a system is close to a critical state (which will lead to the latter situation) or not (then, many efforts to change the system will be in vain). In fact, complex systems often counteract the action. In chemical systems, this is known as Le Chatelier's principle.[4] Therefore, if it is necessary to change a system, the right strategy is to drive it to a critical point first. Then, it will be easy to drive it into a new state, but the potential problem is that the resulting state is often hard to predict.

Regarding such predictions, classical time series analysis will normally provide bad forecasts. The problem of opinion polls to anticipate election results when the mood in the population is changing, is well-known. In many cases, the expectations of a large number of individuals, as expressed by the stock prices at real or virtual stock markets, is more indicative than results of classical extrapolation. Therefore, auction-based mechanisms have been proposed as a new prediction tool. Recently, there are even techniques to

[4] Specifically, Le Chatelier's principle says: "If a chemical system at equilibrium experiences a change in concentration, temperature, or total pressure, the equilibrium will shift in order to minimize that change."

forecast the future with small groups [27]. This, however, requires to correct for individual biases by fitting certain personality parameters. These reflect, for example, the degree of risk aversion.

2.2 Guided Self-Organization is Better than Control

The previous section questions the classical control approach, which is, for example, used to control machines. But it is also frequently applied to business and societies, when decision-makers attempt to regulate all details by legislation, administrative procedures, project definitions, etc. These procedures are very complicated and time-consuming, sensitive to gaps, prone to failures, and they often go along with unanticipated side effects and costs. However, a complex system cannot be controlled like a bus, i.e. steering it somewhere may drive it to some unexpected state.

Biological systems are very differently designed. They do not specify all procedures in detail. Otherwise cells would be much too small to contain all construction plans in their genetic code, and the brain would be too small to perform its incredible tasks. Rather than trying to control all details of the system behavior, biology makes use of the self-organization of complex systems rather than "fighting" it. It *guides* self-organization, while forceful control would destroy it [28].

Detailed control would require a large amount of energy, and would need further resources to put and keep the components of an artificial system together. That means, overriding the self-organization in the system is costly and inefficient. Instead, one could use self-organization principles as part of the management plan. But this requires a better understanding of the natural behavior of complex systems like companies and societies.

2.3 Self-Organized Networks and Hierarchies

Hierarchies are a classical way to control systems. However, strict hierarchies are only optimal under certain conditions. Particularly, they require a high reliability of the nodes (the staff members) and the links (their exchange).

Experimental results on the problem solving performance of groups [29] show that small groups can find solutions to difficult problems faster than any of their constituting individuals, because groups profit from complementary knowledge and ideas. The actual performance, however, sensitively depends on the organization of information flows, i.e. on who can communicate with whom. If communication is unidirectional, for example, this can reduce performance. However, it may also be inefficient if everybody can talk to everyone else. This is, because the number of potential (bidirectional) communicative links grows like $N(N-1)/2$, where N denotes the number of group members. The number of communicative or group-dynamical constellations even grows as $(3^N - 2^{N+1} + 1)/2$. Consequently, the number of possible information flows explodes with the group size, which may easily overwhelm the communication and information processing capacity of individuals. This explains the slow speed of group decision making, i.e. the inefficiency of large committees. It is also responsible for the fact that, after some transient time, (communication) activities in large (discussion) groups often concentrate on a few members only, which is due to a self-organized information bundling and differentiation (role formation) process. A similar effect is even observed in insect societies such as bee hives: When a critical colony size is exceeded, a few members develop hyperactivity, while most colony members become lazy [30].

This illustrates the tendency of bundling and compressing information flows, which is most pronounced in strict hierarchies. But the performance of strictly hierarchical organizations (see Fig. 6) is vulnerable for the following reasons:

- Hierarchical organizations are are not robust with respect to failure of nodes (due ot illness of staff members, holidays, quitting the job) or links (due to difficult personal relationships).
- They often do not connect interrelated activities in different departments well.
- Important information may get lost due to the filtering of information implied by the bundling process.
- Important information may arrive late, as it takes time to be communicated over various hierarchical levels.

Therefore, hierarchical networks with short-cuts are expected to be superior to strictly hierarchical networks [31, 32, 33]. They can profit from alternative information paths and "small-world" effects [34].

Note that the spontaneous formation of hierarchical structures is not untypical in social systems: Individuals form groups, which form companies,

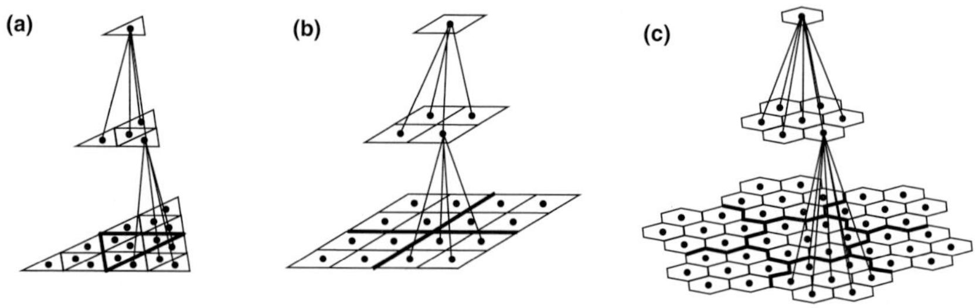

Fig. 6. Illustration of different kinds of hierarchical organization (after [31]). As there are no alternative communication links, strict hierarchies are vulnerable to the failure of nodes or links

organizations, and parties, which make up a society or nation. A similar situation can be found in biology, where organelles form cells, cells form organs, and organs form bodies. Another example is well-known from physics, where elementary particles form nuclei, which combine to atoms with electrons. The atoms form chemical molecules, which organize themselves as solids. These make up cellestial bodies, which form solar systems, which again establish galaxies.

Obviously, the non-linear interactions between the different elements of the system give rise to a formation of different *levels*, which are *hierarchically* ordered one below another. While changes on the lowest hierarchical level are fastest, changes on the highest level are slow.

On the lowest level, we find the strongest interactions among its elements. This is obviously the reason for the fast changes on the lowest hierarchical level. If the interactions are attractive, *bonds* will arise. These cause the elements to behave no longer completely individually, but to form *units* representing the elements of the next level. Since the attractive interactions are more or less 'saturated' by the bonds, the interactions *within* these units are stronger than the interactions *between* them. The relatively weak *residual interactions* between the formed units induce their relatively slow dynamics [35].

In summary, a general interdependence between the interaction strength, the changing rate, and the formation of hierarchical levels can be found, and the existence of different hierarchical levels implies a "separation of time scales".

The management of organizations, production processes, companies, and political changes seems to be quite different today: The highest hierarchy levels appear to take a *strong* influence on the system on a relatively *short* time scale. This does not only require a large amount of resources (administrative overhead). It also makes it difficult for the lower, less central levels of organization to adjust themselves to a changing environment. This complicates large-scale coordination in the system and makes it more costly. Strong interference in the system may even destroy self-organization in the system instead of using its potentials. Therefore, the re-structuring of companies can easily fail, in particularly if it is applied too often. A good example is given in Ref. [36].

Governments would be advised to focus their activities on coordination functions, and on adaptations that are relevant for long time scales, i.e. applicable for 100 years or so. Otherwise the individuals will not be able to adjust to the boundary conditions set by the government. If the government tries to adjust to the population and the people try to adjust to the socio-economic conditions on the same time scale of months or years, the control attempts are expected to cause a potentially chaotic dynamics and a failure of control.

Anyway, detailed regulations hardly ever reach more fairness. They rather reduce flexibility, and make the anyway required processes inefficient, slow, complicated, and expensive. As a consequence, many people will not be able to utilize their rights without external help, while a specialized minority will be able to profit from the regulations or exploit them.

2.4 Faster Is Often Slower

Another common mistake is to push team members to their limits and have machines run at maximum speed. In many cases, this will not maximize productivity and throughput, but rather frustration. Most systems require some spare capacity to run smoothly. This is well illustrated by queuing systems: If the arrival rate reaches the service rate, the average waiting time will grow enormously. The same applies to the variation of the waiting time. Jamming and full buffers will be an unfavorable, but likely side effect. And there will be little reserves in case of additional demand.

The situation becomes even more difficult by dynamic interaction effects, when a system is driven to its limits. In traffic systems, for example, this leads to a "capacity drop". Such a capacity drop occurs often unexpectedly and is a sign of inefficiencies due to dynamical friction or obstruction effects. It results from increasing coordination problems when sufficient space or time are lacking. The consequence is often a "faster-is-slower effect" [38] (see Fig. 7). This effect has been observed in many traffic, production, and logistic systems. Consequently, it is often not good if everybody is doing his or her best. It is more important to adjust to the other activities and processes in order to reach a harmonic and well coordinated overall dynamics. Otherwise, more and more conflicts, inefficiencies and mistakes will ruin the overall performance.

2.5 The Role of Fluctuations and Heterogeneity

Let us finally discuss the role of fluctuations and heterogeneity. Fluctuations are often considered unfavorable, as they are thought to produce disorder. They can also trigger instabilities and breakdowns, as is known from traffic flows. But in some systems, fluctuations can also have positive effects.

While a large fluctuation strength, in fact, tends to distroy order, medium fluctuation levels may even cause a *noise-induced ordering* (see Fig. 8). An eventual increase in the degree of order in the system is particularly expected if the system tends to be trapped in local minima ("frustrated states"). Only by means of fluctuations, it is possible to escape these traps and to eventually find better solutions.

Fluctuations are also needed to develop different behavioral roles under initially identical conditions. This eventually leads to a differentiation and specialization (heterogeneity), which often helps to reach a better group performance [40] (see Fig. 9).

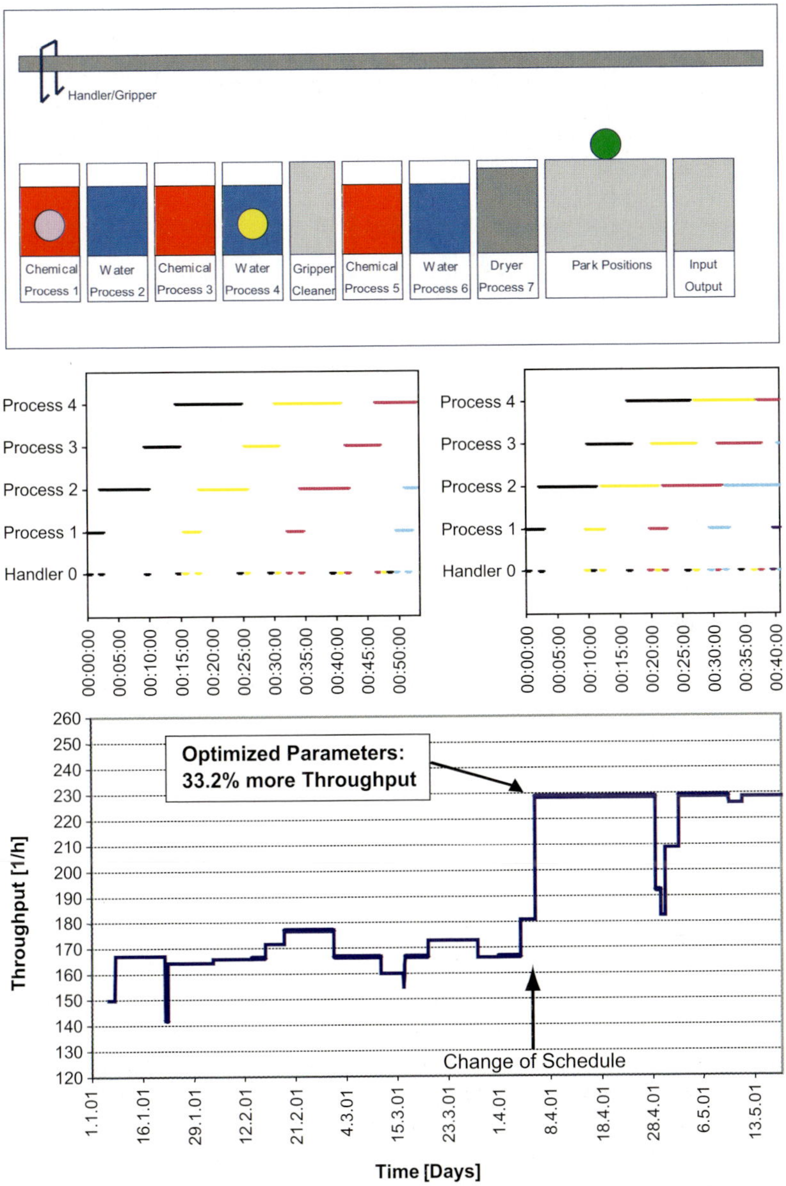

Fig. 7. Top: Schematic representation of the successive processes of a wet bench, i.e a particular supply chain in semiconductor production. **Middle**: The Gantt diagramms illustrate the treatment times of the first four of several more processes, where we have used the same colors for processes belonging to the same run, i.e. the same set of wafers. The left diagram shows the original schedule, while the right one shows an optimized schedule based on the "slower-is-faster effect". **Bottom**: The increase in the throughput of a wet bench by switching from the original production schedule to the optimized one was found to be 33%, in some cases even higher (after [37])

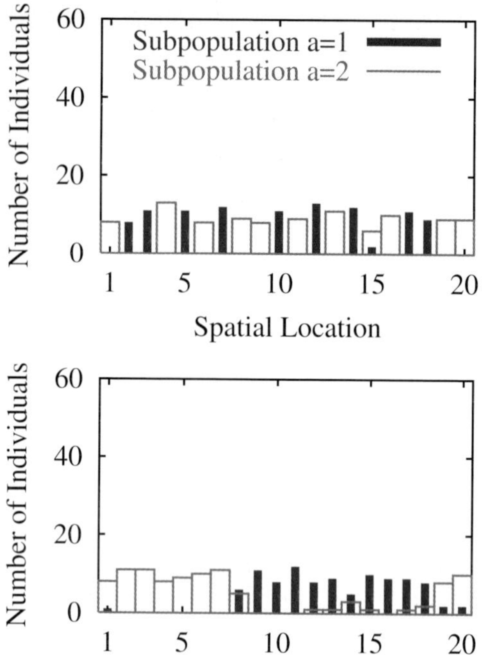

Fig. 8. Illustration of frequency distributions of behaviors in space (after [39]). **Top**: Separation of oppositely moving pedestrians perpendicularly to their walking direction for a low fluctuation strength. **Bottom**: Noise-induced ordering for medium fluctuation levels leads to a clear separation into two spatial areas. This reduces frictional effects and increases the efficiency of motion

Furthermore, the speed of evolution also profits from variety and fluctuations ("mutations"). Uniformity, i.e. if everybody behaves and thinks the same, will lead to a poor adaptation to changing environmental or market conditions. In contrast, a large variety of different approaches (i.e. a heterogeneous population) will imply a large innovation rate [41]. The innovation rate is actually expected to be proportional to the variance of individual solutions. Therefore, strong norms, "monocultures", and the application of identical strategies all over the world due to the trend towards globalization implies dangers.

This trend is re-inforced by "herding effects" [13]. Whenever the future is hard to predict, people tend to orient at the behavior of others. This may easily lead to wrong collective decisions, even of highly intelligent people. This danger can be only reduced by supporting and maintaining a plurality of opinions and solutions.

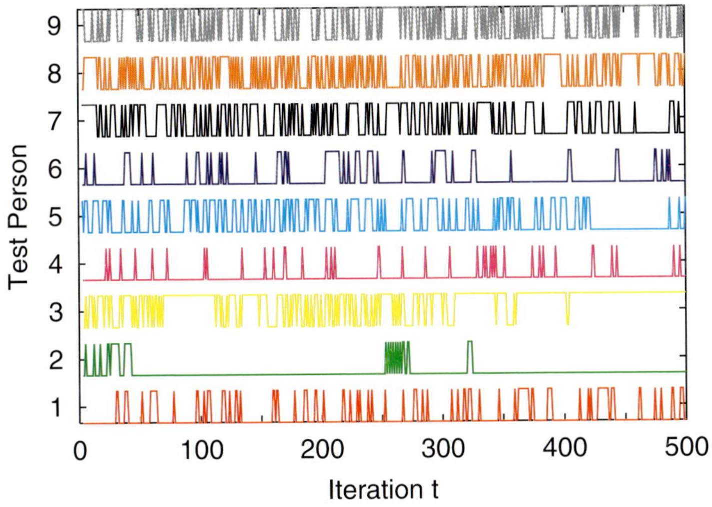

Fig. 9. Typical individual decision changes of 9 test persons in a route choice experiment with two alternative routes. Note that we find almost similar or opposite behaviors after some time. The test persons develop a few kinds of complementary strategies ("roles") in favour of a good group performance (after [40])

3 Summary and Outlook

In this contribution, we have given a short overview of some properties and particularities of complex systems. Many of their behaviors may occur unexpectedly (due to "catastrophies" or phase transitions), and they are often counter-intuitive, e.g. due to feedback loops and side effects. Therefore, the response of complex systems to control attempts can be very different from the intended or predicted one.

Complex behavior in space and time is found for many multi-component systems with non-linear interactions. Typical examples are companies, organizations, administrations, or societies. This has serious implications regarding suitable control approaches. In fact, most control attempts are destined to fail. It would, however, be the wrong conclusion that one would just have to apply more force to get control over the system. This would destroy the self-organization in the system, on which social systems are based.

Obtaining a better understanding of to make use of the natural tendencies and behaviors at work. A management that supports and guides the natural self-organization in the system would perform much more efficiently than an artificially constructed system that requires continuous forcing. Companies and countries that manage to successfully apply the principle of self-organization will be the future winners of the on-going global competition.

In conclusion, we are currently facing a paradigm shift in the management of complex systems, and investments into complexity research will be of competitive advantage.

References

1. Haken, H.: Synergetics. Springer, Berlin Heidelberg New York (1977)
2. Ausiello, G., Crescenzi, P., Gambosi G., et al: Complexity and Approximation – Combinatorial optimization problems and their approximability properties. Springer Berlin Heidelberg New York (1999)
3. Strogatz, S.H.: Nonlinear Dynamics and Chaos: With Applications to Physics, Biology, Chemistry and Engineering. Perseus (2001)
4. Kuramoto, Y.: Chemical Oscillations, Waves, and Turbulence. Springer Berlin Heidelberg New York (1984)
5. Pikovsky, A., Rosenblum, M. and Kurths, J.: Synchronization: A Universal Concept in Nonlinear Sciences. Cambridge University Press (2003)
6. Manrubia, S.C., Mikhailov, A.S. and Zanette, D.H.: Emergence of Dynamical Order. Synchronization Phenomena in Complex systems World Scientific, Singapore (2004)
7. Helbing, D.: Traffic and related self-driven many-particle systems. Reviews of Modern Physics **73** (2001) 1067
8. Bonabeau, E., Dorigo, M. and Theraulaz, G.: Swarm Intelligence: From Natural to Artificial Systems. Santa Fe Institute Studies in the Sciences of Complexity Proceedings (1999)
9. Kesting, A., Schönhof, M., Lämmer, S., Treiber, M. and D. Helbing, Decentralized approaches to adaptive traffic control, see pp. 179ff in this book.
10. Helbing, D. and Lämmer, S.: Verfahren zur Koordination konkurrierender Prozesse oder zur Steuerung des Transports von mobilen Einheiten innerhalb eines Netzwerkes [Method to Coordinate Competing Processes or to Control the Transport of Mobile Units within a Network]. Pending patent DE 10 2005 023 742.8 (2005)
11. Axelrod, R.: The Evolution of Cooperation. Basic Books (1985)
12. von Neumann, J., Morgenstern, O., Rubinstein, A. and Kuhn H.W.: Theory of Games and Economic Behavior. Princeton University Press (2004)
13. Schelling, T.C.: The Strategy of Conflict. Harvard University Press (2006)
14. Helbing, D., Schönhof, M., Stark, H.-U. and Holyst, J.A.: How individuals learn to take turns: Emergence of alternating cooperation in a congestion game and the prisoner's dilemma. Advances in Complex Systems **8** (2005) 87
15. Glance, N.S. and Huberman, B.A.: The dynamics of social dilemmas. Scientific American **270** (1994) 76
16. Hardin, G.: The Tragedy of the Commons. Science **162** (1968) 1243
17. Zeeman, E.C.: Catastrophe Theory. Addison-Wesley London (1977)
18. Stanley, H.E.: Introduction to Phase Transitions and Critical Phenomena. Oxford University Press (1971)
19. Schroeder, M.: Fractals, Chaos, Power Laws: Minutes from an Infinite Paradise. Freeman (1992)

20. Helbing, D., Ammoser, H. and Kühnert, C.: Disasters as extreme events and the importance of network interactions for disaster response management. In: Albeverio, S., Jentsch, V. and Kantz, H.(eds.): The Unimaginable and Unpredictable: Extreme Events in Nature and Society. Springer Berlin Heidelberg New York (2005) 319–348
21. Bak, P., Tang, C. and Wiesenfeld, K.: Self-organized criticality: An explanation of $1/f$ noise, Phys. Rev. Lett. **59** (1987) 381
22. Bak, P.: How Nature Works: The Science of Self-Organized Criticality. Copernicus New York (1996)
23. Aleksiejuk, A. and Hołyst, J.A.: A simple model of bank bankruptcies. Physica A **299**(1-2), 198 (2001)
24. Aleksiejuk, A., Hołyst, J.A. and Kossinets, G.: Self-organized criticality in a model of collective bank bankruptcies. International Journal of Modern Physics C **13** (2002) 333
25. Tubbs, S.L.: A Systems Approach to Small Group Interaction. McGraw-Hill Boston (2003)
26. Arrow, H., McGrath, J.E. and Berdahl, J.L.: Small Groups as Complex Systems: Formation, Coordination, Development, and Adaptation. Sage (2000)
27. Chen, K.-Y., Fine, L.R. and Huberman, B.A: Predicting the Future. Information Systems Frontiers **5** (2003) 47
28. Mikhailov, A.S: Artificial life: an engineering perspective. In: Friedrich, R. and Wunderlin, A. (eds.): Evolution of Dynamical Structures in Complex Systems. Springer Berlin Heidelberg New York (1992) 301–312
29. Ulschak, F-L.: Small Group Problem Solving: An Aid to Organizational Effectiveness. Addison-Wesley Reading Mass. (1981)
30. Gautrais, J., Theraulaz, G., Deneubourg, J.-L. and Anderson, C.: Emergent polyethism as a consequence of increased colony size in insect societies. Journal of Theoretical Biology **215** (2002) 363
31. Helbing, D., Ammoser, H. and Kühnert, C.: Information flows in hierarchical networks and the capability of organizations to successfully respond to failures, crises, and disasters. Physica A **363** (2006) 141
32. Adamic, L.A. and Adar, E.: Friends and neighbors on the web, preprint http://www.cs.man.ac.uk/˜rizos/web10.pdf
33. Stauffer, D. and de Oliveira, P.M.C.: Optimization of hierarchical structures of information flow. International Journal of Modern Physics C **17** (2006) 1367
34. Watts, D.J. and Strogatz, S.H.: Collective dynamics of smallworld networks. Nature **393** (1998) 440
35. Helbing, D.: Quantitative Sociodynamics. Stochastic Methods and Models of Social Interaction Processes Kluwer Academic, Dordrecht (1995)
36. Christen, M., Bongard, G., Pausits, A., Stoop, N. and Stoop, R.: Managing autonomy and control in economic systems.
37. Fasold, D.: Optimierung logistischer Prozessketten am Beispiel einer Nassätzanlage in der Halbleiterproduktion. MA thesis, TU Dresden (2001)
38. Helbing, D., Seidel, T., Lämmer, S. and Peters, K.: Self-organization principles in supply networks and production systems. In: Chakrabarti, B.K., Chakraborti and A., Chatterjee, A. (eds.): Econophysics and Sociophysics - Trends and Perspectives Wiley Weinheim (2006) 535–558
39. Helbing, D. and Platkowski, T.: Self-organization in space and induced by fluctuations. International Journal of Chaos Theory and Applications **5** 47–62. (2000)

40. Helbing, D.: Dynamic decision behavior and optimal guidance through information services: Models and experiments. In: Schreckenberg, M. and Selten, R. (eds.): Human Behaviour and Traffic Networks Springer Heidelberg Berlin New York (2004) 47–95
41. Helbing, D., Treiber, M. and Saam, N.J.: Analytical investigation of innovation dynamics considering stochasticity in the evaluation of fitness. Physical Review E **71** (2005) 067101

Part I

Markets and Business

Market Segmentation: The Network Approach

Insights into a giant Online Market

Jörg Reichardt, Stefan Bornholdt

Institute for Theoretical Physics, University of Bremen, Otto-Hahn-Allee 1, 28203 Bremen, Germany reichardt|bornholdt@itp.uni-bremen.de

1 Introduction

The internet has changed the way a large number of people communicate, work and do business. A small number of enterprises have had enormous economical success in a very short time span, others have gone bankrupt only shortly after they were established and many are struggling for survival. The sheer ease by which new players may enter the market and establish a business is one of the reasons why the dot-com economy is so dynamic. One of the largest internet enterprises grounds its success in facilitating just the very establishment of online businesses: eBay. By providing one large online platform over which private persons and small firms can trade goods, eBay brings together supply and demand for almost everything one can think of. From large industrial machinery, over gourmet food to every day cosmetics, there is virtually nothing one cannot find on eBay. Today, eBay figures more than 150 million registered users world wide, only 4% are active though, having bid or listed items in the past year [1]. Another aspect of eBay's success is its transparency. Buyers and sellers leave mutual feedback about how satisfied they were with the quality of the articles sold, but also the speed of payment. Ebay keeps track of this feedback data and users hence can establish an online reputation. At the same time, this makes the market fully transparent as the trading history of every user is disclosed to everyone on the internet. Details about individual transactions are available for three months after the end of the auction. Obviously, this opens the way for extended both cross-sectional, and longitudinal studies of the structure of this giant market place.

Figure 1 explains the principle after which eBay operates. Users may offer goods through the eBay platform and set a deadline when their auction ends. Articles are listed under a certain taxonomic product category by the seller and are searchable platform wide. Users with a particular demand either browse through the articles listed in an appropriate category or search for articles directly. Until the end of the auction they may bid on the article. The user with the highest bid at the end of the auction wins (so called hard-close)

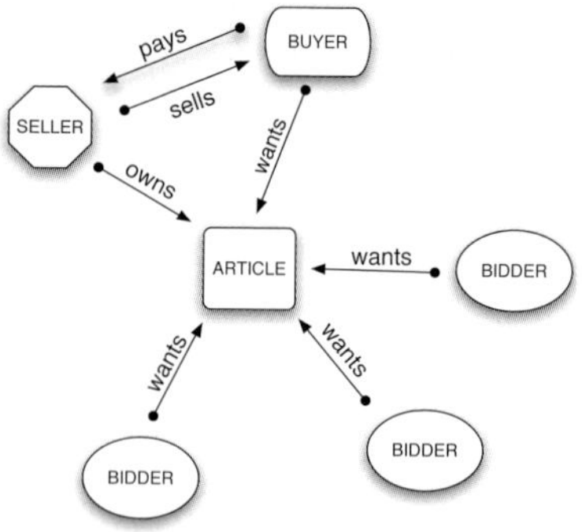

Fig. 1. Structure of a single auction. Users express their common interest in a particular article by bidding. The user with the highest bid wins the auction and exchanges money and the article with the seller. EBay earns a fee with every transaction. Users of the auction site, *i.e.* bidders, buyers or sellers, may change their role in a different auction of another article

and buys the article. In every new auction, users may assume different new roles as sellers, bidders or buyers. The market can be represented as a large graph with the users and/or articles as the nodes and the links denoting their interactions as shown in Figure 1.

A number of researchers have presented statistical studies of trading [2] or analysis of bidding strategies and auction ending rules [3, 4]. In this contribution we focus on the market segmentation of the eBay auction site. Our approach is based on the hypothesis that at a certain level of abstraction the population of consumers can be separated into relatively clear cut and homogenous sub-groups corresponding to certain customer milieus or market segments [5]. We further hypothesize that customers of the same type are described by a common pattern in their consumer interests which leads to a higher probability of bidding for the same article [6].

In particular, we perform a cluster analysis [7, 8] of the bidding behavior of almost one million users. Groups of eBay users with common interest or demand are found exploiting only the information of which users competed in the same auction. The classification is based on a very sparse and very high dimensional data set [9] with only slightly more than 3 auctions per bidder on average (out of 1.6 million possible auctions) and we can say that in principle every article exists only once. Conventional analysis techniques such as correspondence analysis [5, 10] have to use of a similarity measure between articles in order to coarse grain the data, such as exploiting the annotation of articles into product categories. However, this bears several pitfalls. First, the annotations are defined by the seller who lists the article such that the article can be found effectively, hence, the categorization is mainly a taxonomy. Using this to coarse grain the data would introduce a bias in the analysis.

Second, eBay categories differ largely in size, both when counting the number of articles in the category as the number of sub-categories and one would have to correct for this, which again may introduce a bias. Third, using the category taxonomy for coarse graining induces a hierarchy in the data, as all articles below the cut in the taxonomy tree are subsumed. Fourth and most importantly, it is not clear at which level in the category tree a coarse graining should be performed and if this level should be the same for all of the 32 available main categories. The analysis we present is based on the level of individual articles and independent of taxonomic categories and hence does not suffer from any bias introduced by coarse graining the data. It allows both hierarchical and overlapping cluster structures and we find evidence for both. The product categories are only used to interpret the results of our study, *i.e.* provide interest profiles of the user groups found in terms of this taxonomy.

By clustering users directly according to a common demand spectrum, we also circumvent problems of conventional basket analysis done by frequent item sets [11, 12, 13, 14, 15]. The latter asks which articles are frequently demanded by a single person. This analysis is performed for all articles and averaging over the entire population of consumers. The result are sets of items of common demand which may then be bundled together and marketed together to the whole population of customers. This analysis yields information about articles. In a market of diverse user demands, they represent the common denominator of user interests. The same is true for cluster analysis of eBay categories [4]. Clustering customers directly, however, reveals information about people and their diverse and maybe very special interests. In order to perform targeted marketing on a particular customer milieu, this kind of information is indispensable.

We obtain a classification for bidders observed during a short time span before Christmas 2004 according to their bidding activity. In September 2005, we checked what users classified at Christmas according to their bids had bought between June and September. A striking resemblance of the classification from Christmas and the buying pattern during the summer was found with only slight adjustments in the relative strengths of user interests within each profile.

The proposed network cluster analysis at the level of individual articles is able to precisely discover meaningful groups of users with common interest and these interest profiles remain relatively stable over a long period of time for each user group. We conclude with a discussion of the implications of these findings for the growth potential of online vendors.

2 Dataset

A dataset consisting of over 1.59 million auctions was obtained from the German eBay site `www.ebay.de` ending during the pre-Christmas season between December 6th and 20th 2004. Considering only articles with locations in Germany, we recorded the user-id of seller, buyer, and all bidders competing

Table 1. Divisions of users into their roles on eBay as observed between Dec. 6th and 20th 2004. Numbers in millions

auctions observed:	1.59
users acting as buyer:	0.95
users acting as seller:	0.37
users acting as bidder:	1.91
users acting as seller and bidder:	0.14
users acting as seller and buyers:	0.08

in each auction, as well as the individual bids and the product category in which the article was listed (excluding articles listed in the real estate category which was in a beta testing phase at the time). Since auctions last between 7 and 10 days depending on the choice of the seller, we thus cover a bidding period of up to 25 days. We believe the pre-Christmas time is a suitable time for analysis for the following reasons: First, traffic is very high. In fact, there was a broad advertising campaign in Germany advertising to shop for Christmas presents on eBay. Second, we only considered auctions and expect that users are unlikely to bid for articles for which they cannot assess a fair price. Third, if users shop for presents, then we can gain some information about their family background, e.g. people shopping for toys will most likely have a child themselves or among their closer relatives. Our findings indicate that this is indeed the case. Table 1 summarizes the dataset in its basic parameters.

3 User Activity and User Networks

The activity of the users is measured via the probability mass distributions of the number of articles sold $p(s)$, bought (auctions won) $p(w)$, and bid on $p(a)$. Though it is possible to bid multiply in a single auction, we neglect this fact and use "bid" and "take part in an auction" synonymously. Similar to previous studies [2], we find fat tailed distributions of the user activity, *i.e.* we find very broad distributions without a characteristic peak around the mean value or in other works activity at all scales, ranging from only a single bid or article sold to several hundreds of objects, bought, sold or bid for. Due to the short time span observed and a constant growth of the market, one cannot regard these distributions as representing a steady state. Nevertheless, some insight can be obtained.

The distribution of the number of articles sold per seller falls off slowest, followed by the number of articles bid on and the number of articles bought. Here, we may observe the professionalization on the seller side of the market. There are "power-sellers" making a living from selling via eBay, but there are hardly any "power-buyers" professionally buying on eBay. This shows that eBay is more of a selling platform than an actual trading site, where selling and buying activities would be more balanced.

The fat tails of the distribution are striking given the short time span observed. Consider the most active bidder taking part in over 800 auctions! This user seems to follow a gambling strategy bidding only minimal amounts as he/she wins only a few of these auctions. The most successful buyer who won 201 auctions on the other hand took part in only 208 auctions. This hints at a diversity of strategies employed by users of the online auction site. Curiously, the article most desired and attracting 39 different bidders was a ride in a red Coca-Cola-Truck. The fat tails of the distribution also show that there is no "typical" user activity, rather, one observes activities at all scales.

From the original data a number of market networks can be constructed, such as the network of users connected by actual transactions, or the network of sellers that are connected if they have sold to (or received bids from) the same user. Then, the links in the network would represent competition or a possibility for cooperation, depending on the portfolio of articles offered by these sellers.

Here, we focus only on the bidder network based on single articles. Two bidders are linked if they have competed in an auction. Since all users who bid in a single auction are connected, this network results from overlaying fully connected cliques of bidders that result from each auction. Such graphs are also known as affiliation networks [16, 17, 18].

Prior to a cluster analysis in this bidder network, we study its general statistical properties looking for indications of cluster structure [19]. We compare the results to a randomized null model (RNM) obtained from reshuffling the original data, *i.e.* keeping the attractiveness of each auction and the activity of each bidder constant, but randomizing which bidders take part in which auction.

The shapes of the cumulative degree distributions agree quite well between the RNM and the original data. However, for higher values of k, the distribution of the empirical data lies below that of the RNM. Since meeting the same competitor twice in different auctions does not lead to an increase in the number of neighbors in the network, this shows that competitors meet more often in the real world than expected from the random null model. A theoretical expectation for the average number of neighbors in the bidder network is given by $\langle k \rangle = 2(\langle b \rangle - 1)\langle a \rangle = 14$ where $\langle b \rangle$ is the average number of bidders per auction and $\langle a \rangle$ is the average number of auctions taken part in by a bidder. This estimate is in excellent agreement with the result from the RNM ($\langle k \rangle = 13.9$), but larger than in the actual data ($\langle k \rangle = 12.9$) confirming our expectation. See Table 2 for a summary of the basic parameters of the empirical data and the RNM.

Comparing the cumulative distribution of the link weights, *i.e.* the number of times two bidders have met in different auctions, we find a much more prominent difference between the data and the RNM. The weights of the links in the bidder network are distributed with a power law tail. Approximately

Table 2. Summary of basic parameters for the bidder network with two bidders linked, if they have competed in an auction. Shown are the actual data, the parameters for a random null model (RNM) obtained by reshuffling the bidders in different auctions and the reduced version of the network used for cluster analysis containing only those bidders having taken part in more than one auction

	data	RNM	reduced
number of nodes:	1.8×10^6	1.8×10^6	0.9×10^6
number of links:	11.6×10^6	12.6×10^6	7.4×10^6
average degree:	12.9	13.9	16.4

6% of all links correspond to pairs of bidders who have met more than once. If there would be no common interest among bidders, practically all links would have weight 1 as is indeed the case for the RNM.

Additionally to the distribution of degrees and link weights, we compare the distribution of the clustering coefficient as a function of the degree of a node. The clustering coefficient $c(k)$ denotes the average link density among the neighbors of a node of degree k. Due to the construction process of the network as an affiliation network, we expect that for large numbers of neighbors k the clustering coefficient $c(k)$ scales as k^{-1} in case of random assignment of bidders to auctions [17]. This is indeed the case for the RNM, but the actual data deviates strongly for bidders with a large number of neighbors and shows higher clustering. This effect can arise from two processes: either bidders with whom one competes in two different auctions also meet independently in a third auction, or that there is an increased probability that one will compete again with a bidder one has already met once in an auction. Both explanations support our assumption of the presence of clusters of users with common interest.

With these comparisons, we have shown that the probability to meet a given bidder twice in an auction is higher for the empirical bidder network than for RNM. Naturally, we attribute this to an overlap in the interests of users and we will proceed by studying this overlap of interests for which we have found indirect evidence already.

4 Market Segmentation

4.1 Network Clustering

The analysis of the user interests in the eBay market is based on the bidder network as constructed in the previous section. The links in this network represent articles the connected bidders (nodes) have a common interest in. Since we are eventually interested in the overlap of user interests, we reduced the full network to only those bidders having taken part in at least two auctions and considered only auctions with a final price below 1,000 Euro, a price range in which we expect users to consider bidding on and buying several items. Though this reduction takes out the least active bidders from the data set, we believe that in order to sensibly classify a user according to its bidding interests during a certain time span, there should be a minimum number of observations of bidding activity. See Table 2 for the basic parameters of this reduced network.

If we now find groups of users (clusters or communities [20, 21, 22]) with a high density of links among themselves and a low density of links to the rest of the network, the total set of links within such a group of users can be interpreted as a unifying common interest of this group. We assign a community or cluster index to each bidder as to maximize a well established quality function known as network modularity Q defined by Girvan and Newman (GN) [23]. The definition of Q can also be written as [24]:

$$Q = \frac{1}{M}\sum_{s}^{q}\underbrace{(m_{ss} - \gamma[m_{ss}])}_{c_{ss}} = -\frac{1}{M}\sum_{s<r}^{q}\underbrace{(m_{rs} - \gamma[m_{rs}])}_{a_{rs}}. \tag{1}$$

Here, the first sum runs over all group indices s, while the second over all pairs of different group indices $s > r$, m_{ss} is the number of internal links in group s, *i.e.* connecting bidders with the same group index s. Accordingly, $[m_{ss}]$ is an expectation value for this quantity assuming a random assignment of bidders into groups of the same size and degree distribution and is given by $[m_{ss}] = K_s^2/4M$. By K_s we denote the total number of links emanating from members of group s, while M is the total number of links in the network. Equivalently, m_{rs} is the number of links between members of group r and s and $[m_{rs}]$ is the corresponding expectation values given by $[m_{rs}] = K_sK_r/2M$ [24]. Q is maximal, when the sum of cohesions c_{ss}, defined as the difference between the actual and expected number of within group links, is maximal. Equivalently, Q is maximal when there are many links less between different groups than expected for a random assignment of nodes into communities, *i.e.* the sum of adhesions a_{rs} is minimal. Note that any assignment of bidders into groups which maximizes Q will be characterized by both, maximum cohesion of groups, and minimal adhesion between groups. If Q is maximal, every node is classified in that group to which it has the largest adhesion, otherwise it could be moved to a different group to increase Q. Additionally to the original definition of Q by GN, we have introduced a parameter γ which allows to adjust the relative influence of actually present and expected links in the definition (setting $\gamma = 1$ recovers the original definition of GN). At any (local) maximum of Q, the cohesion c_{ss} of a group is never negative and the adhesion between different groups a_{rs} is never positive. Hence, values of γ smaller or greater than one lead to larger or smaller communities, respectively. Comparing classifications obtained at different values of γ allows the detection of hierarchy and overlap in the cluster structure. See Ref. [25, 24] for examples and further details of this variation.

The technical details of how the bidders can be assigned into groups such that Q is maximized using simulated annealing are given in Refs. [24, 21, 26, 27]. Note that the optimization routine exploits the sparsity of the data as it only operates on the links of the network. We allowed for maximally $q = 500$ different groups of bidders in our analysis which gives a sufficient level of detail.

The left part of Figure 2 compares the results obtained with $\gamma = 0.5$ and $\gamma = 1$. Shown are the adjacency matrices A_{ij} of the largest connected component of the bidder network. A black pixel at position (i, j) and (j, i) is shown on an 889828×889828 square if bidders i and j have competed in an auction and hence $A_{ij} = 1$, otherwise the pixel is left white corresponding to $A_{ij} = 0$. The rows and columns are ordered such that bidders who are classified as being in the same group are next to each other. The internal order of bidders within groups is random. The order of the groups was chosen to optimally show the correspondence between the ordering resulting from the $\gamma = 0.5$ and the $\gamma = 1$ ordering. In this representation, link densities correspond to pixel densities and thus to grey levels in Figure 2. Information about the exact size and link density contrast of the clusters is given in Table 3. Note the high contrast between internal and external link density.

At the top of Figure 2, we show the adjacency matrix ordered according to an optimal assignment of bidders into groups with $\gamma = 0.5$. Clearly, a small number of major clusters of bidders and a large number of smaller clusters are identified, strongly connected internally and well separated from one another. The largest 8 clusters are marked with letters A through H. Of all bidders in the network, 85% are classified in these 8 clusters. At the bottom of Figure 2, we show the same adjacency matrix, but now rows and columns are ordered according to an optimal assignment of bidders into groups with $\gamma = 1$. As expected, we find a larger number of smaller, denser clusters, the largest of which are numbered 1 through 13. In order to analyse whether the network has a hierarchical or overlapping cluster structure, we define a consensus ordering of the bidders from the $\gamma = 0.5$ and $\gamma = 1$ ordering by reshuffling the internal order of the $\gamma = 0.5$ clusters according to the $\gamma = 1$ clustering. Remember the orderings for the two values of γ were obtained independently. If the network possesses a hierarchical structure in the sense that the clusters obtained at higher values of γ lie completely within those obtained at lower values of γ, then the consensus ordering would not differ from the ordering at $\gamma = 1$. If, however, clusters at lower values of γ overlap and this overlap forms its proper cluster at higher values of γ, the network is not entirely hierarchical. These aspects will become immediately clear by looking at the middle part of Figure 2. For clarity, we have marked the borders of the $\gamma = 0.5$ clustering. Clusters 1 and 2 fall entirely within cluster A giving an example of a cluster hierarchy. Cluster 3, however, is split by the consensus ordering into one part A3 belonging to A, and B3 belonging to B (see arrows in figure). It is now clear that clusters A and B actually have some overlap which was not visible in the $\gamma = 0.5$ ordering due to the random order of all bidders with the

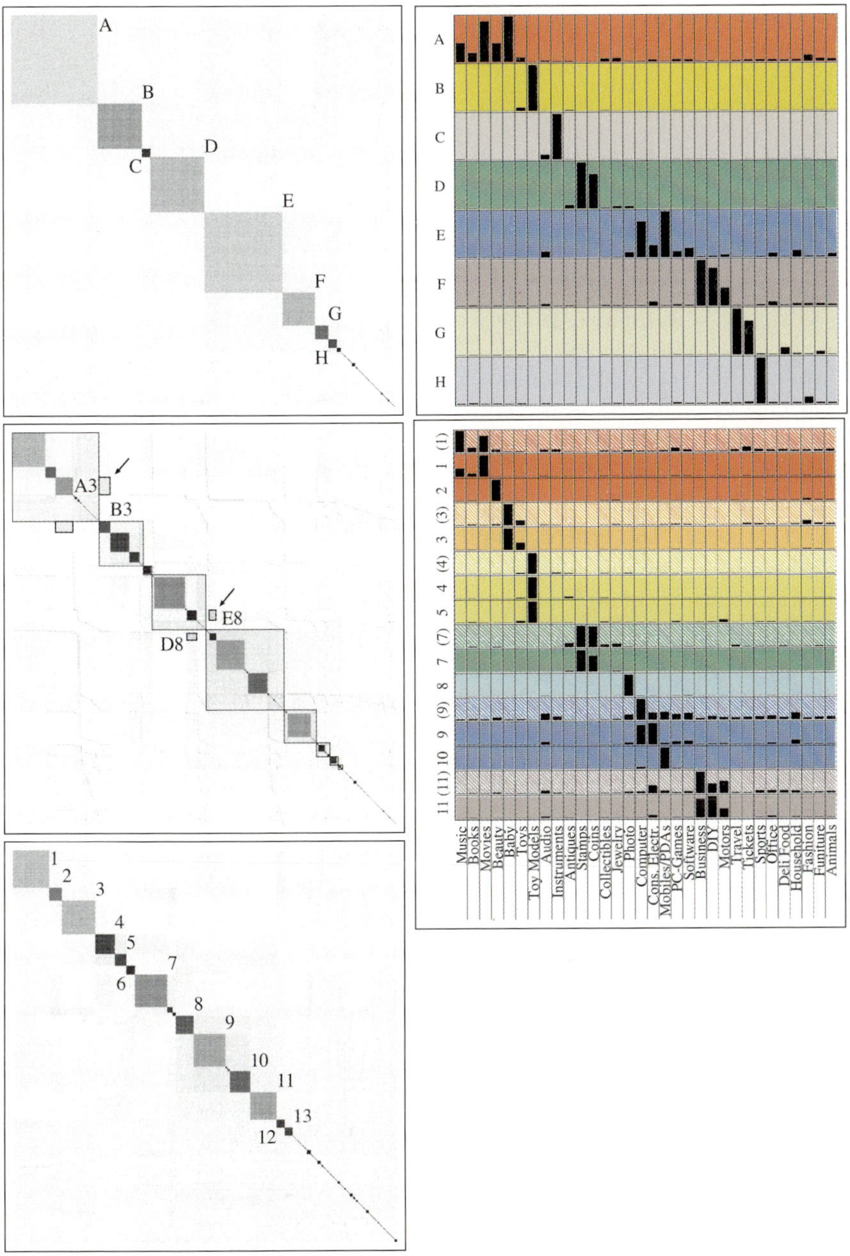

Fig. 2. Left: N×N adjacency matrix of the bidder network in three different orderings. A pixel in row i, column j corresponds to an auction in which bidder i and j have competed. Shown are $N = 889,828$ bidders (nodes) and $M = 7,373,008$ pairwise competitions (links). Grey levels correspond directly to link density in this network and hence to the probability of competing in an auction. Top: $\gamma = 0.5$ ordering, bottom: $\gamma = 1$ ordering and middle: consensus ordering of top and bottom. **Right:** Odds ratios of bidding in one of the 32 main eBay product categories for classified users (normalized to maximum value for each cluster). Top: from $\gamma = 0.5$ classification, bottom: from $\gamma = 1$ classification. Spectra with a dashed background (cluster id in parenthesis) show customer purchases $6-9$ months after original classification. See text for details

same cluster index. This overlap is concentrated in cluster 3, parts of which belong stronger to either A or B. Clusters 4 and 5 then fall again completely within cluster B. Clusters 6 and C are practically identical. Cluster D has a number of sub-clusters, the largest of which is 7 and overlaps with cluster E through cluster 8 as before (see arrows again). Group E has two more sub-groups 9 and 10 while clusters 11, 12 and 13 fall entirely within clusters F, G and H, respectively. More details about hierarchical and overlapping cluster structures including some toy examples can be found in [24].

4.2 Cluster Validation, Interpretation and Time Development

To validate the statistical significance and to rule out the possibility the observed cluster structure is merely a product of the clustering algorithm or the particular method of constructing the network from overlapping cliques of bidders, we need to compare the results to those obtained for appropriate random null models [28].

Luckily for us, the type of quality function of (1) is well known to statistical physicists. The modularity Q can be interpreted as the negative of the ground state energy of a physical system called spin glass [29]. It is long known that an intimate relation exists between finding the optimal solution of hard optimization problems such as the Travelling Salesman, Graph Partitioning or, as in our case, graph clustering, and finding the ground state of a spin glass [29]. In particular, with methods from the statistical physics of spin glasses, it is possible to calculate expectation values for the optimal solution of hard optimization problems [30, 31]. Using such results, we can also show that purely random graphs will cluster into equal sized communities and calculate that a random graph with the same number of nodes and links, *i.e.* disregarding the scale free degree distribution and the affiliation network structure of the graph, would yield only $Q = 0.23$ [24]. Additionally, we repeated the maximization of Q also for the RNM version of the bidder network, again taking into account those bidders who took part in at least two auctions. We find a value of $Q = 0.28$ at $\gamma = 1$.

Both values are significantly less than the value of $Q = 0.64$ for the empirical data. Furthermore, as expected, the RNM shows all equal sized clusters, while the real network clearly possesses major and minor clusters. Though this analysis does not yet give a quantitative measure of the statistical significance of the cluster analysis it certainly indicates a strong deviation from the RNM. However, more refined calculations in principle also allow the calculation of the cluster detection accuracy. Instead, here the analysis of the temporal development of the user interests below will provide additional validation.

Until now we have only found groups of bidders that have an increased probability to meet other members of their groups in the auctions they take part in. The eBay product categories are now used in order to find an *interpretation* for the common interests that lead to the emergence of the cluster structure of the bidder network. Since cluster sizes vary and the number of articles in the individual categories is very diverse, we calculate the odds ratios (OR) for bidding in one of the 32 main categories. This odds ratio is defined as

$$OR_{Cs} = \frac{P(\text{bidding in } C|\text{member of cluster } s)}{P(\text{bidding in } C|\text{not member of cluster } s)}, \tag{2}$$

i.e. the ratio of the odds of bidding in category C, given a bidder is member of group s vs. the odds of bidding in category C given the bidder is member of any group $r \neq s$. The right hand side of Figure 2 shows a graphical representation of the odds ratios for clusters A through H and most of the clusters 1 through 13. All spectra are normalized to their maximum value. The exact non normalized numerical values can be found in Table 4. Clusters from the $\gamma = 1$ assignment are more specific with less entries in the category spectrum and larger ORs.

Cluster A unites bidders interested in articles listed in the baby, beauty, fashion, books, movies and music category. Cluster 1 then represents a more specifically content oriented user group mainly interested in books, movies and music. As we have seen, cluster 1 is an almost complete sub-cluster of A. Cluster 2 is also a complete sub-cluster of A and encompasses bidders mainly interested in cosmetics and fashion.

Cluster B contains two sub-clusters 4 and 5, both annotated in the toy model category. Closer inspection, however, reveals that cluster 4 is mainly characterized by its interest in model railways while the bidders in cluster 5 have a passion for model cars, radio controlled models, slot cars and the like. Note the advantage of clustering based on single articles. The clusters we find with one simple unbiased method combine top level categories as in the case of cluster 1 or can only be described be resorting to sub-categories as in the case of clusters 4 and 5. From the left part of Figure 2, we had observed that cluster 3 is responsible for a large part of the overlap between clusters A and B. We see that users in this group 3 have their main interests in the baby and toy category. The overlap of cluster A and B is hence mediated via the toy category. Members of cluster A and B mainly meet in toy auctions. The interpretation of the other clusters is then equally straightforward.

Bidders in clusters C and the practically identical cluster 6 take interest in audio equipment and instruments. Cluster D represents bidders with an inclination to collecting, their bids being placed in the antiques, jewelry, stamps and coins category (cluster 7). The bidders in cluster E are mainly shopping for technological gadgets, computers, consumer electronics, software, mobile phones, PDAs etc. (clusters 9 and 10). Their overlapping interest with bidders from cluster D is in items from the photo category (cluster 8). In groups F and 11, we find predominantly practically oriented users who place their bids mainly in the categories of automotive spare parts, business and industry (where a lot of tools and machinery are auctioned) and do-it-yourself. Finally, in groups G and 12 we find event oriented customers with strong bidding activity in the tickets and travel category and in group H and 13, we find people bidding on sports equipment.

Let us now focus on the time development of the user interests. The data for our analysis was collected during only a relatively short time span (25 days) and we base our results on an extremely sparse data set. Remember that every bidder in the network took part in only 3 auctions on average. Is it really possible to predict meaningful patterns of consumer interest from such sparse data? One could further argue that the few most active bidders account for a large portion of the bids, thus holding the network together and "defining" the clusters of interest, because they also contribute a large number of links. In order to address this question, we revisited the data set in the beginning of September 2005, more than nine months after our original study. From the 6 largest clusters of the $\gamma = 1$ ordering, we uniformly and randomly sampled $10,000$ users each. Note that this removes possible bias towards very active users, they are now represented in the data according to their proportion in the population. Then we looked at the trading history of these users as far back as eBay permits - 90 days. For these $60,000$ users, we determined the product categories of the articles they had bought between June and September. Again, we calculated the odds ratios, this time of *buying*, *i.e.* winning an auction, from a particular category and with the new sample of users as basic population. The results are shown on the right hand side of Figure 2 with a dashed background and the cluster id from which the users were sampled in parenthesis. The stability of the interest profiles is quite remarkable. The main interests have remained unchanged as compared to the initial study though in some cases the spectrum has become more diverse. For instance the content oriented bidders of cluster 1 now also show increased buying activity in the PC-games and tickets category. At the same time the main interest has shifted from movies to music. The largest number of product categories with increased odds of bidding in this category is found for cluster 9, the members of which are the most technology affine users anyway and which would be expected to satisfy a very broad range of consumer needs from online vendors. The members of cluster 7 (the collectors) and cluster 4 (the toy model builders) are much more conservative and almost do not change their profile at all. Without second hand data about the age structure of the bidders classified, we can only speculate that these clusters are formed by older customers who tend to stick to particular categories.

5 Conclusion

Employing a recently developed network clustering technique, we have presented a detailed study of the user behavior on the online auction site `www.ebay.de` during the pre-Christmas season of 2004. Fat tailed distributions of user activity in terms of the number of articles sold, bought, and bid on were found. The attractiveness of articles, measured in terms of the number of bidders participating in an auction, shows an exponentially decaying distribution. Focussing on the bidding behavior, we constructed a network of bidders from their competition for single articles. Nodes in the network correspond to bidders and links to the fact that these bidders have expressed a common interest in at least one article. Studying the general statistical properties and comparing to appropriate random models, we find clear indications for a non trivial cluster structure. This cluster structure, its hierarchy and overlap was studied using a community detection algorithm. Our analysis did not need the definition of any kind of similarity measure between articles or product categories. Rather, we solely used the taxonomic information about articles provided by eBay to interpret our results. We can classify 85% of the users into only a small number of well separated, large clusters, all of which have a distinct profile of only a few main interests as revealed by annotating the articles in the taxonomy of product categories. Some of the clusters show sub-clusters or overlap with other clusters. The interest profiles we identified are remarkably stable. Sampling randomly from the clusters and checking, what these users bought during a three month period in the summer 2005, we found that the profiles of articles bought were almost identical to those from the classification 6 months earlier.

This is striking because virtually everything is offered on eBay and one would expect users to satisfy a much broader range of shopping interests. However, it appears that the major clusters mainly correspond to people's favorite spare time activities. We believe the apparent stability of users' buying and bidding behavior reflects the permanence of their interests which is also stabilized by their social environment and activities. The clear signature in the market data may stem from the fact that users tend to buy online only articles where they have some experience and expertise in. Users seem hesitant to bid on articles from categories in which they have not previously bid in. This may be due to the fact that inexperienced users cannot judge what is a fair price for an article in an auction and they have difficulty assessing to what extent the article offered really suits their needs. At the same time, user interests are reinforced by online recommender systems [32, 33], which suggest similar articles to those already bought by the user. Here lies enormous growth potential for the auction site by extending the way in which people use the site, *i.e.* to facilitate buying goods the user has no experience with and the introduction of "eBay Express" or the "Buy it now" option are developments in this direction. It would be interesting to study how the interest profiles differ between auctioned items and those acquired via the "Buy it now" option.

The analysis presented also shows a way of how to detect possible routes for extending the way people use the online auction site. As we have seen, people mainly interested in collectors items such as stamps or coins tend to buy from the photo equipment category where they compete with those users interested primarily in electronics equipment. Hence, it may make sense to try to also market technological gadgets to users classified as collectors and vice versa.

Clearly, the present investigation can yield valuable information for the design of targeted marketing campaigns. We note that the proposed clustering algorithms may be implemented easily in a distributed way and allows an online monitoring of cluster structures which further allows to study the growth or shrinkage of individual interest groups while the trading progresses. However, such analysis would necessarily have to be installed at the back end of an online vendor. Careful analysis of possible time patterns may also be used to detect trends in the market at an early stage.

One of the major questions that remain is whether the interest profiles found change on a larger time scale as users get older or whether the profiles remain relatively stable and users undergo several transitions of interest over time. Our everyday experience provides evidence for both, but to what extent the two mechanisms are present could only be quantified through a long time study on data sets available form online vendors.

On the other hand, the profiles found and their temporal stability corroborate the hypothesis that the presence of latent interest profiles in the society per se leads to the emergence of user groups with common interest. Transparent markets such as online auction sites in which users act independently and anonymously are perfect starting points to tap into this collective behavior and methods from physics can provide the tools leading to its understanding.

A Cluster Parameters

Table 3. Summary of basic parameters for the major communities found in the bidder network (annotated as in Figure 3). N denotes the number of bidders in the cluster, $\langle k_{in} \rangle$ and $\langle k_{out} \rangle$ the average numbers of neighbors within the cluster an in the rest of the network, respectively. By p_{in} and p_{out} we denote the internal and external link density, respectively. The average link density in the network is $\langle p \rangle = 1.9 \times 10^{-5}$

Cluster	N	$\langle k_{in} \rangle$	$\langle k_{out} \rangle$	p_{in}	p_{out}
A	200630	10.2	3.4	5.1E-05	5.0E-06
1	84699	10.3	4.0	1.2E-04	5.0E-06
2	29323	9.0	5.2	3.1E-04	6.0E-06
3	76182	10.1	4.1	1.3E-04	5.0E-06
B	102188	18.6	3.9	1.8E-04	5.0E-06
4	44830	24.6	4.2	5.5E-04	5.0E-06
5	26325	14.2	5.2	5.4E-04	6.0E-06
C	19915	14.1	4.3	7.1E-04	5.0E-06
6	20020	14.5	4.3	7.3E-04	5.0E-06
D	124702	16.5	3.8	1.3E-04	5.0E-06
7	74913	17.2	4.1	2.3E-04	5.0E-06
8	41359	16.8	5.9	4.1E-04	7.0E-06
E	183313	15.4	4.2	8.4E-05	6.0E-06
9	73722	13.4	6.5	1.8E-04	8.0E-06
10	47937	17.5	5.9	3.7E-04	7.0E-06
F	74657	10.5	4.9	1.4E-04	6.0E-06
11	62115	11.1	5.0	1.8E-04	6.0E-06
G	31337	11.0	6.0	3.5E-04	7.0E-06
12	18835	11.8	6.1	6.3E-04	7.0E-06
H	19620	10.0	4.4	5.1E-04	5.0E-06
13	18286	9.9	4.4	5.4E-04	5.0E-06

Ausgesondert

Table 4. Odds Ratios for *bidding* in one of the 32 main categories during the pre-Christmas season 2004. Shown are only values above 1 signifying an increased interest in articles from this category. Values larger or equal 5 are set in bold font. Note how the clusters from the $\gamma = 1$ clustering $(1 - 13)$ are more specfic than those from the $\gamma = 0.5$ clustering (A-H), e.g. there are less categories with an OR larger 1 and those that are deviate stronger from 1. Also note how the overlap between cluster A and B is mediated via the toy category and the overlap between cluster D and E via the photo category. For the six largest clusters of the $\gamma = 1$ clustering, we show the OR of *buying* from the corresponding category during the summer 2005 as explained in the text

Category	A	1	2	3	B	4	5	C	D	7	8	E	9	10	F	11	G	H
Music	**5.6**	**8.6 (12.0)**						1.5										
Books	2.7	3.4 (2.5)		(1.1)														
Movies	**12.1**	**22.1 (9.2)**																
Beauty	**5.6**	(1.5)	**29.3**	(1.3)									(1.7)					
Baby	**13.7**			**41.0 (20.0)**														
Toys	1.3			**14.2** (4.9)	**5.0**	(1.6)	1.1											
Toy Models					**70.2**	**21.9 (28.0)**	**10.2**											
Audio		(1.3)						**16.2**			1.2	3.4	2.1(3.9)		1.0	1.4 (2.3)		
Instruments		(1.5)				(1.9)		**144.3**					(2.2)					
Antiques					1.2	1.8 (1.1)			**6.6**	**6.3** (3.7)								
Stamps									**88.6**	**118.0 (18.0)**								
Coins									**66.7**	**88.3 (18.0)**								
Collectibles									1.6	1.7 (2.6)								
Jewelry	1.1		1.3						4.5	**5.3** (3.1)								
Photo						(1.2)			4.4		**47.9**	2.8	1.5 (3.2)			(1.4)		
Computer												**19.6**	**14.7 (12.0)**	3.1		(1.4)		
Cons. Electr.												**6.7**	**16.1** (4.3)		2.0	2.5 **(5.9)**		
Mobile												**25.2**	1.3 (4.7)	**33.0**		(2.1)		
Games		(2.2)		(1.1)								3.3	2.3 (3.4)	1.4				
Software		(1.2)									1.5	**5.1**	2.7 (4.0)			(1.5)		
Business													(1.1)		**21.5**	**16.9 (15.0)**		
DIY													(2.2)		**17.9**	**19.1 (7.0)**		
Motors							1.4					1.5	(1.6)		8.6	**8.2 (8.9)**		
Travel		(1.2)								(1.7)			(1.4)				**58.2**	
Tickets		(2.7)		(1.7)									(1.7)				**42.9**	
Sports		(1.3)		(1.3)									(2.0)			(1.9)	1.4	**16.0**
Office		(1.1)										2.3	1.1(2.4)		1.9	1.6 (2.0)		
Deli Food		(1.2)											(1.5)			(2.0)	**9.0**	
Household												3.7	3.3 (4.1)		1.1	(1.8)	1.8	
Fashion	1.9		2.7	1.7 (4.1)									(1.2)				1.1	2.3
Furniture		(1.0)	1.1	(1.6)					1.3				(1.6)				4.2	
Animals		(1.2)	1.0	(1.3)								2.0	(1.9)			(1.4)		

References

1. Special report on ebay. The Economist **375** (2005) 8430
2. Yang, I., Jeong, J., Kahng, B. and Barabási, A.–L.: Emerging behavior in electronic bidding. Phys. Rev. E **68** (2003) 016102
3. Ockenfels, A. and Roth, A.E.: Late and multiple bidding in second price internet auctions: Theory and evidence concerning different rules for ending an auction. Am. Econ. Rev. **92** (2002) 1093
4. Yang, I. and Kahng, B.: Bidding process in online auctions and winning strategy: rate equation approach. physics (2005) 0511073
5. Wasserman, S. and Faust, K.: Social Network Analysis. Cambridge University Press, Cambrigde (1994)
6. Newman, M.E.J.: Mixing patterns in networks. Phys. Rev. E. **67** (2003) 026126
7. Jain, A.K., Murty, M.N. and Flynn, P.J.: Data clustering: A review. ACM Computing Surveys **31** (1999) 264–323
8. Arabie, P. and Hubert, L.J.: Combinatorial data analysis. Anual Review of Psychology **43** (1992) 169–203
9. Steinbach, M., Ertöz, L. and Kumar, V.: New Vistas in Statistical Physics – Applications in Econo-physics, Bioinformatics, and Pattern Recognition, chapter Challenges of clustering high dimensional data. Springer-Verlag (2003)
10. Greenacre, M.J.: Correspondence Analysis in Practice. Academic Press (1993)
11. Agrawak, R. and Skrikant, R.: Fast algorithms for mining association rules. In: 20th VLDB Conf., Santiago, Chile (1994) 487–499
12. Hipp, J., Güntzer, J. and Nakhaeizadeh, G.: Algorithms for association rule mining - a general survey and comparison. SIGKDD Explor. Newsl. **293** (2000) 58–64
13. Han, E.H., Karypis, G., Kumar, V. and Mobasher, B.: Hypergraph based clustering in high-dimensional data sets: A summary of results. Data Engineering Bulletin **21** (1998) 15–22
14. Sarwar, B., Karypis, G., Konstan, J. and Riedl, J.: Item-based collaborative filtering recommendation algorithms. In: WWW10, Hong Kong (2001) 285–295
15. Sarwar, B., Karypis, g., Konstan, J. and Riedl, J.: Analysis of recommendation algorithms for e-commerce. In: EC'00: Proceedings of the 2nd ACM conference on Electronic commerce, Minneapolis, Minnesota, USA ACM Press (2000) 158–167
16. Strogatz, S.H.: Exploring complex networks. Nature **410** (2001) 268–276
17. Newman, M.E.J.: Properties of highly clustered networks. Phys. Rev. E. **68** (2003) 026121
18. M. E. J. Newman. Random graphs with arbitrary degree distributions and their applications. Phys. Rev. E **64** (2001) 026118
19. Newman, M.E.J., Watts, D.J. and Strogatz, S.H.: Random graph models of social networks. Proc. Natl. Acad. Sci. USA **99** (2002) 2566–2572
20. Newman, M.E.J. and Girvan, M.: Community structure in social and biological networks. Proc. Natl. Acad. Sci. USA **99** (2003) 7821–7826
21. Newman, M.E.J.: Modularity and community structure in networks. Proc. Natl. Acad. Sci. USA **103** (2006) 8577–8582
22. Danon, L., Dutch, J., Arenas, A. and Diaz-Guilera, A.: Comparing community structure indentification. J. Stat. Mech. (2005) P09008
23. Newman, M.E.J. and Girvan, M.: Finding and evaluating community structure in networks. Phys. Rev. E **69** (2004) 026113

24. Reichardt, J. and Bornholdt, S.: Statistical mechanics of community detection. Phys. Rev. E **74** (2006) 016110
25. Reichardt, J. and Bornholdt, S.: Detecting fuzzy community structures in complex networks with a Potts model. Phys. Rev. Lett. **93** (2004) 218701
26. Guimera, R. and Amaral, L.A.N.: Functional cartography of complex metabolic networks. Nature **433** (2005) 895–900
27. Guimera, R. and Amaral, L.A.N.: Cartographyof complex networks: modules and universal roles. JSTAT (2005) P02001
28. Guimera, R., Sales-Pardo, M. and Amaral, L.A.N.: Modularity from fluctuations in random graphs and complex networks. Phys. Rev. E. **70** (2004) 025101(R)
29. Mezard, M., Parisi, G. and Virasoro, M.A.: Spin Glass Theory and Beyond. World Scientific (1987)
30. Reichardt, J. and Bornholdt, S.: When are networks truly modular? Physica D **224** (2006) 20–26
31. Reichardt, J. and Bornholdt, S.: Graph partitioning and modularity of graphs with arbitrary degree distribution. arXiv:cond-mat/0606295 (2007)
32. Resnick, P. and Varian, H.R.: Recommender systems. Commun. ACM **40** (1997) 56–58
33. Konstan, J.A., Miller, B.N., Maltz, D., Herlocker, J.L., Gordon, L.R. and Riedl, J.: Grouplens: applying collaborative filtering to usenet news. Commun. ACM **40** (1997) 77–87

Managing Autonomy and Control in Economic Systems

Markus Christen[1], Georges Bongard[2], Attila Pausits[3], Norbert Stoop[4], Ruedi Stoop[5]

[1] University Research Priority Program Ethics, University Zürich, Switzerland
markus@ini.phys.ethz.ch
[2] Swisscom IT Services AG, 3050 Bern, Switzerland
Georges.Bongard@swisscom.com
[3] Center for Telematics, Donau Universität Krems, Austria
attila.pausits@donau-uni.ac.at
[4] Department of Physics, ETH Zürich, Switzerland / Vitus GmbH, Fehraltorf
norbert@ini.phys.ethz.ch
[5] Institute of Neuroinformatics, University/ETH Zürich, Switzerland
ruedi@ini.phys.ethz.ch

1 The Management of Economic Systems

The Problem. Management is the process in which specified persons in an economic system guide, direct and influence people's activities and processes, with the aim of efficiently reaching predefined goals. This general notion of "management" is applicable to both microeconomic (e.g. companies) and macroeconomic (e.g. a national economy) systems, although the modalities, under which goals are formulated and managers are selected, certainly differ. Companies may aim at developing new markets based on the company owner's or director's decision, whereas a national economy defines its goals by means of a political process – but both systems rely on organizational structures (e.g. an organizational chart or a legal system) and designated persons (e.g. a team leader or the head of the central bank) in order to influence human, financial, material, intellectual or intangible resources so that the specified goals can be achieved. The embodiment of this process of management is basically a *control task.*

In modern western societies, the design of this control task has to be balanced with a core value that is deeply interweaved with our concept of state and social organization: *autonomy*, the idea of self-government and a person's or an organization's ability to make independent choices. Looking back upon the history of economic reasoning [21] and management [34], wee see that in the western world the principles of autonomy have continuously increased their influence and changed the paradigms of control. Although the embodiment of autonomy and control on both the microeconomic and the macroeconomic level certainly is controversial in the fields of political philosophy [7], national economics [35] and management theory [34], we suggest to understand *management* as the problem of finding appropriate definitions and implementations of autonomy and control, on both the microeconomic and the macroeconomic scale.

A Historical Illustration – Autonomy in Management. We briefly sketch the increased importance of autonomy in the management of economic systems on the microeconomic scale. In western societies, the industrial revolution brought about the emergence of large-scale business with its need of professional managers. Although military, church and governmental organizations provided models of management, the specific economic focus of businesses led in the late 19th century to the development of so-called *scientific management*, that focussed on worker and machine relationships. The task was to economize time, human energy, and other productive resources. The most prominent management model of that time was formulated by Frederick Taylor [28] and Henri Fayol [10]. Their model applied a *strict control regime* upon procedures and methods on each job of the production chain, which led to a tremendous increase in productivity, up to a factor four in the examples provided by Taylor. This optimization of the "human motor" [22] by means of a strict control regime was for a long time the paradigmatic view towards management in the industrialized production society.

This early concept of management was later challenged. Not only the conflict of this model with basic notions of humanity – brilliantly exhibited by Charlie Chaplin's movie "Modern Times" (1936) – was the cause for this change. Also the need of a post-industrialized "knowledge society" to exploit the creative potential of its members in order to be able to faster adopt the production processes, was incompatible with such a strict control regime. Consequently, new behavioral [23], systemic [8] and context-related management approaches were developed – to name just a few. This diversification of the theory of management can be seen as the result of balancing autonomy versus control of individual units within economic systems.

Outline of this Contribution. We intend to demonstrate the interplay of autonomy and control on both the microeconomic and the macroeconomic level, by means of two case studies. We first define our notions of "autonomy", "control" and "economic system". On the microeconomic level, we interpret *reorganizations* as a tool for setting up new control regimes upon productive teams (business units) of large companies. We analyze the influence of a reorganization upon the information-flow network of a business unit, as this network basically underlies the productive power in knowledge-based companies. On the macroeconomic level, we feel that the ability to formulate and communicate a *sufficiently simple control optimality* is a core problem in order to formulate and enforce an economic policy within democratic societies. We investigate this problem merely from a modeling perspective, by demonstrating the effect of a simple control mechanism, the properties of which have recently been fully analyzed [26], in the context of a basic model of an economic system. For both levels of economics, we will demonstrate the superiority of simple control mechanisms that allow for a maximal autonomy within the boundaries set by the control.

2 Autonomy and Control

The Nature of Autonomy. The concept of "autonomy" in the sense of self-governance was originally formulated by the antique Greek city states, aspiring independence from the Persian empire. In the age of Enlightenment, however, autonomy was seen as the ability for self-governance in combination with the commitment of responsibility [14]. In this way, autonomy became a fundamental concept of moral philosophy. Contemporary philosophers differ in their notion of autonomy, depending on whether the individual person (personal autonomy [29]), morality (moral autonomy [24]) or political systems (social autonomy [7]) are the focus of discussion. Although this context-dependency led to a rich differentiation of the concept, it is undisputed that – in the western tradition – autonomy is a basic moral and political value, affecting how individuals interact, and the rights they are provided with. Thus, the discussion on the nature of autonomy is not a mere philosophical debate, but has a large impact on how society is organized. A prominent example is medicine where the bioethical "principle of autonomy" [2] reflects a fundamental change in the relation between doctors and patients, on both the social and the legal sides.

In an economic context, several aspects reflect the importance of this *paradigm of autonomy* (see Fig. 1): On the macroeconomic level, the idea of a "free market" – i.e. the organizational principle that supply and demand of economic goods should be unregulated except for the country's competition policy – may serve as the most prominent example. Implicitly, the concept of a free market assumes that the agents of a market (individuals, organized structures like companies etc.) know best of their needs and goals, and how to satisfy and achieve them. Free trade – a principle that addresses the interaction of national economies – is a second, prominent characteristic of this paradigm. The need of a regulatory framework that protects the practical implementation of these principles – e.g. by property rights – is a third, important characteristic. We are well aware of the fact that these embodiments of the autonomy paradigm are subject to intense discussions, which basically reflects that a control problem lurks behind (see the next paragraph). Nevertheless, we consider free market and free trade in combination with a regulatory framework protecting these types of interactions as the main specifications of autonomy on the macroeconomic scale.

On the microeconomic scale, the autonomy paradigm is basically reflected by personal autonomy. This includes "free will" (the power and ability of

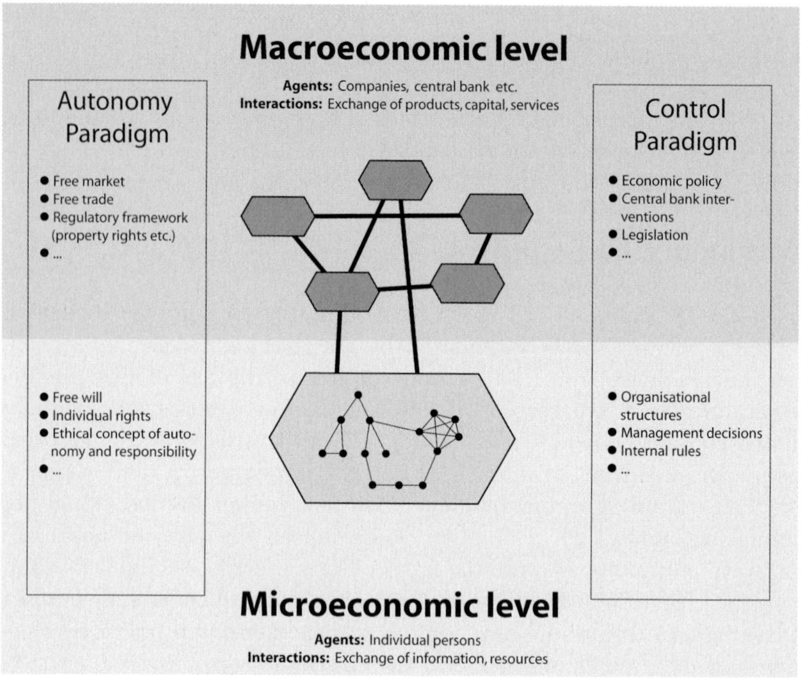

Fig. 1. Autonomy and control paradigms (not concluding lists) on both the microeconomic and the macroeconomic level of social organization

making free choices unconstrained by external agencies), the ethical concept of autonomy (the feature of the person by virtue of which he or she is morally obligated), and a set of basic personal rights (usually implemented on a constitutional level). We will not dwell into the various discussions that accompany these terms. We rather exhibit the basic ingredients that are used in economic systems to implement autonomy on the microeconomic scales (in particular by companies). This non-exhaustive list includes:

- *Interaction autonomy*: Employees are able to freely exchange information in order to solve business-related problems within the boundaries given by the organizational chart.
- *Profit-center organization*: Sub-units within a company may act autonomously in terms of client-relations, budgeting and accounting.
- *Global budgeting*: Central control is only enforced by strategic allocation of financial resources, whereas the allocation within the unit is led to local management.

New Public Management (NPM, [30]) is a prominent example of this new management philosophy. However, NPM is not unchallenged [25], where the deeper reasons for this are again related to the control problem. We thus stress that the requirement to control is the main aspect that shapes the implementation of autonomy in social systems.

The Nature of Control. We suggest to define control as the deliberate use of constraints in order to influence the dynamics of a system such that defined goals are reached. This working definition inspired by control theory [36] certainly has its pitfalls when applied to social system, but it nevertheless outlines major ingredients of the general concept of control. First, control is *deliberate*, i.e. based on some written or verbally manifested decision that includes the goal of the control, naming the system that has to be controlled and the means that are used or acceptable for control. Second, control affects the *dynamics* of a system, i.e. those system variables that have been chosen as relevant for measuring the fulfillment of the control goal. In a macroeconomic system, this could be the money supply controlled by the central bank, in a microeconomic system the output of a production unit. Third, control is implemented in the process characterizing the system dynamics by a *constraint* that requires a certain control effort, as, without the control, the "natural" dynamics would be different. However, we suggest a "liberal" understanding of constraint, which includes, for example, the choice of a certain training program in order to encourage certain goals. Fourth, in order to be able to implement a *goal-oriented* control, a certain degree of "predictive understanding" of the system is required. In economic systems, this requirement is notoriously hard to fulfill, which leads to undesired effects of control due to incomplete system knowledge. We will demonstrate that already in microeconomic systems, the understanding of the system dealt with might focus on wrong aspects leading to undesired control results. The dynamics of macroeconomic systems is even harder to predict. Therefore, we will concentrate our analysis on the application and consequences of control, applied to a model of a chaotic macroeconomics.

We will also not discuss the question of how to find a (good) control goal. In economic systems, however, it is convenient to relate this goal to an *efficient use* of relevant resources – time, capital or material – in relation to the product or service provided. At the macroeconomic level, the same basic parameters are the objects of discussion. Examples are the reduction of working hours, the foreign trade deficit problem or the efficient use of commodities due to ecological reasoning. The control tools appropriate on the two economic levels however, differ. The classic control tools on the macroeconomic level are legislations in order to guarantee certain minimal constraints for the system (e.g. minimum wages), short-term economic policies in order to address current problems (e.g. prize limiters) or monetary interventions of the central bank.

On the microeconomic level, control can be much more sophisticated, which actually powers a tremendous industry of management consultants. This discussion we will basically avoid; we only sketch the main aspects of embodied control. The first control tool of managers is to *constraint the interaction* of employees through a specified organization chart. This important aspect is the main focus of our first case study in the next section. The second control tool is the amount of allowed resources like capital, material, space, or time of collaborators (*input control*). The third tool is real-time control (*process control*) – either by detailed rule-sets (in the tradition of Taylor's scientific management) or by several means of supervision (which could be perverted into a kind of "big brother" control). Finally, control can be implemented by measuring specific results of the sub-unit (*output control*) and followed by a re-arrangement of processes, resources etc., within the unit. In practical life, control by management is usually implemented as a mixture of all four instances of control.

Microeconomic Systems – Business Units. Large companies perform *business processes* within specified organizational units – business units – in order to create products or services that are supplied to the market. The business processes are mapped to the units such that the organizational structure of the business unit optimizes production. The adaptation of the organizational structure of companies through reorganizations is a widely used control tool used by management. The measure to evaluate the effectiveness of a reorganization is efficiency in terms of the time needed to perform business processes [31]. Customarily, the *organizational structure* of the business unit is grasped by the organization chart, where several different forms can be distinguished (e.g. line organization or matrix organization [17]). The coaction of the steps associated with a business process is described by the *operational structure* – e.g. in the form of a flow chart. As a business unit usually performs more than one business process, it is the task of the manager to find a organizational structure that can be mapped in an optimal way to the different operational structures. Thus, we define a *reorganization* as the adaptation of the organizational structure to a new set of operational structures.

Reorganizations result in new constraints for the interaction of the employees of a business unit. In knowledge-based companies, these interactions basically consists of information transfer – text, e-mail, program-code etc. – that form the edges of a *social network* [33]. The structure of this network is crucially dependent of the interaction autonomy the employees have. The structure is also informal in the sense that the personality of the individuals involved lead to implicit optimizations of the operational structure that may not be recognized by members of the senior management. This makes reorganizations a challenging task, as they may influence the social network of business units in an unforeseen way [27]. For example, a person A could have important knowledge in order that person B can perform a specific step of a business process, leading to information exchange (i.e. an edge) between A and B. A reorganization could transfer B to another unit with a less satisfying work, such that B refuses to further collaborate informally with A (as it is "no more his or her business"). In our case study, we will demonstrate how this effect can be quantified in terms of "robustness" of the network and how it leads to unforeseen, negative results due to a wrong control approach.

Macroeconomic Systems – Economic Cycles. Economic booms and bouts affect modern societies strongly, with a direct impact on individuals biographies. In western economies, cycles have been an ubiquitous and undesired observation. Among the most remarkable,Kitchin cycles emerged [16]. These macroscopic variables are the target of control when the economic system is object of investigation. Until the 1970s, as the legacy of John Maynard Keynes [15], cycles were regarded as primarily due to variations in demand (company investments and household consumption). As a consequence, economic analysis focused on monetary and fiscal measures to offset demand shocks. During the 1970s, it became obvious that stabilization policies based on this theory failed. Shocks on the supply side, in the form of rising oil prices and declining productivity growth, emerged as equally crucial for the generation of cycles. In a paper published in 1982, Finn Kydland and Edward Prescott [18] offered new approaches to the control of macroeconomic developments. One of their conclusions was that the control should be kept constant throughout a cycle, in order to minimize negative effects.

Cycles and crises may be inherent to the principles on which our economics is based. However, if they could be predicted and their origin understood, they might be engineered to take a softer course. An extreme form of this control approach was taken in the centrally planned economies in the former socialist countries. This approach failed, as it was not able to aggregate sufficient and reliable information about supply capacities and demand needs, which is necessary for efficient control. Furthermore, in the western tradition, the strong and ubiquitous control approach of socialistic economies is not compatible with our basic notion of autonomy on the personal as well as on the social level. In order to deal with the control problem of macroeconomic systems in western democratic societies, it is necessary to be able to communicate a sufficiently simple optimality policy. For obtaining it, the understanding of the response to control in simple economical models may provide important guidelines. As the number of variables that govern a macroeconomic system is vast, one is confronted with a hard prediction problem. We suggest, that the prediction problem of macroeconomics is related to the one in chaotic processes, where strategies for overcoming it have been developed. Although the question of to what extent real economies are classified as chaotic can readily be disputed, low-dimensional chaotic models yield insight into the mechanisms that govern the response of economics to control policies.

3 Control in Toy Models

3.1 Microeconomic Level: Reorganization and Robustness

Definitions We use the social network paradigm in order to understand the effect of reorganizations. Thus, we describe a *business unit* as a network, where the nodes represent employees and the edges represent information transfer. The main concepts are defined as follows:

- *Business process* P: A sequence $\{p_1 \dots p_n\}$ of n processing steps associated with a specific product.
- *Business unit* $B(k)$: A social network of k nodes associated with a class $\{P\}$ of business processes. The part of $B(k)$ that performs processing step p_i is called *process unit* B_{p_i}.
- *Process operating expense* $E_P = \sum_i e(p_i)$: The sum of the times $e(p_i)$, associated which each p_i, needed to perform P.
- *Process runtime* T_P: The total time from initiation to completion of P.
- *Robustness* $R(l)$: Defined as $R(l) = 1 - I_P(l)$, where $I_P(l)$ is the probability of process interruption in dependence of the relative fraction of node outage of a business unit B (l/k, where l is the number of nodes that turned out).

Note that we distinguish E_P and T_P, because employees can be absent (due to illness etc.), possibly leading to an interruption of P if no redundance is implemented in the network. T_P is an estimate of the efficiency of P. We have $T_P \geq E_P$, as a temporary outage of a B_{p_i} increases T_P.

Defining Robustness. Robustness refers to the ability of a network to avoid malfunctioning when a fraction of its constituents is damaged [4]. One differs between *static robustness*, the influence of deleting nodes without redistribution of information flow, and *dynamical robustness*, which takes the latter into account. In our case study, dynamic robustness basically reflects the degree of interaction autonomy the employees have. Thus, it is related to the informal networks that are formed in the business unit within the boundaries given by the organizational chart. Dynamical robustness is usually warranted only *within* a process unit B_{p_i} that is formed by a sub-set of employees of the business unit. We therefore calculate the robustness of our business unit as the static robustness of the network of process units. This probability is calculated using the hypergeometric distribution (see appendix).

Robustness alone does not account for the *relevance* of process interruption for process runtime. For example, longer downtime of a node may have cumulative effects on runtime. To model this effect in a simple way, we weight the probability of process interruption by a factor that accounts for the additional time that prolongs process runtime. By changing this weighting factor we can analyze the parameter space spanned by this factor and the relative fraction of node outage.

To demonstrate our approach, we investigate a reorganization in a toy example (Fig. 2). Here, the management intends to concentrate a business unit $B(k)$ on its core business, by reducing the number of business processes P from three to two and by releasing one employee in each B_{p_i}. We assume that $e(p_i)$ for a specific P is reduced from 3 to 2, as the number of p_i that have to be processed in parallel by each B_{p_i} decreased, which reduces friction losses. Thus E_P decreases from 9 to 6. However, when the decrease of robustness of the social network (Fig. 2.c) is taken into account, depending on the weight of process interruption, the reorganized $B(k)$ may be less efficient than before (Fig. 2.d). In the bright region of parameter space, the organizational structure

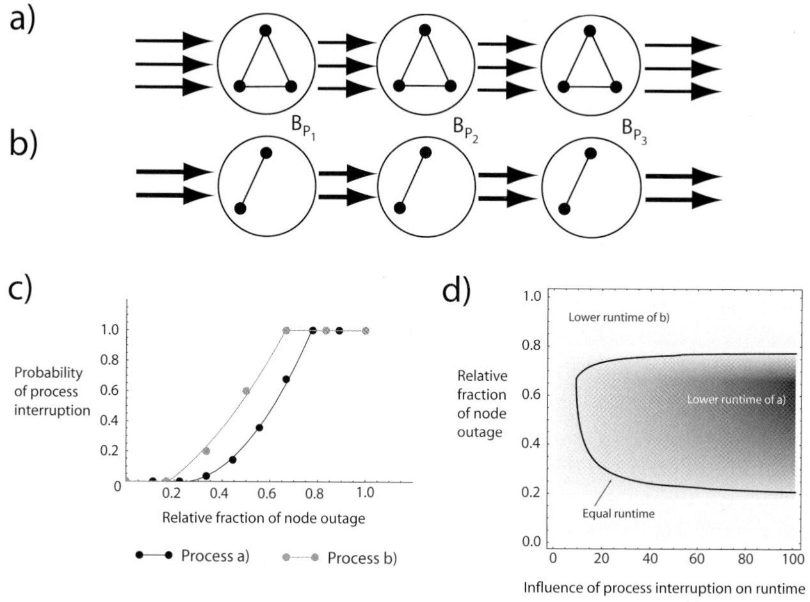

Fig. 2. a) Three business processes P incorporated in a business unit B consisting of three process units B_{P_i} of three employees each. **b)** B after reorganization: the number of business processes has been reduced from three to two and in each B_{P_i} an employee has been released. **c)** Robustness before and after reorganization for a single P. The increase of interruption probability (= decrease of R) is approximated by quadratic fit-functions. **d)** Process runtime in dependence of the weight of process interruption and the relative fraction of node outage (the darker region identifies the parameter space where P in the organizational structure of a) is processed faster compared to the structure of b)

b) is more efficient in performing a business process P, whereas in the darker region, the structure a) is superior to b).

3.2 Macroeconomic Level: Controlling the Dynamics

Logistic Economy. When exponential growth is possible, real economies have little problem. But when the limits of the economic systems are reached, their prediction becomes difficult. From the mathematical point of view, this is due to the nonlinearities that are required to keep the system within the boundaries. Economies naturally tend towards the recruitment of all available resources – which can be interpreted as a unrestricted exhaustion of the autonomy paradigm. This drives the macroeconomy towards the boundaries and fosters a natural tendency of the system to evolve towards maximally developed nonlinearities. We can describe economics in a simplified and abstract way in terms of a parameter indicating the degree of globalization of resources (nonlinearity parameter a), a dynamical parameter x expressing a generalized consumption, and a noise parameter r modeling short-term fluctuations, which are often of local or external origin. Thus, the evolution of this simple model of economics takes place on three timescales: a slow one which modifies parameter a, an intermediate term variable x that is assumed to be deterministic, and momentary perturbations that are included in x in the form of noise. The underlying deterministic system is defined by the property that for states far from full exploitation of the resources, the consumption can grow almost linearly. Close to maximal exploitation, the next consumption is required to be small, to let the system recover. Over a large parameter range of small a (local economics), this behavior, however, is avoided and a state of quasi-constant consumption emerges. A most simple and generic setting for modeling this dynamics is provided by the iterated logistic map (see Fig. 3), whose mathematical properties are explained in the appendix.

A simple illustration of this type of economics is whaling in the Northern Atlantic Ocean. If the whaling fleet is small (captured by $a \ll 1$), the annual catch x_n will be small and affect the whale population little, so x_n will stay at a quasi-fixed point. An increase of a will raise the average catch $\bar{x}$. Larger ships will start venturing to the whole of the Atlantic Ocean. At

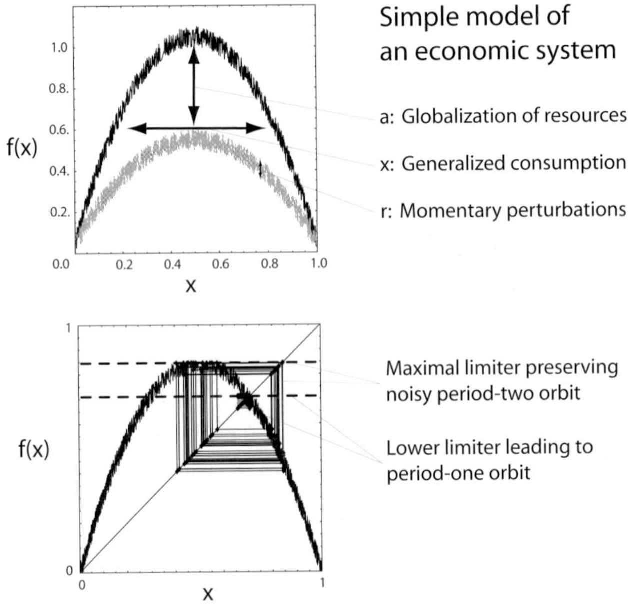

Fig. 3. a) Illustrating the three parameters of a logistic economy (for mathematical details please consult the appendix). **b**) Hard limiter control for the noisy logistic map. Placement of the limiter around the maximum of the map preserves the natural noisy period-two orbit (*black*). For lower placement, a modified period-one behavior is obtained (*grey*)

the point when we start to exploit a considerable part of the whole system ($a \to 4$), the fixed-point behavior naturally ceases to hold. After a situation of almost complete exploitation ($x_n \approx a/4$), the system needs an extended time to recover. Novel technologies may annihilate the constraints that originally defined the confinement to the unit interval. The universality underlying the above-discussed route to ever more complex dynamical behavior, however, implies that under the new constraints, the whole process will repeat, leading to a cascade of such processes.

Controlling Chaos. The logistic map is a generic example of a chaotical system. From a dynamical system's point-of-view, chaos is composed of an infinite number of unstable periodic orbits of diverging periodicities. In order to exploit this reservoir of characteristic system behavior, methods to stabilize (or control) such orbits using only small control signals have been developed and applied mostly in electronic system. These practical applications often require that the orbits are quickly targeted and stabilized. For the classical Ott-Grebogi-Yorke [20] and for feedback control, this is a problem. Recently, Corron and coworkers [9] introduced a new control approach (termed control by simple limiters) and suggested that it could overcome the limitations of the previous methods. The general procedure can be summarized as follows: An external load is added to the system, which limits the phase space that can be explored. As a result, orbits with points in the forbidden area are eliminated. The authors also observed that modified systems tend to replace previously chaotic with periodic behavior. The mathematical properties of this type of chaos control has recently been fully described (see appendix for more details).

A variety of economic models are based on the logistic approach, as it generically implements the dynamic effects of shortage of goods – thus catching the core problem of economy [3]. Therefore, our approach intends to analyze the effect of simple limiter control upon such models. In economic terms, simple limiter control is realized for example by prize limiters (minimum or maximum price levels). Such an economic policy is indeed simple and must not necessarily contradict the paradigm of autonomy. Benefits and pitfalls of this type of control are discussed in the next section.

4 Empirical and Modeling Results

4.1 Inefficiency of Hierarchic Business Information Networks

Characteristics of the Microeconomic Case Study. We investigated the IT division of a Swiss telecommunication company that develops products for telecommunication and insurance companies [5]. The division had 1400 employees before, and 1250 employees after reorganization. We focused on an business unit within this division that performed four classes of P: project management, application development, operations and maintenance of IT services. Before reorganization, $B(k)$ (23 employees) was organized as profit center and acted autonomously on the marked in terms of client relations, budgeting and accounting (mean turnover: 16.5 Mio. CHF, profit: 2 Mio. CHF). This large degree of autonomy of the unit led to a small-world social network [32] with a low characteristic path length $L = 2.05$ and a high clustering coefficient $C = 0.92$. The business unit contained three hierarchical levels and two persons supporting the unit managers in administrative needs (Fig. 4.a).

In 2002/2003, the whole division was subject of a reorganization. The reorganization intended to separate the different P more clearly and to map them on more precisely defined B_{p_i} in order to increase efficiency in terms of E and T_P. Furthermore, the reorganization aimed to increase the control on the beforehand autonomously acting business units, as competition between the units within the division sometimes led to the situation, that external customers obtained different tenders for the same product and could choose for the best solution within the division. Thus, after the reorganization, the information flow within the business unit was much stronger restricted, leading to a hierarchical network characterized by a characteristic path length that was more than doubled ($L = 4.72$) and a strongly reduced clustering coefficient ($C = 0.07$) (Fig. 4.b). Furthermore, more unit managers and additional

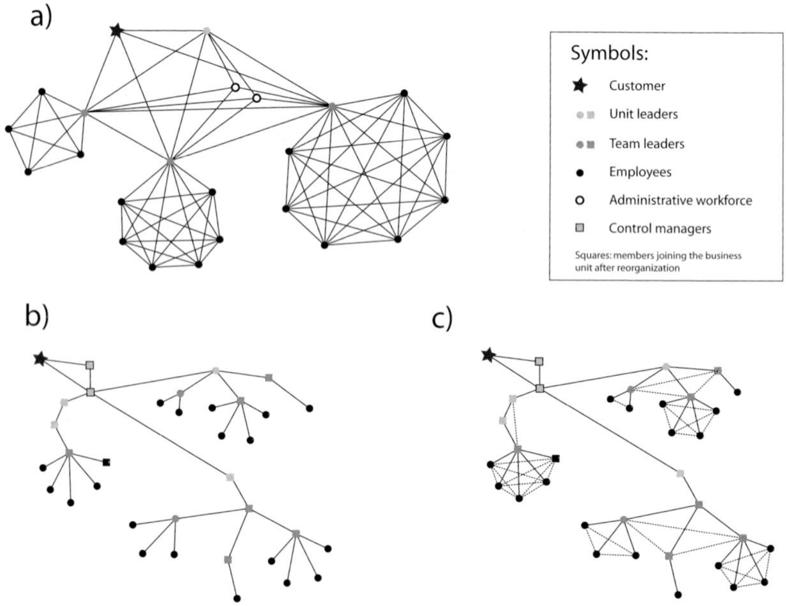

Fig. 4. Information flow network of a business unit before **(a)** and after **(b/c)** reorganization. Quadratic nodes in b) and c) indicate new members joining the business unit after reorganization. Dashed lines in c) indicate informal information transfer emerging within the teams

control managers taking care of customer relation were included, such that the business unit contained 34 employees after reorganization (the administrators have been released). After some time, informal information transfer emerged within the teams, leading to an increase of the clustering coefficient ($C = 0.73$), but not strongly affecting the characteristic path length ($L = 4.33$) (Fig. 4.c).

Day-to-day experiences of the employees of $B(k)$ aroused the suspicion that T_P increased significantly after reorganization. This phenomenon was investigated for several classes of P performed by the unit by determining E and T_P empirically [5]. Although both parameters could not be measured precisely due to comparability issues, valuable estimations could be gained. In the following, we focus on project management processes, where the most trusted results have been obtained.

Explaining Inefficiency as a Result of Decreased Robustness. Project management processes are performed within the business unit according to general procedures. In the IT company we investigated, the procedure emerged out of the so-called Hermes-method – a standard procedure that has been implemented in the late 1970s in the large public enterprizes of Switzerland [13]. This widely distributed standard has been used by the business unit before and after reorganization, so that basic comparability is given. One task of the unit was to develop tender offers for large IT projects. Whereas realization and implementation of such projects largely depend on their individual character, the tender phase was much more uniform, allowing to compare process operating expense and process runtime before and after reorganization (the details of the measurement process are outlined in Ref. [5]).

We find that, in the mean, E_P slightly decreased after reorganization, whereas T_P considerably increased (Table 1) – confirming the general impression of the employees. This observation becomes explicable when determining the change in robustness of the social network (Fig. 5). Before reorganization, basically two B_{p_i} with in total 11 employees were involved in the process. After reorganization, the project management process was separated into a system engineering branch and an application development branch, where five B_{p_i} with in total 16 employees were involved. As Fig. 5 demonstrates, the project management process after reorganization is much less robust compared to the process before reorganization. Interestingly, even the larger number of employees involved in the process after reorganization (16 instead of 11) does

Table 1. Mean process operating expense E and process runtime T_P before and after reorganization for the project management process

	E_P	T_P
Before reorganization	88 hours	19 days
After reorganization	86 hours	35 days

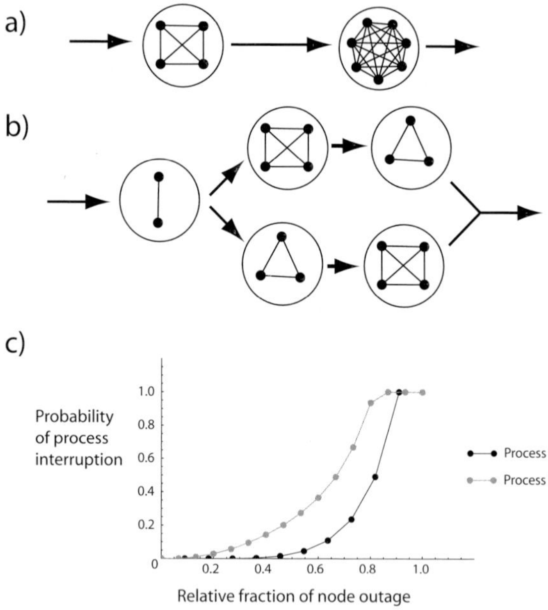

Fig. 5. Project management process before (**a**) and after (**b**) reorganization of the business unit performing a project management process. (**c**) Decreased robustness of the network after reorganization (*grey*) that elucidates the empirically measured decrease in process runtime

not increase the robustness – independent whether the relative fraction of node outage or the absolut number of turned out nodes is taken as reference.

4.2 Efficiency of Hard Limiter Control

Effects of Simple Control. We investigate the effect of controlling the simple model of economics introduced in the previous section by placing a limiting value on x that the system is not allowed to cross (hard limiter control). In Fig. 6, three time series generated from this model are displayed. For the first series, the system was tuned so as to generate a noisy, superstable period-four orbit. For the second series, a limiter at the highest cycle point was inserted, whereas in the third series the control was on the unstable period-one orbit. It is easily seen that the period-one orbit yields the highest average value. In an analysis that is mathematically more involved, it can also been shown that implementing the limiter at maximal system response is generally a suboptimal solution [6]. We found that the system average is generally optimized by controlling a period-one cycle. In the presence of a substantial amount of noise, only low-order cycles can be controlled.

The *efficiency* of this control approach emerges in two aspects: First, hard limiter control yield higher averages of the generalized consumption x. Second,

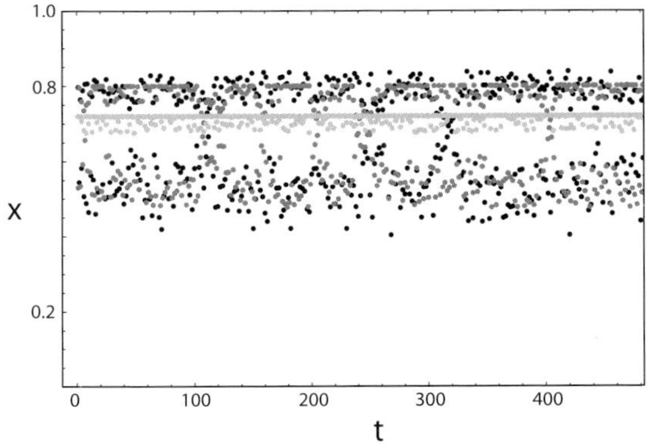

Fig. 6. Time series of a superstable period-four orbit: uncontrolled (*black*), controlled in the maximal cycle point (*dark grey*), and controlled in the unstable period-one orbit (*light grey*). The period-one orbit yields the highest average of x (t: number of iterations)

by means of the applied control, the system predictability increases (i.e. control leads from chaotic to periodic, or from higher periodic to lower periodic orbits). Both aspects are interrelated, as the "simplest" system dynamics (period-one cycle) also yields the highest average. Latter, however, must not necessarily be of positive value from a point of view of political economics. Hea and Westerhoff applied this type of limiter control to a model of commodity market, basically reproducing our result [12]. Both the implementation of a bottom price level (to support producers) or a top price level (to protect consumers) can reduce market price volatility – thus increases the predictability of the system. But as the variable x represents prices in their context, a maximum price limit increases the average price, which can be considered as a unwanted side-effect.

5 Summary and Outlook

We characterized management as the task to balance the autonomy of the constituents of an economic system and the control of the dynamics of the system in order to reach predefined goals. Our case studies provide inside of benefits and pitfalls of this understanding of management on both the microeconomic and the macroeconomic level. The first case study can be characterized as the analysis of a *failure* of management through implementing a new control structure in a business unit by means of a reorganization. We have shown that robustness, determined in terms of how business processes are affected by an outage of nodes in the information flow network, can be a critical parameter that tends to counteract the intended gain in efficiency by reorganizations. The example demonstrates that reorganizations focussing on efficiency by optimizing the division of labor and by increasing the control has the effect that the information flow network looses the small-world property. This property result from the interaction autonomy of the employees within the business unit, which obviously had a beneficial effect on the efficiency of how the processes within the unit have been performed, although the system looked rather complex from an external point of view. In our real-world example, these negative effects have been recognized by the senior management and the reorganization has basically been retracted.

But how should this control problem be solved? One possibility would have been to change the perspective towards the result of internal competition within the division, i.e. by benefitting the successful profit centers and by focusing on those units that were less productive. Alternatively, the concept of a "hard limiter", suggested by the macroeconomic case study, could have been implemented – for example by defining that tender offers are not allowed to surpass a certain minimal amount. This would indeed be a simple control policy, not touching interaction autonomy within the units.

In macroeconomies, the autonomy paradigm leads to a tremendous complexity in number and types of interactions between the agents. Although it has been found that for either very underdeveloped or mature economies, stable fixed-point behavior is predominant [1], at an intermediate level, complex economics emerges that can induce chaotic dynamics – e.g. leading to large fluctuations of entrepreneurs wealth W_n [19]. There, a control task for economic policy could emerge. Our model suggests, that hard limiter control in the form a tax on assets with a sufficiently fast progression could be applied, forcing W_n to remain below a maximal value, $W_{\max}$. With sufficient care, control on a period-one system could be achieved, and excessive economic variations due to chaotic dynamics could be prevented. Political realizability will often require the use of "softer" limiters (in the sense that $W_n > W_{\max}$ is not strictly prohibited), but the main features of hard limiter control will be valid even in these cases.

We emphasize that control mechanisms of limiter type are indeed common in economics. This control, however, inherently generates superstable system behavior, whether the underlying behavior be periodic or chaotic. Political activism may suggest a frequent change of the position of the limiter to be a suitable strategy in order to compensate for the amplified or newly created cyclic behavior. This strategy, however, will only result in ever more erratic system behavior. Our analysis shows that it is advantageous to keep the limiter fixed, adjusting it only over timescales where the system parameter a changes noticeably. In this way, reliable cycles of small periodicity should emerge. Among these cycles, the period-one cycle appears to be the optimal one, from most economic points of view. To recruit this state, a strong initial intervention is necessary and the control should be permanent. Otherwise, a strong relaxation onto the suboptimal natural behavior sets in. In discussions of real economics, these effects will be natural arguments against the proposed control. To overcome such arguments, a sufficiently simple control policy must be formulated in democratic societies.

Unfortunately, the "control trend" in most western societies goes into a different direction, as an increasing body of legislation in combination with a decreasing ability to enforce these rules is observed [25] – a control strategy that we consider as being the contrary of hard limiter control. In this way, a regulatory system interfering in a hardly predictable way on many levels of social organization emerges, that affects beneficiary effects of the autonomy paradigm. Our empirical investigations of a microeconomical problem as well as the theoretical analysis of a macroeconomical model suggest, that a control regime in order to manage the behavior of economic systems should be simple but enforceable. Or course, the central problem of how to find the appropriate goals of management is not addressed by this argument. But as soon as those are defined, simple control mechanisms that allow maximal autonomy within the control's boundaries should be implemented.

Appendix

Calculating the Robustness of Networks

We define the robustness of a business unit as the static robustness of the network of process units. We consider each process unit (r members) in a business unit (k members) separately. We have to calculate the probability that from a number l of nodes that fail in the network those x nodes fail that interrupt the process by means of the hypergeometric distribution. As $x = r$ in our case (i.e. the complete process unit has to fail), the effect of a single B_{p_i} on the probability of process interruption caused by an outage of l nodes is calculated as

$$I_{B_{p_i}}(l) = \frac{\binom{r}{x}\binom{k-r}{l-x}}{\binom{k}{l}} \overset{x=r}{=} \frac{\binom{k-r}{l-r}}{\binom{k}{l}} \tag{1}$$

$I_{B_{p_i}}(l)$ is calculated for all l up to a value where at least one B_{p_i} fails definitely (i.e. the probability of process interruption is one – this depends on the network topology) and for all B_{p_i}. Basically, $I_P(l) = \sum_n I_{B_{p_i}}(l)$ applies – but one has to take into account that specific constellations of node-outage may possibly be counted twice (this, again, depends on the network topology). These cases have to be identified and incorporated when calculating $I_P(l)$. In this way one obtains – for each given $l = 1 \ldots n$ – the probability of process interruption $I_P(l)$ and thus the robustness $R(l)$. Due to the dependence of R on the network topology, no general analytic formula for $R(l)$ can be provided.

Mathematics of Hard Limiter Control

Our economic model contains a parameter indicating the degree of globalization of resources (nonlinearity parameter a), a dynamical parameter x expressing a generalized consumption, and a noise parameter r modeling short-term fluctuations. Whereas in the case of small a such perturbations are stabilized by the system itself, for larger a they lead to ever more long-lived erratic excursions. To incorporate these fluctuations within our model, we perturb x with multiplicative noise, for simplicity chosen uniformly distributed over a finite interval. The size of the noise sampling interval denoted by $\bar{r}$, is a measure for the amount of noise. The most simple and generic setting for modeling this dynamics is provided by the iterated logistic map.

$$f : [0, 1] \rightarrow [0, 1] : \quad x_{n+1} = ax_n(1 - x_n) + r \tag{2}$$

The self-organization towards an ever-growing exploitation of the phase space [0, 1] is reflected in a slow increase of the order parameter a towards $a = 4$. At $a = 4$, it can easily be seen how the nonlinearity keeps the "orbits" x_n away from the boundary: starting with small values, x_n increases almost linearly (with factor a). As soon as x_n approaches the upper phase-space boundary (at $x_n = a/4 = 1$), this is counterbalanced by the factor $1 - x_n$. If a is increased further, large-scale erratic behavior sets in, as the process is no longer confined to the previously invariant unit interval. After a potentially chaotic transient, the system settles into a new area of stability, where the same scenario takes place anew, starting at rescaled small a. We believe that in particular the effects of technical shocks may be adequately described in this framework. On its way towards the globalization of resources ($a \rightarrow 4$), the system undergoes a continued period-doubling bifurcation route, where a stable period-one solution is transformed, over a cascade of stable orbits of increasing orders 2^n (where $n = 2, 3, 4 \ldots$), into a chaotic solution (the Feigenbaum period-doubling cascade [11]). Our model is characteristic for the whole class of systems that are subject to such a process of self-organization.

By introducing a limiter, orbits that sojourn in the forbidden area are eliminated. Modified in this way, the system tends to replace previously chaotic with periodic behavior. By gradually restricting the phase space, it is possible to transfer initially chaotic into ever simpler periodic motion. When the modified system is tuned in such a way that the control mechanism is only marginally effective, the controlled orbit runs in the close neighborhood of an orbit of the uncontrolled system. This control approach was successfully applied in different experimental settings. The properties of the control method are fully described by the one-parameter one-dimensional flat-top map family, implying that orbits are stabilized in exponential time, independent of the periodicity and without the need for targeting. Fine-tuning of the control is limited by superexponential scaling in the control space, where orbits of the uncontrolled system are obtained for a set of zero Lebesgue measure. In higher dimensions, simple limiter control is a highly efficient control method, provided that the proper limiter form and placement are chosen [26].

In applications, the time required to arrive in a close neighborhood of the target orbit is an important characteristic of the control method. With the classical methods, unstable periodic orbits can only be controlled when the system is already in the vicinity of the target orbit. Hard limiter control renders targeting algorithms obsolete, as the control-time problem is equivalent to a strange repeller-escape (control is achieved as soon as the orbit lands on the flat top). As a consequence, the convergence onto the selected orbit is exponential. These properties of 1D hard limiter control systems fully describe the effects generated by the limiter control. Due to the control, only periodic behavior is possible.

References

1. Aghion, P., Bacchetta, P. and Banerjee, A.: J. Monetary Econ. **51** (2004) 1077
2. Beauchamp, T.L. and Childress, J.F.: Principles of Biomedical Ethics, 4th edition. Oxford University Press, Oxford (1994)
3. Benhabib, J. and Day, R.J.: Rev. Econ. Stud. **48** 459 (1981)
4. Boccaletti, S., Latora, V., Moreno, Y., Cavez, M., Hwang, D.–H.: Physics Report **424**(4/5) (2006) 175
5. Bongard, G.: Der Einfluss von Reorganisationen auf Robustheit und Effizienz von Business-Einheiten. MS Thesis, Donau Universität, Krems, Austria (2005)
6. Christen, M., Ott, T., Kern, A., Stoop, N., Stoop, R.: Journal of Statistical Mechanics: theory and experiment (2005) 11013
7. Christman, J. and Anderson, J.: Autonomy and the Challenges to Liberalism. Cambridge University Press, Cambridge (2005)
8. Churchman, C.W.: The Systems Approach. Delacorte Press, New York (1968).
9. Corron, N., Pethel, A. and Hopper, B.: Phys. Rev. Lett. **84** (2000) 3835
10. Fayol, H.: General and Industrial Management Sir Isaac Pitman & Sons, London (1949)
11. Feigenbaum, M.: J. Stat. Phys. **2** (1979) 669
12. Hea, X.–Z. and Westerhoff, F.H.: Journal of Economic Dynamics & Control. **29** (2005) 1577
13. Hermes: an open standard within Swiss public administration (e-print `http://www.hermes.admin.ch/internet/hermes/hermes/index.html?lang=de`).
14. Kant, I.: Grundlegung zur Metaphysik der Sitten. Werke in sechs Bänden, Band 3 (Könemann Verlagsgesellschaft, Köln (1995)
15. Keynes, J.M.: The General Theory of Employment, Interest and Money. Cambridge University Press, Cambridge (1936)
16. Kitchin, J.M.: Rev. Econ. Stat. **5**, (1923) 10
17. Kosiol, E.: Einführung in die Betriebswirtschaftslehre Gabler, Wiesbaden (1968)
18. Kydland, F.E. and Prescott, E.C.: Econometrica **50** (1982) 1345
19. Michetti, E., Caballé, J. and Jarque,X.: Financial development and complex dynamics in emerging markets. Proceedings of the New Economic Windows Conference, Salerno (2004)
20. Ott, E., Grebogi, C. and Yorke, J.A.: Phys. Rev. Lett. **64** (1990) 1196
21. Pribham, K.: A History of Economic Reasoning. The Johns Hopkins University Press, Baltimore (1983)
22. Rabinbach, A.: The Human Motor: Energy, Fatigue, and the Origins of Modernity. Basic Books, New York (1990)

23. Roethlisberger, F.J. and Dickson, W.: The Management and The Worker. Human Relations School, Cambridge MS (1939)
24. Schneewind, J.B.: The Invention of Autonomy Cambridge University Press, Cambridge (1998)
25. Schweizer, R.W., Jeanrenaud, C., Kux, S. and Sitter-Liver, B.: Verwaltung im 21. Jahrhundert. Herausforderungen, Probleme, Lösungswege Universitätsverlag, Freiburg (2003)
26. Stoop, R. and Wagner, C.: Phys. Rev. Lett. **90** 1541011 (2003)
27. Sydow, J.: Management von Netzwerkorganisationen Gabler, Wiesbaden (2003)
28. Taylor, F.W.: The Principles of Scientific Management Harper & Row, New York (1911)
29. Taylor, J.S.: Personal Autonomy. New Essays on Personal Autonomy and Its Role in Contemporary Moral Philosophy Cambridge University Press, Cambridge (2005)
30. Thom, N. and Ritz, A.: Public Management. Innovative Konzepte zur Führung im öffentlichen Sektor. Gabler, Wiesbaden (2000)
31. Thommen, J.–P.: Managementorientierte Betriebswirtschaftslehre. Versus, Zürich (2000)
32. Watts, D.J. and Strogatz, S.H.: Nature **393** (1998) 440
33. Weyer, J.: Soziale Netzwerke. Oldenbourg Verlag, München (2000)
34. Wren, D.A.: The History of Management Thought. John Wiley& Sons, Hoboken NJ (2004)
35. Yergin, D. and Stanislaw, J.: The Commanding Heights: The Battle for the World Economy. Simon & Schuster Inc., New York (1998)
36. Zabczyk, J.: Mathematical Control Theory: An Introduction. Birkhäuser, Boston (1993)

Complexity and the Enterprise: The Illusion of Control

Karl G. Kempf

Decision Technologies Group, Intel Corporation, 5000 W. Chandler Blvd., Chandler, Arizona, USA 85226 karl.g.kempf@intel.com

1 Introduction: A Generic Model of the Enterprise

A generic model of an enterprise is described. Rising complexity in all facets of the model is shown to result in a tendency for all activities of the enterprise to require higher budgets and longer durations over time. However, the market in which the enterprise operates expects lower prices and faster response moving forward. One approach to this dichotomy is to exercise continuous improvement inside the company to reverse the trends toward longer and more expensive activities. Examples of successful efforts in this direction are presented from Intel Corporation. Another approach is to improve demand management in the marketplace including better forecasting. This proves to be a particularly difficult task that is fraught with practical problems, a few of which are presented in detail. It is proposed that, while internal improvement projects are necessary, they are not sufficient for the continued success of the enterprise. Improved forecasting is required, but may be precluded by rising complexity, leading to a dangerous illusion of control.

Essential functions of an enterprise that manufactures goods and resides mid-echelon in a supply chain are represented in Figure 1. Although motivated by Intel Corporation, the diagram is generic. On the one hand it is highly simplified since it ignores such important support functions as information technology, finance, and personnel. On the other hand the core functions included are more than adequate for discussing complexity and adaptive behaviour.

1.1 The Enterprise as a Seller of Goods

The enterprise exists to generate a profit. The most important function of the enterprise as shown in the upper right of Figure 1 is selling goods for more on average than the aggregate cost of producing, marketing, and distributing them. If it fails to succeed in this function over time it ceases to exist. In 2006 Intel Corporation enjoyed net revenues of over $35 billion and net income of

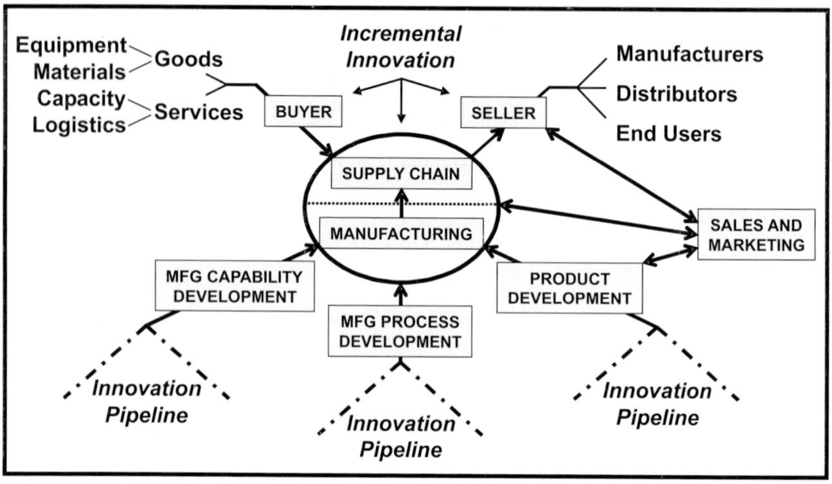

Fig. 1. Relationships between the major functions in the enterprise model. (NOTE: Neglects such important functions as finance, personnel, information technology)

over $5 billion realized through the sale of hundreds of distinct products to thousands of different customers around the world.

1.2 Integrated Areas of Development

There are three basic development areas that enable and support the production of goods for sale. Product development is shown on the lower right of Figure 1 and represents the continuous improvement and expansion of offerings into the market. The lower middle of Figure 1 shows manufacturing process development and includes progress on ever more effective and efficient production methods. Manufacturing capability development is shown on the lower left of Figure 1 and concerns the design and construction of new manufacturing facilities to execute the manufacturing processes that produce the products. In the successful enterprise, these three development processes must be highly integrated. This is true even if one or more of the processes is outsourced. Intel's reputation as a high-technology company rests to a large extent on its demonstrated expertise in these areas.

1.3 Manufacturing and Supply Chain Execution

Goods are produced through the execution of a hierarchy of processes shown in the middle of Figure 1. Topping the stack is the process of forecasting the type and volume of products that the market will purchase over multiple future timeframes. Strategic forecasts are used to plan expansion, contraction, and rearrangement of the production capabilities of the enterprise. Plans for material acquisition and release into factories, stockpiling of intermediate and finished goods, and transportation of products to customers are generated using tactical forecasts . At the bottom of the stack is manufacturing execution at Intel factories turning supply chain plans into shippable products 24 hours per day 365 days per year.

1.4 The Enterprise as a Buyer of Goods and Services

Thousands of suppliers provide tens of thousands of goods and services at a cost of several billion dollars each year to support Intel's efforts. Goods include the materials to build and the equipment to outfit factories as well as the raw materials and equipment spares to build products. Some service subcontractors provide the transportation modalities needed to acquire equipment and raw materials and to distribute products. Others provide manufacturing capacity buffers used to hedge uncertainty in demand.

1.5 Sales and Marketing

Sales and marketing spans many of these core functions. Looking out a year or more into the future, sales and marketing personnel assist product development in determining the appropriate features and functions for new products. In the time frame of quarters and months, demand must be forecast to direct most of the ongoing activities in acquiring goods and services as well as running the supply chain and manufacturing, and sales and marketing is active here too. Finally week to week and day to day Intel sales and marketing interfaces directly with a plethora of customers to service their needs before, during, and after the sale.

1.6 The Role of Innovation

All products become commodities over time. The life blood of the enterprise is innovation across all of the functions shown in Figure 1. With the sophistication of its products and the level of technology required to produce them, Intel maintains long term innovation pipelines for the development of products, manufacturing processes, and facilities to extend its technology leadership. These formal pipelines act as funnels with many (sometimes radical) ideas under initial investigation being reduced over a multi-year incubation period to just a few implementations resulting ultimately in new and better products. In parallel, usually with less formality, incremental innovations are needed in the buying and selling functions including supply chain and manufacturing planning and execution to support Intel's market leadership. This continuous improvement is aimed at decreasing product costs and increasing responsiveness to customers. As these are operational functions proceeding on a daily basis, they are sensitive to disruption and so innovation is usually (but not always) realized incrementally.

1.7 The Enterprise Visualized as a Machine

In the enterprise represented in Figure 1, the material flow is from left to right and the financial flow is from right to left in the top of the diagram. Ideas are implemented up the development pipelines in the bottom of the diagram. Information flows in every direction. Radical and incremental innovation is mandatory and profit is the goal. It is easy for the personnel that operate this kind of an enterprise to build mental analogies based on machines. The gears mesh and drive the enterprise forward. The technologists provide the fuel and the business managers steer the course to success.

2 One Generic Problem Set for the Enterprise

Over time, almost everything in Figure 1 has become more complex for many enterprises including Intel. This can be attributed to many things including a) steady progress in science and engineering, b) stronger competition in the international marketplace with an expanded set of business techniques, c) revolutionary improvements in communications supporting broad and rapid access to information, and d) ever more sophisticated consumers seeking customized products quickly delivered to their doorstep, to mention but a few.

This rising complexity inherently results in time and cost pressure on all the core functions of the enterprise. Development efforts can take longer and cost more to complete since added complexity increases the probability of unforeseen problems, errors, and delays. Planning and executing supply chain and manufacturing activities tend to consume more time and budget as they rise in number, interactions, and complexity. Arranging the acquisition of goods and services that are vital to the enterprise can consume more time and budget as the number of suppliers and transactions rise along with the complexity of the relationships.

The marketplace also exerts time and cost pressures, but unfortunately for the enterprise, in the opposite direction. In many cases competition in the marketplace drives shorter times for the development and delivery of the next generation of products. In most cases consumers expect more features, functions, and reliability from lower cost products generation after generation.

Forecasting and managing what the market wants along any of these axes over time becomes particularly difficult in the face of these complexities. This difficulty generates a disturbing scenario. From a control theoretic perspective, if your ability to look forward is diminishing at the same time your speed of response is degrading, you will suffer serious consequences. The question is not whether you will loose control, but rather when it will occur and if you will be able to recognize it when it happens.

One response to this frightening scenario is to continuously work to manage the internal complexities in such a way as to reverse the trends toward activities consuming more time and budget. Another response is to devise improved ways to manage demand in the marketplace including forecasting. Both of these responses represent specific types of innovations generically included in Figure 1. The following sections contain descriptions of some of the complexities Intel and other companies face, efforts that have been successful to varying degrees in managing internal complexity to contain durations and costs with specific examples from Intel, and approaches to the management and forecasting of demand. It will be obvious that progress on internal improvements has far outpaced progress on demand management. It will be proposed that this asymmetry could lead to the illusion that enterprises are further away from loosing control than is actually the case.

3 Complexities in the Enterprise with Examples from Intel

The complexities described here cover a broad spectrum. The origins of some are relatively obvious but they are nonetheless very impactful to the enterprise. Others are subtle in root causes but just as important (although few in the enterprise may recognize them as such). Examples are drawn from both direct Intel experiences and the basic physics of development and operations in any enterprise. Coverage is not encyclopaedic but rather illustrative.

3.1 Product and Manufacturing Process Complexity

Intel has had the luxury of being guided by Moore's Law in the understanding of the complexity of its technology treadmill [26]. The guidance is that the number of transistors that can be placed onto a given area of silicon will double roughly every 1.5 to 2 years (or that a given number of transistors will occupy half the area of silicon in that time). Smaller transistors are faster transistors and so lead to products able to exhibit higher performance and broader functionality over time. This progression is enabled by incremental downscaling of the physics and chemistry involved in the manufacturing process. Figure 2 shows the product and process becoming more complex in lockstep since Intel invented the microprocessor.

Early microprocessors were built from a few thousand transistors while current generation devices contain a few billion as shown on the left axis of Figure 2. The right axis shows the capability of the supporting process technology represented as the minimum feature size achievable. This dramatic increase in the complexity of the product and the process has kept the related development processes under constant pressure to contain durations and costs.

3.2 Manufacturing Capability and Operations Complexity

The steady decrease in the minimum feature size of the manufacturing process also impacts other core functions of the enterprise. For example, over the

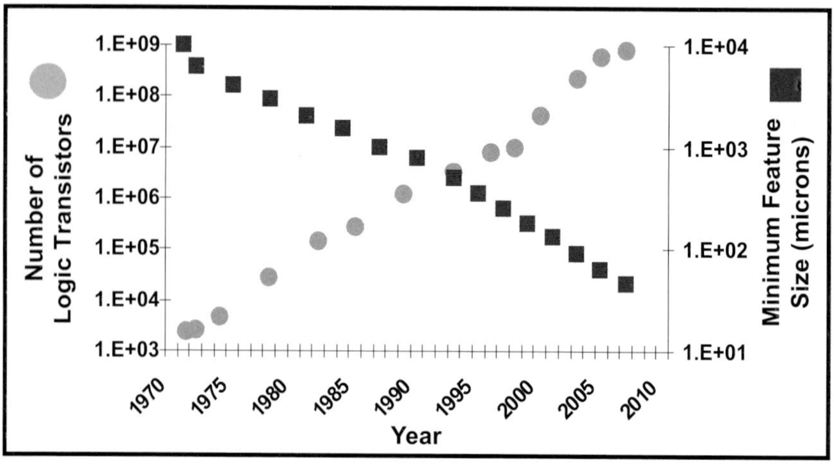

Fig. 2. The product and process complexity aspects of Moore's Law

time period represented in Figure 2, Intel factories have become increasingly complex. Three sequential types of complexity can be recognized over the factory lifecycle. First is the development of the manufacturing capability. In the current environment, it costs over $3 billion US dollars to construct and outfit a new high volume semiconductor fabrication facility. Obviously with an investment of this magnitude minimizing the time from green field to operational readiness is especially important. However management of the construction and fitup is complex based on the thousands of personnel involved, hundreds of pieces of equipment to be installed, and tens of thousands of tasks to be choreographed over the roughly one year project duration. Second is the transfer of the complex manufacturing process from the facility where it was developed to the factories where it will be run in high volume. There are a practically infinite number of complications that could occur given emphasis on both speed of completion and quality of the result during this transfer. Fortunately Intel has developed the COPY EXACTLY! (CE!) process to manage this transfer [25]. Third is the minute to minuute operation of the factory including managing thousands of lots of wafers as work in progress, hundreds of machines requiring preventive and unscheduled maintenance, and a complex automation system controlling material movement and data collection. These dramatic increases in the complexity of manufacturing capability realization and operations have kept high volume manufacturing under constant pressure to contain durations and costs.

3.3 Manufacturing and Supply Chain Execution Complexity

Stochasticity in supply and demand adds further complex to the enterprise. On the supply side, few manufacturing processes have fixed throughput or throughput time. Among the primary drivers of variability in throughput time is equipment reliability. Raw material released into the factory at a particular point in time will result in finished product at some time in the future that can only be represented as a distribution because of random equipment problems. A driver of throughput variability is the quality of the manufacturing process and the fact that no realizable process can be perfect. A subset of the raw material will be converted into high quality finished goods, but again the percentage can only be represented as a distribution.

Perhaps more concerning are the inescapable and in many cases uncontrollable stochastic processes involved in morphing demand from the customers of the enterprise. Surprise moves by competitors can change a customer's mind even after orders have been placed. Changes in the tastes of the customer's customers can lead to requests to adjust orders in the pipeline. Problems in the production facilities of the customer can force them to alter their plans leading to altering orders. In extreme cases, products shipped to a customer can be returned to the enterprise. Practically speaking, given the dynamics of the marketplace, even firm orders are simply forecasts until shipped, received, and invoices paid.

For an international company with a wide variety of products like Intel, basic supply and demand stochasticity is magnified. Supply stochasticity applies to each individual product but in a different way depending on its complexity and maturity. Serving an international market usually involves a broad network of factories as shown in Figure 3. Each product is manufactured in multiple factories and each factory manufactures multiple products. This arrangement mitigates risk and satisfies diverse markets, but each factory contributes a variant on the stochasticity (contained in Intel's case to large degree by CE! [25]).

3.4 Buying and Selling Complexity

Selling into an international market raises complex logistics and legal issues. Providing excellent service to customers scattered around the world from a disperse but finite number of manufacturing facilities presents a complex transportation problem in its own right. But when customs and taxation protocols vary from border to border, complexity escalates even higher.

Buying goods and services in the international marketplace involves additional complexity. Not only does the reliability of suppliers vary from country to country, but so do contract and arbitration laws. While a variety of financial approaches have been developed to share risk between buyers and sellers in dynamic markets, negotiating and executing such arrangements internationally can be challenging.

3.5 Complexity due to Structure

The enterprise is comprised of a number of entities arranged in structures on a number of scales with self similarities. The organizational chart of Intel

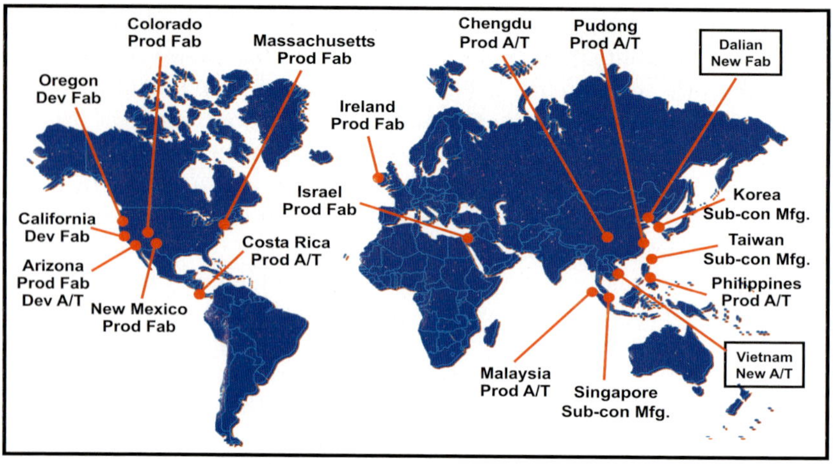

Fig. 3. Intel worldwide production facilities (including new facilities being built in China and Vietnam)

Corporation shows roughly 80,000 employees structured in layers with recurring team structures, reporting relationships, and performance metrics. The thousands of products Intel manufactures aggregate bottom to top in recurring structures based on performance, price, and intended application. The same can be shown for factory equipment, purchased goods, and so on.

Global properties aggregate on a number of spatial and temporal scales. For example, the minute to minute decisions made at a particular machine in a factory having to do with production and maintenance aggregate over all machines and over days and weeks and quarters to quantify the performance of that factory. The same is true over all the factories in a product line to quantify that line's performance, and over all product lines resulting in the performance of the enterprise.

3.6 Complexity due to Nonlinearity

There are many well known nonlinearities in the enterprise described generically in Section 1. At the lowest level of manufacturing execution, there is a nonlinear relationship between throughput and work in progress (due to congestion, throughput rises at a decreasing rate as work in progress rises). In supply chain execution, there is the bullwhip effect in which fluctuations in orders and inventories increase dramatically as information moves upstream relative to the flow of materials. In the buying and selling activities of the enterprise there is a nonlinear relationship between price and volume (price falls at a decreasing rate as volume rises due to economies of scale and market saturation). Furthermore, complex behavior has been shown in surprisingly simple enterprise models. Re-entrant manufacturing systems with as few as two machines and two products have been show to exhibit complex performance with repeat patterns measured in years as well as sensitive dependence on initial conditions [4, 12]. Switched flow manufacturing systems have been shown to produce nonlinear phenomena including deterministic chaos [6, 29]. The same is true of simple supply chains with as few as four decision makers (retailer, wholesaler, distributor, manufacturer) [27].

3.7 Complexity due to Decision Making

A distinguishing feature of artificial systems is their reliance on decision making. Our understanding of the quality of decision making in the enterprise setting has changed dramatically over time. Initially it was asserted that "... by directing that industry in such a manner as its produce may be of the greatest value, he intends only his own gain, and he is in this, as in many other cases, led by an invisible hand ..." implying impressive efficiency and effectiveness in decision making [33]. Although it took nearly 200 years, Simon has shown that decision making in enterprises can at best be described as "satisficing" [32]. This is due to bounds on the rationality of the decision maker based on incomplete knowledge of available alternatives and inability to predict the outcome of pursuing an alternative based on limited computational power and inherent uncertainty in the external world. Additional study by Kahneman of the psychology of intuitive beliefs and choices have demonstrated a number of additional mechanisms that bound the rationality of human decision makers [16] including a variety of biases (e.g. anchoring, framing, optimism) and logical fallacies (e.g. base rate, conjunction, planning). The application of chaos theory to the study of organizations like those that manage enterprises has been illuminating [37, 38].

3.8 Complexity due to Self Reorganization

An enterprise has the ability to collect and internalize data and information. It also has the ability to rationalize the data to fit the enterprise's mental model of the world. In the best of cases when the rationalization is congruent with objective reality, learning takes place. More often the rationalization is used to proactively and reactively act upon itself and its ecosystem. Acting upon itself frequently results in some form of self organization, or rather self reorganization. This is done with the goal of increasing efficiency and effectivity. Unfortunately this is not always the result due to the strictly bounded rationality of the decision makers and the complexity of the relationships between the parts of the system (that is, the enterprise and all of its pieces as well as the enterprise and its multi-component external ecosystem).

3.9 The Resulting Enterprise Behaviour

With complexity due to inherent structure and nonlinearity as well as bounds on rationality in decision making and self reorganization, it is not surprising that enterprises exhibit very complex trajectories. There is seldom any approach to equilibrium and the enterprise never periodically returns to some standard base state. Small inputs sometimes produce dramatically large results. Sometimes a small change to a product design dramatically increases its popularity in the market, sometimes it does not, and sometimes even a large change has no effect. The same apparent action sometimes yields a different result. A sales promotion that achieved record results sometimes does not work in another time period or another geography, and sometimes it does. In all cases, behavious emerges and can be logically rationialized after the fact, but not predicted in advance. There is ample evidence that it is more appropriate to view the enterprise as an artificial complex adaptive system than as a machine: artificial since the enterprise is the result of human manipulation as distinct from an entity occurring in nature, complex as described in Section 3., and adaptive but in a different sense than natural systems since decision making is a major and inherent component of the behavior of the enterprise.

4 Progress on Containing Internal Durations and Costs

Many enterprises including Intel continue to develop methods to combat the impact of rising complexity on the duration and cost of their internal activities. Here we draw concrete examples from Intel over a range of time scales from minutes to years and a range of tasks from equipment maintenance to product development. Coverage is not encyclopaedic but rather illustrative.

4.1 Minutes and Hours in Factory Operations

Given the cost of constructing its facilities and the value of the products manufactured in them, Intel operates its factories around the clock every day of the year. Multiple decisions are required minute to minute to maximize throughput (supporting minimizing cost) while minimizing throughput time. Continuously improving these decisions is required in the face of continuously rising complexity in products, manufacturing processes, and supply chain needs. (For an overview of Intel manufacturing see [36].)

The initial incremental innovation in this area was the application of Goldratt's Theory of Constraints (ToC) [18]. Intel employed the basic ToC concepts modifying them to suit its needs including managing a) work in progress and machine loading, b) preventive and unscheduled maintenance, c) training and shiftly work assignments for floor personnel, and d) the timing of material release into factories. This approach was used in factories primarily concerned with process development as well as those focused on high volume manufacturing, and spanned fabrication and assembly-test factories. In all cases factory throughput went up by 10% to 20% while inventory (and consequently throughput time) went down with no capital outlay or increase in operating expense. Simply getting the same people to make better and more integrated decisions about the same work in progress and equipment was the key.

This initial success has been continuously improved in many directions. Extending the basic ideas of ToC about bottleneck equipment, we have become better at identifying [2, 17] and acting on [35] the dynamic set of subproblems that exist on the factory floor at any point in time given shifting bottlenecks, prioritized materials based on market demand, and related issues. This progress on managing manufacturing dynamics has occurred at two levels. Looking top down at the factory as a whole, we have explored a spectrum of control algorithms. Among the simplest is real time contol of the push-pull point [28] and among the most complex is the application of Model Predictive Control (MPC) [40] adapted from the continuous process industry. Looking bottom up at the individual tools that populate the factory, we have explored algorithms for optimizing the performance of machines with setups, machines that batch [19], and machines that have complex internal processing substeps [9].

4.2 Hours and Days in Supply Chain Execution

Assembly-test factories provide the final differentiation points for our products at a level more abstract than the minute to minute operation of the factory floor (Figure 4). Approximately seven days prior to shipment to customers, decisions are made concerning whether to put microprocessor "die" fabricated in upstream factories into packages known as substrates for application in server, desktop, or laptop computers. Roughly one day prior to shipment, decisions are made about the "finishing" process that sets the final performance configuration and surface markings of products to satisfy demand.

The initial breakthrough here was an extension of the control perspective used so effectively on the factory floor as described in Section 4.1 [24]. More recently this has been expanded to again utilize the power of model predictive control (MPC) techniques [5, 20]. At the beginning of the assembly segment is an inventory holding position known as "assembly die inventory" (ADI). At the end of the test segment is "semi-finished goods inventory" (SFGI) as an inventory position. Finishing of SFGI material leads to the shipping warehouse and dock. Control decisions at ADI and SFGI relate to how much material to release into the line to satisfy demand and maintain downstream inventory at appropriate levels. (Setting the target levels for inventories is discussed in Section 4.3.)

A simple mass-balance description of the control algorithm at any inventory position includes a) estimating how much material will be removed by downstream processes in the next few time periods (demand from the shipping warehouse, finishing from SFGI, assembly-test from ADI), measuring how much material is currently present in each inventory position, and estimating how much material will flow in from upstream processes in the next few time periods (die from fabrication factories into ADI, product from assembly-test

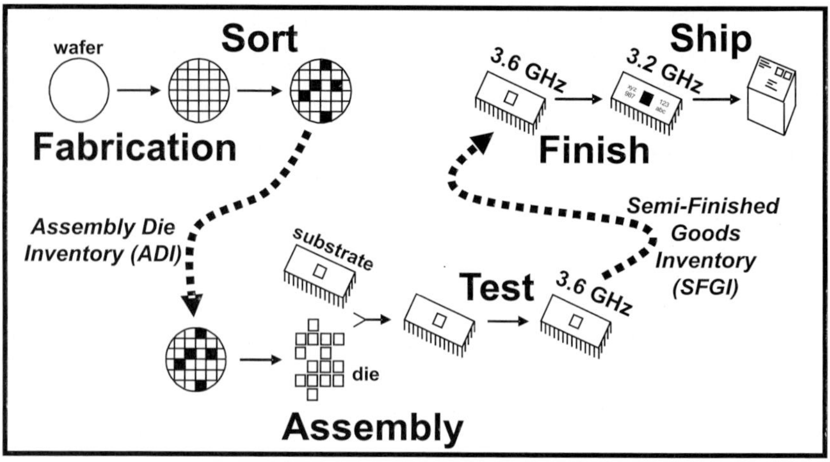

Fig. 4. The semiconductor manufacturing flow

into SFGI, and finished product from finishing into the shipping warehouse). There are two considerations that make this a non-trivial computation. One is that each product is made in multiple factories and each factory makes multiple products, so the algorithm must consider multiple geographically disperse ADI, SFGI, and shipping warehouse combinations to maximize demand satisfaction at minimal cost. The other consideration is that each factory exhibits individual stochasticity that changes in unique ways over time and this is what the model component of the MPC controller manages (Figure 5).

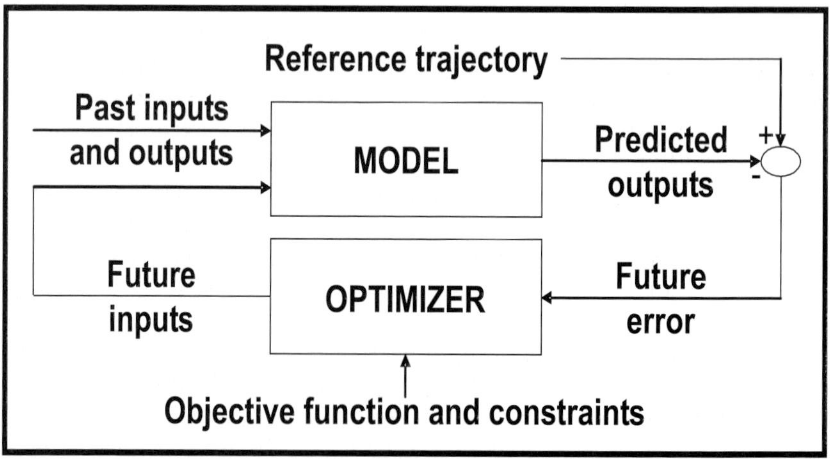

Fig. 5. A high level block diagram of a model predictive controller

Initial simulation trials of this technique have been promising [34]. Historical performance of a single Intel assembly-test factory was retrieved for a typical 6 month period on a small set of products. This included wafers shipped into ADI and product shipped out as well as contents at ADI, SFGI, and the shipping warehouse, all on a daily basis. A discrete simulation matching the factory throughput and throughput time was used to execute commands from an MPC algorithm shown in Figure 5. It was dictated that the simulation accept the same inputs into ADI and provide the same outputs from the shipping warehouse as the actual factory, but could independently decide how to run all intermediate steps. This included material release from ADI and SFGI as well as instructions for final performance configuration in finish. At the end of the simulation, the MPC algorithm had matched shipped outputs but withdrawn 62% less material from ADI, held 55% less finished product in the shipping warehouse, and required 180% less configuration exiting SFGI than the actual factory. Similar results have been obtained in the field on the much more difficult multi-factory multi-product problem. Our current research concerns improving the modelling component of the MPC algorithm for supply side prediction [1].

4.3 Weeks and Months in Manufacturing and Supply Chain Planning

Given the multi-month throughput time required for microprocessor fabrication and the one week throughput time of assembly-test factories, as well as the day to day dynamics of the market and its ever-changing demand for Intel products, deciding weekly and daily material releases into factories is critical to Intel's success. Too much production overflows inventory positions and risks scrapping products that were expensive to manufacture. Too little production risks leaving valuable demand unsatisfied with short term and long term consequences in addition to allowing expensive capacity to sit idle. Continuously improving these decisions is required in the face of continuously rising complexity in products, manufacturing, and supply chain needs. (For an overview of Intel's supply chain see [21].)

With its long standing prowess in product design, manufacturing process development, and high volume manufacturing execution, Intel grew its business for over three decades planning on spreadsheets. Considering the scale of the planning problem this was obviously a process that relied heavily on human glue and tribal knowledge. At the turn of the last century with complexity rising rapidly, this situation had to be dramatically improved. Mathematical optimization proved to be the foundation of a radically improved planning system [3, 20].

Computing with Intel's fastest current microprocessors, planning coordinated material releases into fabrication and assembly-test factories could be done very much more quickly with fewer planners and produced superior plans. Planning groups reported productivity gains of 5X to 15X when moving from spreadsheets to optimizers. Planning throughput time was cut by days even though many more business scenarios were being explored before final plans were agreed. It was shown repeatedly that new planners could be brought up to speed in 50% less time compared to the old system. The optimization tools support improved demand satisfaction and more efficient use of Intel's manufacturing capacity.

Our current work is focused on better managing the stochasticity of both supply and demand described in Section 3.3 as well as the non-linearities mentioned in Section 3.6 into our planning tools. Standard optimization approaches to production planning require a manufacturing throughput time as input to compute factory loading as output. But factory throughput time is a function of loading and so standard approaches are hampered by a mathematical circularity that we intend to avoid though improvements in our core algorithms [22]. Standard approaches to computing inventory targets indicate that demands, demand variabilities, and service levels be available as inputs to produce safety stock targets as outputs. These targets are then utilized as inputs to production planning algorithms. But safety stocks are put in place to buffer production plans against variabilty. In practice, our planning tools are used to play out multiple business scenarios before deciding the production plan to be implemented. Thus there is a logical circularity in standard approaches that we intend to avoid through improved integration of our safety stock and production planning algorithms [41].

4.4 Months and Quarters of Procurement

The procurement of equipment and materials necessary for production supports manufacturing and the supply chain. If the planning tools are used to look out a few months or quarters for loading Intel's factories to satisfy demand, then the plan produced serves to also specify materials to order from suppliers to support the plan. If the planning process looks out multiple quarters, the result may indicate capacity changes needed for Intel and its material suppliers to prepare for future demand. Continuous improvement in this area is vital to Intel's profitability and long term success. (For an overview of Intel's approach to managing procurement risk see [23].)

A natural adjunct to the optimization tool developed for production planning was a set of similar tools for procuring the packages (called substrates) that are joined with die in the early steps of assembly-test manufacturing [31]. The core of the old procurement process contained a very inefficient repetitive request-respond interaction between Intel and its many suppliers. On the Intel side of this negotiation were rough capacity estimates for the suppliers heuristically based on past cycles. The first step in the optimization approach focused on building a confidential capacity model for each supplier that Intel could use to formulate its material orders as a win-win for each supplier and Intel. Based on these models that the supplier regularly updates, three optimization tools were built as shown in Figure 6.

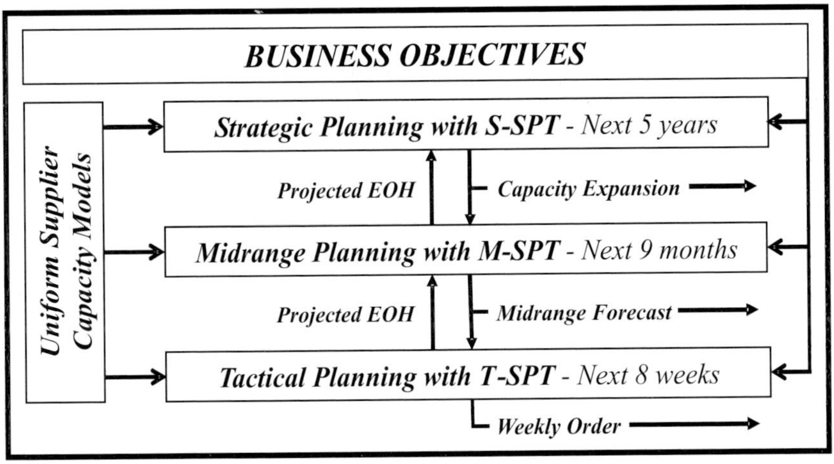

Fig. 6. The materials procurement process as a set of three integrated tools

Looking forward, the tools generate a) weekly orders to support assembly-test manufacturing plans for the next 8 weeks (the Tactical Substrate Planning Tool or T-SPT), b) forewarning of modifications or rearrangements to the suppliers capacities that will be needed over the next 9 months to mesh with Intel's plans (the Midrange Substrate Planning Tool or M-SPT), and strategic advice about substrate roadmaps and long term demand forecasts to be used for capacity expansion at the suppliers (the Strategic Substrate Planning Tool or S-SPT). In practice, the tactical tool has reduced the duration of the business process by 10x and reduced the time spent by Intel personnel by 6x while increasing the number of business scenarios consider by 4x. Issue resolution with each supplier has been reduced from an average of 2 per week per supplier with the old methodology to less than 1 per month with the new approach. Even larger benefits have been gained with the mid-range tool.

Purchasing production equipment is a very different problem than purchasing materials like substrates. With equipment, the volumes are much smaller (10s as opposed to 10s of millions), the unit cost is much higher ($10's of millions per unit versus $10's), and the lead times are much longer (quarters instead of weeks). The risk in equipment purchase is simple to explain but difficult to manage [8].

Assume that Figure 7 shows the situation for a piece of equipment that costs $25M per instance. Using a variety of performance parameters about the tool and the forecasted demand for the product, it has been computed that k tools are required. It is relatively easy to order tools 1 and 2 tool since Intel knows when it will start production and the tools will be used for the lifecycle of the product (typically 2 years). Placing orders for the j-th and k-th tools is much more difficult, especially because they may be used for only a

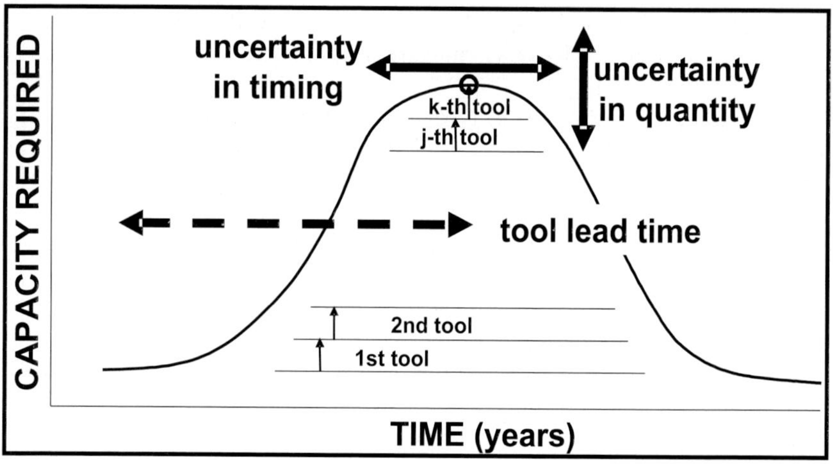

Fig. 7. Uncertainty in the timing and quantity of expensive tools with long lead times

short time. Since the height in capacity and the position in time of the peak demand is simply a forecast, there is considerable uncertainty in the ordering process. If j and k are ordered based on forecasted demand but not needed once actual demand is realized, substantial expense has been incurred with no benefit. If j and k are needed but not ordered, substantial revenue is lost with possible longer term ramifications to the reputation of the enterprise. Similar arguments can be made for purchasing the right quantity of tools but ordering too early or too late. Notice that in the worst case the lead time of the tool from the supplier can be as much as half the life cycle of the product. For this set of situations we have developed a variant of options contracts [39]. Intel's need is to delay the purchase decision for as long as possible and then take delivery of the tool(s) as quickly as possible, and we have begun to succeed in accomplishing this as a win-win with our equipment suppliers.

4.5 Quarters and Years in Capacity Development

New factory construction and outfitting is triggered by some combination of forecasted market demand, new product development, and new manufacturing process development. With the relentless improvement of manufacturing processes yielding ever smaller feature sizes, older factories are simply not able to be re-outfitted for new processes mainly because of particle size contamination issues. An older factory has older generation water purification, air management, and chemical delivery systems that are not suitable for the latest generation processes with more stingent cleanliness requirements based on their reduced feature sizes. And it is the latest generation process that enables the improvement of products to be smaller in size and larger in functionality.

The risk however is too high to introduce a new product and a new process at the same time. New products are initially manufactured on the currently stable high volume manufacturing process. Likewise, the new process is initially used to manufacture the currently stable high performance products. This is the nature of the technology treadmill that Intel has driven for the past 30+ years. But it is the forecasted market demand that is the deciding factor. To spend over $3 billion dollars to build a new fabrication facility, senior management of the enterprise must be convinced of the revenue opportunity in the market.

Once the decision is taken to proceed, two very important processes are initiated. One is the design process where the equipment set for the factory is selected and positioned within the factory envelope. The concepts of ToC described in Section 4.1 are used to determine the appropriate numbers of each equipment type to purchase for efficient operation [18]. On the one hand, ToC instructs us that a manufacturing line where the capacities of each tool set are closely balanced is the most difficult to run. This is because the constraining tool set where all operational policies are anchored changes identity rapidly to whichever tool has experienced the latest unscheduled breakdown. On the other hand, financial considerations clearly show that a line that is too unbalanced is the most expensive to build and operate. While one tool set (usually the most expensive in the factory) will be stablely constraining and so make the factory easier to operate, other tools will be grossly underutilized driving up product cost. Even with the proper tradeoffs in tool selection, the layout of the tools on the floor is critical for flow of material as well as operations and maintenance staff productivity [13]. In complex factories, the number of tools selected and the manner that they are positioned are not independent. Our improved techniques for solving this difficult problem have lead to a 5% to 10% decrease in our capital costs relative to previous methods while delivering the same or improved throughput and throughput times in the resulting factories.

Another very important process is the construction and fitup of the factory. Having taken the decision to build the factory to intercept the perceived market opportunity, speed is critical in moving from a green field to a factory producing saleable product. Applying ToC concepts to this project management problem (often referred to as "critical chain theory" to distinguish it from "critical path theory") has been beneficial for Intel [18]. Although improvement results in the fitup stage have considerable variation depending on the complexity of the type of tool being installed, after multiple trials the expectation is 15% to 40% duration reduction compared to previous project management methods. Recently we have been exploring more advanced techniques to better manage the variability in the contsruction and fitup process to push these expectations even higher [42].

4.6 Years in Product and Manufacturing Process Development

Although at Intel the two longest duration processes in Figure 1 are intimately related as mentioned in Section 1.2, they have very different characteristics. The manufacturing process development effort is based on such technologies as device physics and chemical engineering. It is an unchallenged assumption that smaller faster transistors are a competitive advantage as long as the manufacturing process is affordable and the speed of implementation is appropriate for the intended market place. Product design is concerned with circuit design, logic and memory layout, and platform architecture. The assumption here is that the bundle of advanced integrated features and functions will be the basis of competition. Early introduction of a product that support a significant useage model (e.g. home information and entertainment systems or enterprise servers or hospital information/automation systems) is a typical goal supporting the concept of the enterprise as a seller of goods described in Section 1.1.

The proprietary nature of the continuous improvement of these development processes necessarily limits their description here. However Intel's record of delivering cutting edge products based on industry leading processes indicates that the management of risk in process and product development is a strong focus area [10, 11]. But it must be noted that these two extremely important processes have diametrically opposite dependencies on the market forecast. On the one hand, history indicates that the market continues to be interested in smaller faster transistors so that the process development efforts can be relatively insulated from market forecasts. On the other hand, the markets' taste for features and functions changes more rapidly than the product design can be executed. This means that demand forecasts, and more importantly demand management, is of high interest to product development personnel.

5 Problems with Demand Forecasting

These success stories are important, but to a degree are misleading. In each of the cases cited for Intel successes there is a mathematical formulation that provides a high quality solution to the problem. In many cases the solutions are integrated. Efforts continue to improve the formulations and provide further integration.

But there is a common feature in every one of the formulations described. They all require information about demand as input. Whether the question is how to prioritize equipment maintenance on the factory floor to make production goals this week, or what features and functions to include in a new product to be introduced in two years time, or anything in between, the question is the same. How many will the customers buy and when will they buy them and how much will they pay? Predicting the future has always been a difficult task, and as complexity rises so does prediction difficulty nonlinearly. The abstract constraint for every formulation aimed at managing duration and cost of internal activities is the ability of the enterprise to manage demand. Ultimately an accurate demand forecast ongoing bounds the ability of the enterprise to have the right product (in terms of features and functions and use case) in the right volume at the right price at the right time. An errorful demand forecast can quickly negate any improvement in process or capacity development, manufacturing or supply chain operation, and goods or services procurement. A consistently accurate demand forecast acts as a positive amplifier for all other activities of the enterprise.

Since forecasting is so vitally important, there is a particularly broad literature on the topic suggesting a wide variety of general and domain specific approaches that can not be reviewed here in detail. From the combination of methods employed by Intel, only two will be discussed. The most effective over the years has been the approach suggested by Alan Kay: "The best way to predict the future is to invent it". Intel's technology treadmill has validated this method over and over across more than three decades. Another method that Intel uses, arguably the most widely studied in the literature and the most widely practiced by large enterprises, is analysis of historical data. While this approach is probably necessary, it is clearly not sufficient to appropriately feed the mathematical formulations described in Section 4 in the face of the complexities enumerated in Section 3.

5.1 Demand Forecasting Thought Experiment #1

Consider an enterprise in a particular situation during month M (Figure 8). The enterprise manufactures two products P1 and P2 each with a manufacturing lead time of 1 month that are used by its customers as a component in their products. Months ago in M-6, outside the capacity lead time for organizing production, Corporate Sales and Marketing (CSM) requested that the Corporate Manufacturing Organization (CMO) prepare itself to supply 150K units of P1 and 150K units of P2 per month. CMO responded by preparing to supply 160K units of each product. Although CMO committed only 300K total units to CSM, they held a 10K capacity buffer on each product since in the past a) given the complexity of the forecasting process, CSM had miscalled demand and CMO had been forced to stretch to cover the shortfall, and b) given the complexity of the manufacturing process, CMO occasionally experienced production problems leading to lost output. CMO has an additional degree of freedom since its capacity is partially flexible. In the 320K total units, it could actually produce as many as 180K units of P1 (but then only 140K units of P2) or 190K units of P2 (but then only 130K units of P1). CSM has been given a basic model of this flexibility at the 300K total unit level that CMO has committed.

Based on their global econometric models from months M-6 through M-2, CSM has determined that the Total Available Market (TAM) for P1 is 300K units and they believe that they should be able to capture a 60% Share of Market (SOM) against their weaker competitor. Therefore the CSM goal for P1 is set at 180K units for month M. Similarly the TAM for P2 has been determined to be 300K units with a possibility SOM of 40% against a stronger competitor for a goal of 120K units for month M. The setting of these goals was heavily influenced by the CMO committed production capacity of 300K total units and the flexibility model between P1 and P2. This plan for month M was passed to CMO at the end of month M-2. (Without loss of generality, this thought experiment assumes no initial safety stock for simplicity of explanation.)

The enterprise sells its products in geographies G1 and G2. Customers Ca and Cb are located in G1 and place their orders with the enterprise's G1 CSM representative. The enterprise also has a CSM representative in G2 who takes orders from customers Cx and Cy. The CSM representatives are in close contact with their respective customers, communicating at least on a weekly basis given the dynamic nature of the end user market.

In geography G1, the CSM representative took orders early in month M-1 from Ca and Cb for the coming month M. Ca placed orders for 40K units of P1s and 40K units of P2s while Cb placed orders for 80K units of P1s and 60K units of P2s. The G1 representative therefore had total orders for 120K units of P1s and 100K units of P2s to pass on to headquarters. But this representative had seen the CSM global estimates of 180K units for P1 and 120K units for P2 and did not feel comfortable passing along such large orders, especially since he felt that Cb was being overly optimistic as they had been so often in the past. Therefore the orders that were passed up to headquarters from G1 were 100K P1 and 80K P2 for month M based on Cb being judged down by 20K units on each of its orders.

The orders that the CSM representative in G2 took early in month M-1 were treated in a slightly different manner. Cx ordered 70K units of P1 and 30K units of P2 while Cy ordered 60K units of P1 and 80K units of P2. The total G2 orders were 130K units of P1 and 110K units of P2. With knowledge of the CSM global estimates for the two products, and with sales commissions clearly in her mind, the G2 representative passed the customers orders quantities directly up to headquarters.

The CSM personnel at headquarters combine the judged geography data early in month M-1. This totaled orders for month M of 230K units for P1 and 190K units for P2, well over the committed capacity of CMO as well as the econometric predictions of CSM. Senior personnel decide to simply proportionally commit the planned use of CMO capacity to the orders to maintain a level playing field among the customers. Any other distribution would leave the enterprise vulnerable to accusations of favoritism. This meant that in the middle of month M-1 the month M planned production was committed to the customers through the CSM Geography representatives as shown in Table 1.

On the first day of month M as the committed production quantities began shipping, customer Ca contacted the CSM representative in G1 to ask for an

Table 1. Initial orders in forecasting thought experiment #1

customer	product	requested	judged	committed
Ca	P1	40K	40K	30K
Ca	P2	40K	40K	25K
Cb	P1	80K	**60K**	45K
Cb	P2	60K	**40K**	25K
Cx	P1	70K	70K	55K
Cx	P2	30K	30K	20K
Cy	P1	60K	60K	50K
Cy	P2	80K	80K	50K
totals		460K	420K	300K

additional 10K units of product P1. This request was passed to headquarters. A few days later, customer Cy contacted the CSM representative in G2 to ask to decrease its order for product P1 by 25K units. As this was within the terms of Cy's contract, the representative agreed but delayed passing this message to headquarters in case customer Cx asked for an increased quantity of product P1. This was based on the fact that Cy initially ordered 70K units of P1, but only 55K units were committed by the enterprise. In the middle of month M, customer Cb contacted the G1 representative asking to cancel its entire committed 25K units of product P2 based on the surprise retraction of a very large order that required the P2s as a component. This was a relief to CMO who encountered production problems and would be unavoidably short of P2s by 10K units by the end of month M. Finally at the end of month M, customer Cx contacted the G2 representative to ship back 10K unused units of P2, and based on the terms of its contract, the appropriate arrangements were made for the return.

This is an extremely simplified but representative version of a typical month for a large manufacturing enterprise under conditions of initial orders exceeding capacity. (An equally interesting version could be constructed for conditions of capacity exceeding initial orders.) This scenario raises many questions. There was enough burst capacity to satisfy more demand, but this does not seem to have been used. A cancellation in one geography could have been used to satisfy demand in the other geography, but was not. These are useful questions concerning execution policy and demand fulfillment, but the focus of this thought experiment is demand forecasting and identification. In the process of historical data collection for use in future demand management computations, the important question is "what was the demand in month M?" A number of candidates must be considered as shown in Table 2.

Candidates 1 and 2 are closely related and might be considered as demand. Candidate 1 is simply the initial orders placed by the customers. Candidate 2 includes the judgment of the enterprise's representatives in the field closest to the customers. But it would be difficult to consider these as an accurate

Table 2. Candidates for recording historical demand

	source	**P1**	**P2**	**P1+P2**
1	customer orders	250	210	460
2	geography judged orders	230	190	420
3	CMO manufacturing actual capacity	180	140	320
4	CMO manufacturing realized capacity	180	130	310
5	CMO econometric models	180	120	300
6	CMO manufacturing committed capacity	180	120	300
7	CSM committed orders	180	120	300
8	CSM committed + post commit requests	165	85	250
9	shipped materials	155	95	250
10	shipped materials - returns	155	85	240

demand signal since the customers knew that the enterprise was limited on capacity from discussions with the field representatives and understood from past experiences that there was a "proportion available capacity according to customer orders" rule in place. This knowledge probably lead them to inflate their initial orders. It therefore might be appropriate to consider the original orders as an upper bound on demand, although in this case it would be a high bound.

Candidates 3, 4, 5, 6 and 7 all represent some view of the market by the enterprise. Candidates 3 and 4 represent CMO's view. Although they include the highest values in this subset, they are actually based on a doubly pessimistic approach. One component is pessimism about CSM's ability to predict the market, and the other pessimism about CMO's own ability to flawlessly manage its own production process. Candidates 5, 6, and 7 represent CSM's best efforts to manage the uncertainties in the market based on their models, understanding on CMO's capacity, and efforts to maintain equality in the treatment of their customers. The econometric models of CSM and the committed capacity of CMO (also based on CSM's econometric models over a longer time horizon) clearly influence the enterprise's view of demand.

Candidates 8, 9, and 10 are closely related. Candidate 8 represents what the enterprise committed plus adjustments during month M (+10K units of P1 requested by Ca, -25K units of P1 cancelled by Cx, -25K units of P2 cancelled by Cb, -10K units of P2 returned by Cy). Candidates 9 and 10 are slightly different ways to account for the same adjustments. It might be appropriate to consider these results as a lower bound on demand.

Assume that the personnel recording demand data select 155K units of P1, 85K units of P2, and 240K units overall as the official "demand" for month M. Note that these are 62%, 40%, and 52% respectively of what the customers initially ordered and 86%, 71%, and 80% respectively of what the econometric models of the CSM organization originally forecast.

5.2 Demand Forecasting Thought Experiment #2

Purchasing decisions are usually made on a cost versus value basis, especially by large enterprises that are purchasing components to be used on their own production lines. The competing items in the marketplace are evaluated from a cost perspective including unit cost reductions for volume purchases, special promotional pricing, transportation and related service costs in acquiring the items, post-purchase maintenance costs, and so on. This sometimes takes the form of a cradle-to-grave life cycle cost analysis. Similarly the enterprise purchaser considers how useful each of the items available in the marketplace could be in increasing the value of the product it is producing. A number of intangibles come into the decision as well. What is the reputation of each of the potential suppliers? Is there a tactical or strategic advantage to picking one supplier over another, or should the purchase be split between multiple suppliers? The larger perspective of the purchaser directly influences the decision too. What is the market forecast for the purchaser's products? Is the stock market up or down? What is the outlook for the national and international economy? Purchases of important and expensive components are seldom simple.

Consider the enterprise from thought experiment #1 in Section 5.1. The time is M+18 (that is, a year and a half later). The "demand" data base has the data entered in month M along with data from a number of months before and all months after M. A few things have changed from the situation portrayed in Figure 8. Customer Cy has acquired customer Cx in geography G2. A new enterprise Cz has started up in geography G2 and has become a customer, but is much more difficult to deal with than either Cx was or Cy is since they aim to compete on low price in their marketplace. Product P1 has

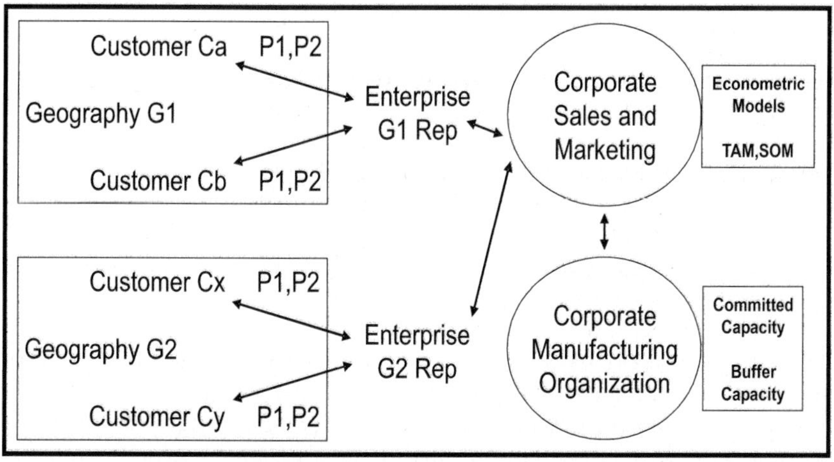

Fig. 8. Participants in forecasting thought experiment #1

been phased out to be replaced by product P3 with expanded functionality. Product P2 which had only been released into the market in month M-4 is now a best seller. The enterprise representative in G1 has taken a different job and has just finished training his replacement. Based on strong enterprise results over the past 6 quarters, CMO has expanded their capacity aggressively and currently have roughly 20% more than the SOM estimated by CSM. The forecasting cycle for M+19 through M+24 is beginning. What analysis technique should be used on the historical data to best support the forecasting process?

As in thought experiment #1, this is a simplified but realistic scenario. On the one hand everything has changed: the customers, the products, the enterprise personnel, and the enterprise capabilities. On the other hand nothing has changed: CSM is running their econometric models, CMO is buffering their capacity, the customers are ordering, and a tactical commitment and a strategic forecast are needed.

Simple projection from the historical data is risky under these conditions. The connection between the data in the "demand" data base and the actual demand at the times it was collected is tenuous on at least two levels. First, the data recorded in fact concerned what was shipped and has only an indirect relation to actual demand as shown in thought experiment #1. Second, little if any of the contextual data that was used by the customers in making their purchasing decisions in the past was captured. It is debatable how much of that data it is possible to capture.

More sophisticated analysis can be devised. Products can be categorized by positions in their life cycles, sizes of intended markets, estimated TAMs and SOMs, and so on. Customers can be sorted by their target markets, overall size, SOM, and others. Analysis can then be driven by finding the historical situation most similar to the current one. But the more complex the situation, the smaller the probability of finding a historical situation that has any relevance to the current one.

5.3 Progress in Practice by Strategic Planning for Uncertainty

In spite of this complexity, since demand uncertainty is the basic constraint in the operation of the overall enterprise system, Intel continues to develop an improved forecasting solution. With the clear realization that predicting the future will never approach perfection, any improvements will be beneficial in the face of rising complexity. The efforts are progressing along three vectors, each related to the operational components described in Section 4.

One way to deal with uncontrollable uncertainty is to characterize its components and then play out a variety of future scenarios with different permutations and combinations of the components. The expectation is not that any of the scenarios will actually occur, but that thinking through possible futures (rather than locking all attention on the one that the enterprise hopes for) will prepare the team to respond more appropriately to inevitable surprises as they occur.

Through the use of case studies we have classified four sources of uncertainty: a) market factors, which includes economic, business, and seasonal cycles, b) product transitions, those of Intel, competitors, and complementors (i.e. software vendors), c) marketing actions, which include pricing, promotions, and advertising, and d) demand management systems, the methods used by Intel to forecast, plan, execute, and measure. Focusing on product transitions, we have developed two new tools for the demand management system that are helping assess and account for uncertainties [7, 15].

The first method, Product Transition Index (PTI), is an assessment tool used to gather information about the product transition of interest. PTI is a model containing eight vectors that explore the pace and success of a transition. A total of 65 factors identified in our research are scored to complete the PTI. The scoring process requires integrating the knowledge of teams across sales, marketing, planning, manufacturing, and engineering.

The second is the application of the idea of Transition Playbooks. The intent of the playbook is to enable strategists and managers to map out the tactics the organization will use to respond to risks that might impact the transition. Developing a playbook for a product transition encourages advance planning and analysis so that the business functions can respond to disruptions quickly and in concert under the stress and time constraints of market dynamics to keep the transition on track.

5.4 Progress in Practice by Improved Tactical Information Integration

With the overall complexity of the enterprise and its ecosystem, it is doubtful that it will be possible for the foreseeable future to collect all relevant tactical information in a manner that will support advanced computation. There may however be ways to improve tactical forecasting using innovative methods for integrating the information that a variety of Intel personnel have.

Analyzing the forecasting process, we learned that four fundamental sources of noise cause difficulty in determining market demand: a) current data, such as orders, work in progress in factories, and inventories, b) market assessments, such as intelligence on how appealing Intel's products (and competing products) might be to the market, c) tactical goals, the success metrics Intel has set for specific products such as sales volume, average price, and share of market, and d) marketing plans, such as which products to promote, how to price them, and how to take advantage of technological and manufacturing capabilities. Nearly all the pitfalls we have discovered in forecasting can be linked to one or a combination of these factors.

One fundamental problem in managing forecasts is the process through which the hard data is created, judged, and passed from group to group. There is a general lack of credibility based on past performance, so each group feels the need to adjust the information based on any number of experiences and heuristics. Groups preparing to publish data are aware of how other groups might judge the data and are prone to adjusting numbers in anticipation of future judgment. Another fundamental problem is data sets themselves do not convey any specific meaning. Meaning can be inferred from how the data compares to expectations or previously published data, but the numbers themselves cannot explain the strategies Intel and its customers are employing or the uncertainties they are facing. Decentralized organizations must transmit information and intelligence from employees who have it to employees who need it to make decisions and plans. Intel has many informal networks that attempt to move such knowledge across the organization, but these networks have many failure modes including turnover of employees in key positions, limited bandwidth of each individual and team, and difficulty systematically discovering the important information to be learned.

To address these problems, we have successfully developed an adaptation of the well-known "Decision Market" (DM) concept [14]. We generate a number of well-specified possible forecasts and our diverse team members buy and sell "shares" in these optional forecasts. At the end of the decision market period, the shares that have the correct answer become valuable and the owners earn rewards while shares in incorrect answers become worthless. When a sufficiently large group of knowledgeable participants estimate a future event, they have an uncanny ability to get that estimate right since the cumulative knowledge of all forecasters is greater than a single forecaster. As the system becomes more complex, no single forecaster can reliably comprehend or integrate all relevant factors. Mathematically, if there are more than two sources of imperfect forecasts having independent information, there exists some optimal blend of predictions that will be better than any single forecast.

Results to this point have been very impressive and we are proliferating the method across multiple product groups. We believe that three factors enable forecast markets to outperform other types of forecasting systems. First, the features of anonymity and incentives work together to draw out good information. Incentives encourage participants to search for the best information they can find and reward trading behavior that is unbiased. Anonymity helps prevent biases created by the presence of formal or informal power, the social norms of group interaction, and expectations of management. Second, the simple mechanism of aggregating data through a market smoothes results over time that is useful for guiding supply as opposed to other methods that sometimes cause thrash in factory operations. Third, increasing the diversity of a pool of participants increases the accuracy of the collective forecast. As long as each additional participant brings some information, adding more, diverse opinions improves the collective judgment. This condition holds true in many cases because good information tends to be positively correlated and sums, while errors are often negatively correlated and cancel.

5.5 Progress in Practice by Control-Relevant Forecasting

Current research is focused on understanding the effects of demand forecast error on the manufacturing and supply chain operational systems [30]. The demand forecast is treated as an external measured disturbance for a multi-degree of freedom feed forward feedback controller of an inventory management system as described in Section 4.2. Because forecast error will be multi-frequency in nature, the effect of error in different frequency regimes has been studied. The closed-loop transfer functions describing forecast error have been derived and the effect of erroneous forecasts studied in the time and frequency domains. A simultaneous perturbation stochastic approximation optimization algorithm has been implemented to develop optimal tuning strategies for a variety of scenarios. By understanding the impact of different types of forecast errors, we hope to better focus our efforts to reduce those errors that are most damaging to the performance of the overall system.

6 The Illusion of Control

From one perspective, enterprises appear to be continuously improving their operational control even as the complexity of their core functions and of their marketplace continues to rise. For example, over the past 20 year Intel Corporation has developed a) improved decision algorithms for designing, building, and operating manufacturing facilities [36], b) optimization methods for acquiring materials, planning material releases into the factories, and sizing inventory buffers [21], and c) financial instruments and related methods to mitigate development and procurement risk and manage business continuity [23]. These efforts have generated billions of dollars for Intel.

From another perspective, enterprises continue to struggle in their efforts to manage their markets. This is particularly apparent in demand forecasting although not surprising given the rising complexity in all facets of the enterprise's activities. The most obvious result is stagnation in the financial growth of the enterprise even though there are continual significant improvements in the operational methods of the company. Operational methods to respond to the market are improving in spite of complexity while forecasting methods to predict the market are declining due to complexity. There is an illusion of control.

Since the operational methods require market forecasts as inputs, there will be an upper limit to the improvements that can be expected to the operational methods. Methods can be continuously improved to design, build, and operate factories, but if the factories were timed and sized on faulty forecasts, little benefit is derived by the enterprise. Methods can be continuously improved to develop new products with advanced features and functions, but if the projected demand for the new products is mistaken little benefit accrues to the corporation. In the best case, the growth of the enterprise stops. In the worst case, with wasted capital expenditures and missed market opportunities, the enterprise fails.

There is at least one plausible explanation for the illusion of exercising improved control over the enterprise as an artificial complex adaptive system. It has to do with the relationship between signal and noise. Decision makers with bounded rationality working in a complex system can easily confuse signal and noise. The result is too often an unnecessary response to the noise. It is possible that the decision support tools that have been put into beneficial enterprise practice have simply helped distinguish the signal and more appropriately react to it while spending less time and effort chasing the noise. In any case, while the operational situation has improved, enterprises still experience repeated failure to accurately forecast. On a scale of 0 to 100, an improvement in operational decision making resulting in a performance gain from 20 to 30 is dramatic, but there is still a long journey remaining. Unfortunately the improvement supports, in fact reinforces, the illusion of control.

One part of dispelling the illusion lies in communications. As the enterprise becomes exponentially more complex over time at every level, people's view of the enterprise and their role in it changes. Under an assumption of bounded individual capacity, increasing complexity forces people to become broader but less deep, or deeper but less broad, or to ignore some tasks while focusing on others. In any case, actually performing the core of their work consumes an ever increasing percentage of their bandwidth. This often means that time spent in high quality communication with other enterprise personnel decreases. Rising complexity would indicate just the opposite should be occurring, that is increased communication. In addition the rise in complexity and the resulting decreased access to the whole of the enterprise results quite naturally in an increased aversion to risk. This may be just the point in the trajectory of the enterprise that a radical innovation might decrease the complexity of the enterprise. But a rising aversion to risk makes the process of adopting a radical innovation that much more difficult. Methods must be developed to better cope with this situation and these methods will have to achieve the difficult cultural goal of dispelling the illusion of control.

As the enterprise and its ecosystem moves along a trajectory from deterministic to complex to chaotic, its theoretical and practical ability to predict or forecast diminishes dramatically. Conventional wisdom instructs that it is better to strive to be agile than to try to be clairvoyant, but this is obviously not useful advise under the current circumstances. While it is certainly necessary to continuously improve operational agility and efficiency (and seductive to do so to the exclusion of other activities), it is clearly not sufficient. Breaking the illusion of control will result in a major shift of emphasis to the continuous improvement of demand management and forecasting. Current complexity theory indicates that this problem will be extremely difficult if not impossible to solve. This is the commercially most important frontier for applied complexity theory and the focus of our current efforts. Finding a process to manage complexity in the enterprise is the grand prize.

References

1. Armbruster, D., Marthaler D., Ringhofer, C., Kempf, K.G. and Jole, T.C.: A continuum model for a re-entrant factory. Operations Research **54** (2006) 933–950
2. Aytug, H., Kempf, K.G., and Uzsoy R.: Measures of subproblem criticality in decomposition algorithms for job shop scheduling. Inter. J. of Production Research **41** (2002) 865–882
3. Bean, J. W., Devpura, A., O'Brien, M. and Shirodkar, S.: Optimizing supply chain planning. Intel Technology Journal **9** (2005) 223–232
4. Beaumariage, T. and Kempf K.G.: The nature and origin of chaos in manufacturing systems. In Proc. of the IEEE/SEMI Advanced Semiconductor Mfg. Conf. (1994) 169–174
5. Braun, M.W., Rivera, D.E., Flores, M.E., Carlyle, W.M., and Kempf, K.G.: A model predictive control framework for robust management of multi-product multi-echelon demand networks. Annual Reviews in Control **27** (2003) 229–245
6. Chase, C., Serrano, J. and Ramadge, P. J.: Periodicity and chaos from switched flow systems. IEEE Transactions on Automation Control **38** (1993) 70–83
7. Erhun, F., Concalves, P. and Hopman J.: The art of managing new product transitions. MIT Sloan Management Review **48** (2007) 73–80
8. Fisher, K., Holland, S, Loop, K., Metcalf, D. Nichols, N. and Ortiz, I.: Managing goods and services acquisition risks. Intel Technology Journal **11** (2007) 115–126
9. Geiger, C., Kempf, K.G. and Uzsoy, R.: A tabu search approach to scheduling an automated wet etch station. Journal of Mfg. Systems **16** (1997) 102–116
10. Golda, J. and Phillippi, C.: Managing new technology risk in the supply chain. Intel Technology Journal **11** (2007) 95–104
11. Goodman, A., Hinman, E.J., Russell, D., and Sama-Rubio, K.: Managing product development risk. Intel Technology Journal **11** (2007) 105–104
12. Hanson, D., Armbruster, D. and Taylor, T.: On the stability of re-entrant manufacturing systems. In Proc. 31st Mathematical Theory of Networks and Systems (1999) 937–940
13. Hilton, C., Mazenko, G.,Solomon, L. and Kempf, K.G.: Assembly floor layout and operation: quantifying the difference. In Proc. IEEE Inter. Symp. Semiconductor Mfg. (1996) IV–3
14. Hopman, J.: Managing uncertainty in planning and forecasting. Intel Technology Journal **9** (2005) 175–184
15. Hopman, J.: Using forecasting markets to manage demand risk. Intel Technology Journal **11** (2007) 127–136
16. Kahneman, D.: Maps of bounded rationality - Nobel Memorial Lecture. (2002)
17. Kempf, K.G.: Intelligently scheduling semiconductor wafer fabrication. In Zweben, M. and Fox M., eds.: Intelligent Scheduling. Morgan Kaufman, San Francisco (1993) 517–544
18. Kempf K.G.: Optimizing performance over the factory life-cycle. Intel Technology Journal **2** (1998) 38–43
19. Kempf, K.G., Uzsoy, R., and Wang, C.: Scheduling a single batch processing machine with secondary resource constraints. Journal of Mfg. Systems **17** (1998) 37–51
20. Kempf, K.G.: Managing supply-demand networks in semiconductor manufacturing. In Armbruster, D., Kaneko, K. and Mikhailov, A., eds.: Networks of Interacting Machines. World Scientific Co., Singapore (2005) 67–100

21. Kempf K.G.: Special Issue: Managing International Supply and Demand at Intel. Intel Technology Journal **9** (2005)
22. Kempf K.G. and Uzsoy, R.: Integrating workload-dependent lead times and safety stocks into mathematical programming models for production planning, to appear. Inter. J. Prod. Res.. (2007) *in press*
23. Kempf K.G.(ed.): Special Issue: The Spectrum of Risk Management in a Technology Company. Intel Technology Journal **11** (2007)
24. Knutson, K., Kempf, K.G., Fowler, J. and Carlyle, M.: Lot-to-order matching for a semiconductor assembly and test facility. IIE Trans. Sched. and Logistics **31** (1999) 1103–1111
25. McDonald, C. J.: The evolution of Intel's COPY EXACTLY! technology transfer method. Intel Technology Journal **2** (1998) 9–14
26. Moore, G. E.: Cramming more components onto integrated circuits. Electronics **38** (1965) 114–117
27. Mosekilde, E., Larsen, E.R. and Sterman, J.D.: Coping with complexity: deterministic chaos in human decision making behavior. In Casti, J.L. and Karlqvist, A., eds.: Beyond Belief: Randomness, Prediction, and Explanation in Science. CRC Press, Boston (1991) 119–229
28. Perdaen, D., Armbruster, D., Kempf, K.G. and Lefeber, E.: Controlling a reentrant manufacturing line via the push-pull point. Inter. J. Prod. Res. (2007) *in press*
29. Rem, B. and Armbruster, D.: Control and synchronization in switched arrival systems. Chaos **13** (2003) 128–137
30. Schwartz, J.D., Rivera, D.E. and Kempf, K.G.: Towards control-relevant forecasting in supply chain management. In Proc. American Control Conference (2005) 202–207
31. Shirodkar, S. and Kempf, K.G.: Supply chain collaboration through shared capacity models. Interfaces **36** (2006) 420–432
32. Simon, H.: Rational decision-making in business organizations - Nobel Memorial Lecture (1978)
33. Smith, A.: An Inquiry into the Nature and Cause of the Wealth of Nations - Book IV. In Cannan, E. eds. Methuen and Co., London (1776) Chapter 2, IV 2.9
34. Smith, K. and Kempf, K.G.: Application of model predictive control and optimization methods to semiconductor manufacturing supply-side inventory replenishment. In Proc. IEEE Inter. Symp. Semiconductor Mfg. (2005) 35–38
35. Smith, S., Keng, N. and Kempf, K.G.: Exploiting local flexibility during execution of pre-computed schedules. In Famili, A., Nau, D.S. and Tong, S.H., eds.: Artificial Intelligence Applications in Manufacturing. MIT Press, Cambridge (1992) 277–292
36. Splinter, M.(ed.): Special Issue: Manufacturing Overview. Intel Technology Journal **2** (1998)
37. Thietart, R. and Forgues, B.: Chaos theory and organization. Org. Science **6** (1995) 19–31
38. Thietart, R. and Forgues, B.: Action, structure and chaos. Org. Science **18** (1997) 119–143
39. Vaidyanathan, V., Metcalf, D. and Martin, D.: Using capacity options to better enable factory ramps. Intel Technology Journal **9** (2005) 185–192

40. Vargas-Villamil, F.D., Rivera, D.E. and Kempf, K.G.: A hierarchical approach to production control of reentrant semiconductor manufacturing lines. IEEE Trans Control Sys Technology **11** (2003) 578–587
41. Willems, S.P., Tian, F. and Kempf, K.G.: An iterative approach to item-level production and inventory planning. European Jour OR. (2007) *submitted*
42. Zafra-Cabeza, A., Riado, M.A., Camacho, E.F., Kempf, K.G., and Rivera D.E. Managing risk in semiconductor manufacturing: a stochastic predictive control approach Control Engineering Practice **15** (2007) 969–984

Part II

Logistics and Production

Benefits and Drawbacks of Simple Models for Complex Production Systems

Oliver Rose

Dresden University of Technology, Institute of Applied Computer Science, 01062 Dresden, Germany oliver.rose@tu-dresden.de

1 Introduction

During the last decade, the role of simulation in improving semiconductor factory operations has been subject to considerable growth. Simulation activities were integrated into the decision-making process of the factory managers of all major semiconductor manufacturers as a powerful tool for understanding the complexity and enhancing the performance of their production facilities. In the past, simulation was mainly used for high level capacity planning. But now its application area also includes solving a variety of operational questions [1].

The main reason for the increased use of simulation in the operational planning lies in the size, complexity, and cost of nowadays semiconductor fabrication facilities (fabs) generated by market and business pressures coupled with the hard limits of physics. Traditionally, many operational decisions in the industry were made based on prior knowledge, experience, and intuition. This is no longer appropriate. There is a need to build a meaningful model of the factory and to perform simulation studies to examine its operational problems. At the moment, there is no other analysis tool available that is capable to support meeting production goals while avoiding unnecessary investments or other costs.

Simulation is used in such areas like capacity planning, scheduling, bottleneck identification, impact of new products or process flows, layout analysis, equipment modeling, factory ramp-up modeling, and operator modeling. Typical performance measures are cycle time, throughput, inventory levels, equipment usage, and cost.

In the early days of operational modeling and simulation in the semiconductor industry, models were often only used once to solve a particular problem. At the moment, there is a trend among all semiconductor manufacturers to develop persistent models that are used over the entire life cycle of a factory from the design phase until it is closed down [2].

Despite the almost ubiquitous use of simulation for operational planning in the semiconductor industry it has to be noted that simulation is not always the right tool. In certain scenarios there is still the need for alternative solution approaches like queuing networks or mathematical programs. At the moment, however, there are no clear guidelines available under which conditions simulation should be replaced by other methods of analysis.

1.1 Semiconductor Manufacturing

Manufacturing semiconductor products is among the most complex production processes known today [3, 4]. This has mainly two reasons: the production process itself and the production planning and control.

Overview of the Production Process

The process of manufacturing semiconductor chips like processors or memory components is divided into two parts, the front end and the back end. The front end factories deal with processing a thin disc of silicon, the wafer. During a series of thin film operations structures are built on this wafer. The front end processing is finished as soon as the wafer is ready to be sewn into pieces, the dies. In back end factories electrical contacts are attached to the dies and they are packaged in plastic or ceramic. After burn-in and final test the semiconductor chips are ready to be sold.

In the following we describe the front end manufacturing process in more detail. The structures on the wafers consist of several layers, up to 50 for memory products. Each layer is formed by about the same sequence of thin film operations.

1. Cleaning
2. Oxidation (to separate the layers electrically)
3. Photo lithography
   a) Coating with light-sensitive varnish
   b) Projection of ultraviolet light or x-rays through a mask to generate pattern on coating
   c) Resist development
4. Etching
5. Deposition or ion implantation
6. Photo resist strip
7. Testing

In total, several hundred of processing steps are required to produce a wafer.

The layout of the front end or wafer fabrication facilities is similar to classical job shops. I.e., tools performing the same operation, for instance, etching, are grouped together in the same area of the shop floor. The wafers are manufactured in lots. Typical lot sizes are 25 or 50 wafers. After finishing a processing step the lot is moved to the next group of processing equipment which might be located in a distant area of the shop floor. Lots are moved either by operators with pushcarts or by an automated material handling system (AMHS) .

Overview of Production Planning and Control

The main purpose of production planning and control is to achieve given performance goals like throughput or on-time delivery. The application of its methods has to lead to a competitive advantage by running the fabrication facility more efficiently. To reduce the complexity of the planning and control task a hierarchically structured set of methods is implemented. A typical way to separate the methods is according to their planning horizons. Usually, three time frames are considered:

- Strategic planning (long-term): factory locations, product lines, marketing strategies, quality improvement strategies, etc.
- Tactical planning (mid-term): staffing schedules, maintenance schedules, advertising, etc.
- Operational planning (short-term): material flow control, lot release, dispatching, scheduling, break-downs, setups, etc.

Of course, the ultimate goal of all planning approaches is to maximize revenue. Unfortunately, it is rather difficult to assess the direct impact on revenue by improvements in operational planning, e.g., the financial benefit of buying a new machine or replacing a certain dispatch rule. As a consequence, financial aspects are replaced by material-flow oriented aspect for the performance assessment of short-term planning methods. It is much easier to measure, for instance, cycle times, throughputs or on-time delivery percentages and to compare planning approaches based on these measurements. There are numerous methods available to control the material flow that are customized to the special requirements of semiconductor manufacturing [5].

1.2 Modeling and Analysis of Material Flows in Semiconductor Wafer Fabrication Facilities

The way of modeling a factory and the analysis methodology are closely related. The selection of a particular modeling technique already limits the range of analytical tools that can be used, and vice versa. There are two categories which are already sufficient to classify the models used to analyze material flows in semiconductor manufacturing. The first category is the amount of stochastic detail that is contained in the model, i.e., whether it is static or dynamic, in other words, deterministic or stochastic. The second category is the level of model details, where the level ranges from very abstract to almost realistic. During the decision process about which type of model suits best for the problem that has to be solved several questions have to be asked.

- Do the analysis tools available for this type of model help to solve the problem?
- Is the level of detail of the result adequate for the problem?
- Is it possible to build the model in time?

- Is it possible to evaluate/analyze the model in time?

In general, the model type decision will be a compromise because the answers to the above questions will not always be positive for a given problem and a particular model type candidate. For instance, static and abstract models usually take small amounts of time for building and evaluation but provide only rough performance estimates. If a detailed answer is not necessary this is definitively the way to go but usually more information is required.

In the following we present a number of analysis techniques that are applied in factory performance analysis , their advantages, and their disadvantages. Here, we do not comment on the time and effort that is necessary to build the models for these methods. Our main interest is evaluation speed and adaptability to real factory problems.

- Spreadsheets
  Model type: static, from abstract to detailed
  + Fast evaluation
  + Comprehensive model
  – Rather simple and coarse model
  – Only basic factory data used
  – No dynamic effects covered
  – Small range of performance measures
- Stochastic petri nets
  Model type: dynamic, from abstract to medium level of detail
  + Fast evaluation (if the model is not too large)
  + Dynamic effects covered
  – Limited to certain probability distribution types
  – Difficult to model certain dispatch rules
- Queuing networks
  Model type: dynamic, from abstract to medium level of detail
  + Reasonable evaluation speed (some evaluation approaches can be rather slow)
  + Dynamic effects covered
  + Large body of literature with factory problem solutions
  – Difficult to model real factory in detail, e.g., real dispatch rules
- Simulation
  Model type: dynamic, from abstract to high level of detail
  + Rich and realistic models
  + Dynamic effects covered
  + Large body of literature with factory problem solutions
  – Slow model evaluation

As a consequence, simulation, or, to be more precise, Discrete-Event Simulation (DES) [6], was and is the main approach to analyze and solve operational problems in many industrial fields.

Before the performance of the factory can be assessed by simulation we need an appropriate simulation model. Depending on the goal of the study models of different complexity can be used ranging from very detailed and close to the real factory to coarse and abstract. For almost all industrial studies full detail models are used. In academic studies both detailed and simple models are being applied.

Typical model components for complete factory models are

- tool set,
- product routes,
- operators, and
- material handling systems.

1.3 Planning and Controlling the Flow of Material

There are several planning and operation control tasks that have to be performed in semiconductor manufacturing. The planning problems are usually almost the same for front-end and back-end factories. First, the company has to make the decision to build a new factory because of increased customer demand for new products that can no longer be fulfilled by the existing factories or by buying these products from a competitor.

The next step is to forecast the product mix for this factory and the required capacity. Based on these forecasts and the process plans of the considered products the planners determine the required tool set taking into account the process and capacity specifications of the tool manufacturers.

With this tool set the layout planning process is started. At this point the material handling system comes into play because it is required to move the product lots from one tool to the next tool. After studying several layout alternatives the final layout of the shop floor including tools and material handling system is found.

In a real factory there are additional planning requirements like supply of clean air, chemicals, electricity, etc. which will not be considered here. As soon as the factory construction is finished and the supply of all required materials is functional the test production starts. This phase is called ramp-up. Tool group by tool group the shop floor is populated and process engineers run test operations to adjust the tool parameters in order to run processes within specifications. During this low-volume test production phase factory simulation models are built that grow with the number of tools in the tool set on the shop floor. These models are used to determine the shortest path from test to full operation of the factory.

When the tool set is complete and operational the usual factory planning and control system has to be established covering forecasting customer demands, planning the product mix, and scheduling the work for the tools.

During the operational phase of the factory additional planning tasks have to be performed. If tools become old and new processes are required for production these machines have to be replaced by better equipment. If the factory capacity is becoming too small faster and/or more tools have to be bought. To reduce the effect of unexpected tool down-times a preventive maintenance framework has to be planned and operated. The operator planning department needs appropriate staffing tools. New scheduling methods and paradigms appear in the literature and come into practice at competitors factories. After its operational phase the factory is ramped down and the process of fading out of products and shutting down tools has to be supported by studies in order to make it smooth and cost-effective.

During the whole life cycle of a semiconductor factory planning and control methods are applied but each phase has its own particular methods and problems. The focus will be on controlling the flow of material in an operational factory because there we find the highest demand for using simulation as a performance assessment tool. Another important field of application for simulation is capacity planning.

1.4 Operational Material Flow Control

The material flow control consists of two parts: the lot release part and the dispatching or scheduling part.

At lot release the intention is to find an optimal release time for a particular lot. Several authors claim that this part is more important than the dispatching part [5]. The reasoning behind that conjecture is that if a lot is started at the optimal point in time into the factory it fits smoothly into the flow of the lots that are already processed inside the factory and only little has to be done to adjust its ranking at particular tool groups and FIFO dispatching is adequate to achieve all performance goals. So far, however, there is no study available that proves this conjecture for the semiconductor industry.

With respect to lot release there are two approaches. The first one is the simple, open-loop-type control which is currently applied in most semiconductor factories. The desired throughput, i.e., the number of wafer starts per week, is given by the planning department. Then the distance between two lot starts is the reciprocal of the throughput transformed into lots per week. This is an open-loop control procedure because no information from within the factory affects the lot release time computation. The procedure is very simple and straightforward to implement.

The second type of lot release methods are closed-loop control approaches. Here information from inside the factory is used to find the optimal release time for a lot. The intention is to obtain a smooth and balanced material flow in the factory by controlling the amount of material that flows into the factory. If the factory is congested the release rate is throttled down. If the factory is below its capacity limit more material will be released to it. The release control methods differ in their definitions of the terms “congested” and “below capacity”. All methods have to define a factory status variable or index that has to be kept in a given window by the release control algorithm. In the following we provide an overview of the closed-loop release control methods for semiconductor manufacturing that can be found in the literature. The majority of the studies deal with front-end factories.

1.5 Practical Problems of Detailed Models

In semiconductor manufacturing, the factory simulation models are usually given as input files for commercial or academic simulation packages. Up to now there is no modeling language available that is accepted among all simulator suppliers. An effort was made to establish the Testbed Data Format for simulation models as quasi-standard but only a limited number of simulation packages are able to read this format.

The modeling capabilities of commercial software are limited to the requirements of the majority of the customers. For the remaining characteristics, it is up to the simulation experts of the semiconductor company to extend the capabilities of the simulations software if this is possible. If the simulator cannot be customized to model a certain characteristic of the real factory, it has to be determined to which extent this lack in modeling capability reduces the usefulness and applicability of the simulator results. It is, however, hard to assess the effects due to lacking capabilities because the only reference which can be used for comparison is the real factory. Even if full modeling capability cannot be achieved, this factory model will be better than having none.

Apart from missing modeling capabilities, the main problems in the industrial environment are to find the correct model parameters and to keep the models up to date. In this respect there are several problem areas.

Tool sets are permanently changing. Due to new products tools are replaced on a regular basis. These tools are usually different from the replaced ones in almost all characteristics. In many cases, the actual characteristics deviate from the manufacturers specifications. As a consequence, it will take a considerable amount of time until the true performance details of the new tool are known. During that time, special care has to be taken when interpreting the simulation results.

The routes and the operations are permanently changing. These changes happen due the same reason as the tool set changes and have similar consequences. It takes time until a new route or even a new operation can be accurately characterized.

We are dealing with the statistical analysis of random measures. It is very difficult to estimate the correct probability distribution type for the breakdown behavior of the tools. Although nowadays all machines are permanently exchanging data with the MES (Manufacturing Execution System) there are still data collection problems due to a fuzzy definition of the term "breakdown". Due to political reasons the unproductive percentage of a tool is intended to be kept as low as possible because engineers and staff who responsible for error-prone equipment are often running into problems with shift managers and the planning department. Therefore, it can happen that short tool failures are not taken into consideration for the breakdown statistic of a tool. Only if a failure lasts long enough or is severe enough such that it becomes obvious that the tool is unproductive, it is added to the statistics collection. The long failures, however, are less frequent than the short ones and, as a consequence, the data set that can be used to estimate is too small to be significant for TTF (Time To Failure) and TTR (Time To Repair) distribution estimation. Due to the permanent replacement of tools there is no time to collect sufficient failure data from the tools.

2 Simple Models for Complex Manufacturing Systems

In this section, we present two studies where a very simple model is used to mimic the behavior of a complex wafer fab. These studies show both the strengths and weaknesses of applying models with a high level of abstraction compared to the real world system.

The first study outlines the construction of the simple model and how it can be used to determine the inventory level trajectory of a wafer fab after a serious machine breakdown [7].

In the second study, we determine the parameters of the simple model from a full-detail simulation model and compare several performance measures of the simple and the complex model [8].

The two above studies indicate that simple simulation models are useful for analyzing the behavior of complex wafer fabs in certain scenarios.

2.1 WIP Evolution After a Bottleneck Work Center Breakdown

In numerous studies (e.g., [9]) the long-term behavior of the fabrication facilities in terms of mean cycle times, average inventory levels, etc. is determined. These studies help to find dispatch rules for achieving given requirements such as a certain probability to meet due dates.

There are fab phenomena, however, that cannot be explained with such classical simulation approaches because only long-term or steady-state performance criteria are taken into consideration.

One of these phenomena is the observation of huge amounts of work in progress (WIP) even weeks after a catastrophic failure of the bottleneck work center, i.e., all machines of the work center that constrains the fab capacity are down for a few days. This particular fab behavior was reported by fab managers of a large memory fab.

In contrast to classical simulation studies, we need to study the evolution of the fab, for instance the WIP over time, after the catastrophic event and not the long-term behavior of the fab. To this end, we apply simulation techniques that were used to compare the performance of routing algorithms in communication networks after a node breakdown [10]. The major disadvantage of such techniques is the fact that instead of tens of simulation runs for classical studies, hundreds of them are required for studies of the system behavior over time.

Hence, we first have to develop a simple fab model that shows the behavior of a complete fab model with respect to our problem. To carry out the study with the complete fab model is not possible due to the enormous run length of hundreds of simulation replications. We carry out several experiments that show how the simulation results change due to modifications in the model and in the dispatching rules. It turns out that the model is capable of reproducing the building up of inventory after catastrophic failure. In addition, we show that due-date based dispatch rules such as critical ratio lead to a worse fab performance in terms of WIP level and cycle time variations than FIFO dispatching for the period after the catastrophic failure. This is very much in contrast to the steady-state results where due-date based dispatching clearly outperforms FIFO dispatching [11].

Simple Factory Model

Due to the layered nature of semiconductors, the wafers visit sequences of machines several times, i.e., they proceed through the fab in cycles. Memory chips may have up to 30 layers. This cyclic visiting sequence of machines is responsible for a large part of the logistic problems of wafer fabs because lots with different due date requirements compete for the machines. If due-date based dispatching is applied, lots that are closer to their due dates are preferred at the cost of waiting time for the other lots. The consequences of this permanent reordering of lot priorities will become apparent in Section 2.1.

To make a simulation study feasible with respect to running time, we require a fab model that shows the aforementioned behavior, but is considerably less complex in terms of the number of machines. Figure 1 shows the proposed factory model. It consists of a bottleneck work center, a delay unit, and a control unit. The bottleneck work center determines the fab performance to a large extent [4] and is therefore modeled in detail considering the number of machines, processing times, and dispatch rules. The rest of the machines are modeled as a delay unit. Each time a lot leaves the bottleneck work center it is delayed for a random amount of time before it either leaves the fab or it requests a bottleneck machine once more. The control unit decides whether the required number of layers/cycles have been finished, and directs the lots to the fab exit or back to the bottleneck work center.

For the simulation experiments we considered the following parameters. These parameters are not chosen arbitrarily but in conformance with current wafer fabs.

There are four products that are manufactured by the fab. Each product has a lot start rate of 0.0925 lots/hour where the time between lot starts is constant. The processing times at the bottleneck work center are 0.7, 0.9, 1.1, and 1.3 hours, respectively. The processing times are assumed to be identical

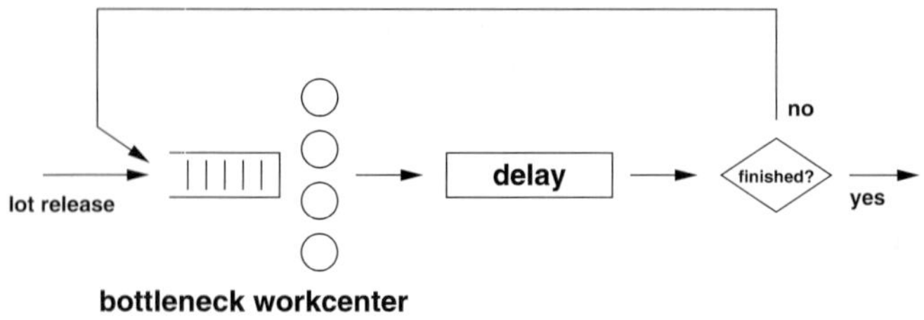

Fig. 1. Factory model

for each layer. All products have 10 layers. This results in a bottleneck work center load of 92.5 % (= 4[products] · 0.0925[lots/hour] · 0.25 · (0.7 + 0.9 + 1.1 + 1.3)[hours] · 10[cycles]/4[machines]). This is a reasonable load for the bottleneck work center if the average percentage of downtimes is assumed to be less than 7.5 %.

The average period of time spent in the delay unit is 125 hours for each product and layer. To facilitate the application of due date based dispatch rules, we require the processing times of all machines. Hence, we have to partition the delay time into a constant amount of 50 hours of processing time and a random amount of time with an average of 75 hours for waiting and other non-processing times. Details on the distributions of the waiting time are given in the next section. We set the lead flow factor to 2.5, i.e., the ratio of cycle time and raw processing time is intended to be 2.5. For the bottleneck tool this is to be achieved by adequate dispatching, whereas the rest of the fab represented by the delay unit is always conforming to the intended flow factor due to defining an average delay time of 125 hours and a processing time of 50 hours. At lot start, a due date of current time + 10 · (125 hours + 2.5 · bottleneck processing time) is assigned to each lot.

The bottleneck work center consists of four machines. We consider the following four dispatch rules.

FIFO (First In First Out) The waiting lots are scheduled in the order of their arrival. This rule is the only one considered that does not lead to a re-ordering of queued lots.

SPTF (Shortest Processing Time First) The lots are scheduled according to their processing times. Lots with the shortest processing time are taken from the queue first.

CR (Critical Ratio) Each time a lot has to be taken from the queue, the following index is assigned to each of the waiting lots:

$$\mathrm{CR} = \frac{\text{due date} - \text{current time}}{\text{remaining processing time}} .$$

The lot with the smallest index value is chosen for processing. As a consequence, lots that are closer to their due dates are preferred.

ST (Slack Time) Compared to CR, the index used for scheduling the lots is based on a difference and not on a ratio:

$$\begin{aligned} \text{ST} &= \text{due date} - \text{current time} \\ &\quad - \text{remaining processing time}\,. \end{aligned}$$

Again, the lot with smallest index is removed from the queue. The ST rule does not increase the priorities as fast as CR when lots are about to miss their due dates.

In the latter three cases, ties are broken by the FIFO rule. In contrast to the first two rules, the latter two rules take into account the due dates of the lots.

The above factory model is used to determine the fab performance after a complete bottleneck work center breakdown. The simulation model was implemented in ARENA [16], and the statistical post-processing algorithms were developed by the author.

As a first step, we determine empirically the start-up phase of the system [6]. Having started with an empty system, the estimated end of the transient phase is at about 1500 hours.

Hence, we schedule the breakdown at 2000 hours of simulated time. All machines of the bottleneck work center become unavailable for processing. After 50 hours of repair time all machines start processing again. The simulation ends after 5000 hours of fab time.

During each simulation replication of 5000 hours, we record WIP changes and cycle times of finished lots. To reduce the amount of data and to facilitate the synchronization of the measurements from different replications, we apply the following method. The simulated time is divided into 10-hour intervals. For each 10-hour interval, we compute the sample mean of the cycle times of the lots that leave the fab during this period. With respect to the WIP, we compute the time-based average of the WIP level during each 10-hour interval, i.e., each WIP level observed during this period is weighted by the percentage of time during which it is kept. For each replication, we obtain a condensed WIP and cycle time sequence of 500 values each (5000 hours/10 hours).

To obtain statistically useful results, each experiment is repeated 500 times. The curves shown in the rest of this section are based on averaging the condensed WIP and cycle time sequences of 500 simulation replications. The 95% confidence intervals of the WIP sequence of the fab with FIFO dispatching are shown in Figure 2. All other experiments lead to approximately the same ratios of confidence interval half-widths and sample means.

Modeling of the Delay

We begin our study with a set of experiments where we intend to determine the effect the delay time model on the behavior of the fab model. First, we

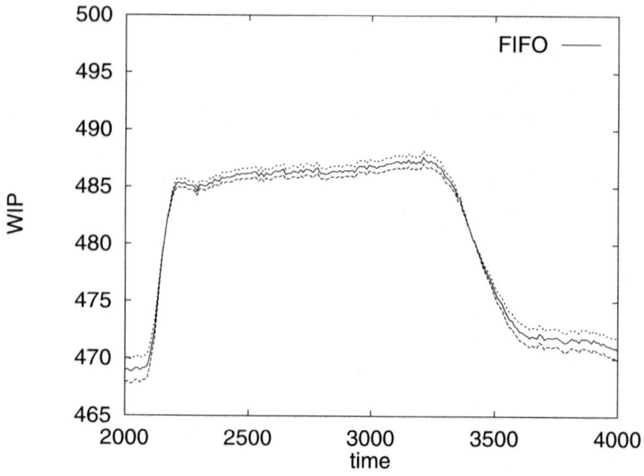

Fig. 2. 95% Confidence intervals of the WIP sequence.

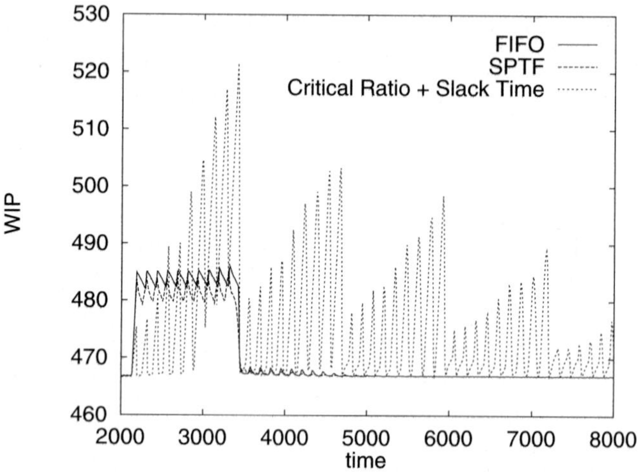

Fig. 3. WIP evolution for constant delay.

consider a delay unit where the delay time is constant, i.e., 125 hours for all products and all layers.

Figure 3 shows the WIP evolution for constant delay under the regime of the four considered dispatching rules at the bottleneck work station. In order to be able to show some interesting effects, the run length of these replications was extended to 8000 hours.

The two dispatch rules that do not consider due dates, FIFO and SPTF, show a fundamentally different behavior from CR and ST. During the breakdown the WIP increases very fast for FIFO and SPTF, stays approximately constant for about 1250 hours, and then drops down to the steady state level immediately. In the case of SPTF, the WIP level is lower than for FIFO. After about 3500 hours, no effects from the breakdown can be observed in the fab. For CR and ST, however, the situation is different. WIP builds up more slowly, but about 500 hours after the breakdown it is becoming significantly higher than the FIFO level. Even after all lots that experienced the catastrophic failure left the fab, this behavior is repeated with decreasing WIP level. It is worth noting that there is almost no difference in the behavior of CR and ST for an intended flow factor of 2.5.

In all cases, the WIP level oscillates at a frequency of about $1/125[\text{hour}^{-1}]$. The reason for this oscillations is the constant delay introduced by the feedback loop to the bottleneck work station. The lots waiting in the bottleneck center queue are processed very fast compared to the feedback delay and show up almost at the same time at the queue again.

Now, we explain the reason for the unexpected behavior of the system if due-date dependent rules are applied. Right after finishing the repair of the bottleneck machines , those lots are preferred by CR that are closest to their due date. Since all lots that saw the bottleneck down have a due date that is closer than all lots that are newly arriving after the end of the failure, all of the blocked lots are always processed ahead of the new lots. New lots will not be able to seize a server until all blocked lots left the queue. Therefore the WIP is building up due to the permanent arrival of new lots. As soon as all lots that experienced the failure have left the fab, the new lots that were blocked by these lots take their role and the phenomenon of increasing WIP repeats itself. Since the bottleneck work center has a spare capacity of 7.5 % the peak level of the WIP is becoming smaller and smaller.

Looking only at the WIP graphs, it is likely to draw the conclusion that SPTF is a reasonable dispatch rule in case of a catastrophic failure since it leads to the smallest WIP level and little WIP level variations. With respect to cycle times, this is no longer the case.

Figure 4 depicts the cycle time evolution for FIFO and SPTF dispatching.

While FIFO cycle times show the same behavior as FIFO WIP levels, the SPTF cycle times show periods where the cycle times are lower than the FIFO ones but also periods where the cycle times are considerably higher. The reason is the priorization of lots with smaller processing times at the bottleneck work center. Thus, the lots with a processing time of 0.7 hours rush through the bottleneck but lots with a processing time of 1.3 hours have to wait until lots of all other products have been processed.

Considering both WIP level and cycle times, there is a clear indication that FIFO dispatching will help to avoid WIP build up and will considerably reduce the cycle time variations.

Up to this point we discussed a fab that is completely deterministic. Of course, this is not the case for a real fab. In particular, the waiting times experienced by a lot are far from being constant. Hence, we consider a new

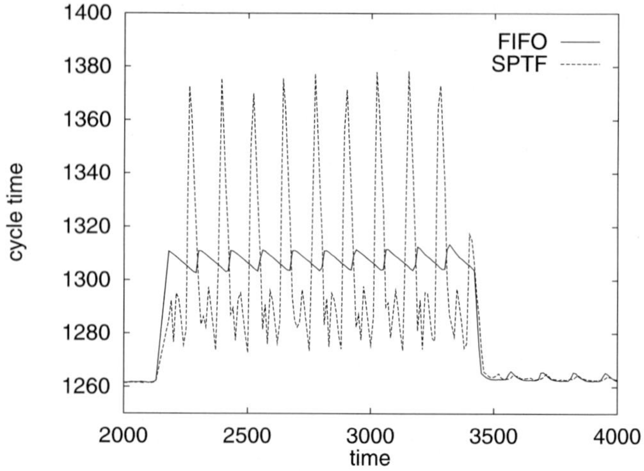

Fig. 4. Cycle time evolution for constant delay.

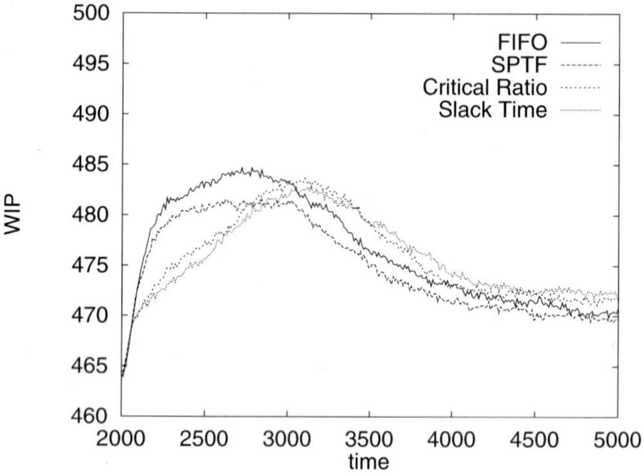

Fig. 5. WIP evolution for shifted exponential delay.

delay model that offers larger variations in the delay times. We choose an Exponential distribution with mean 75 hours shifted by 50 hours, i.e., we assume a constant sum of processing times and an exponentially distributed sum of waiting times and other non-processing times. The variation of this model is higher than that of the Siemens fab.

Figure 5 shows the WIP evolution for shifted Exponential delay.

Neglecting the oscillations, the WIP evolution is similar to that presented for constant delays (cf. Figure 3). For FIFO and SPTF, the WIP stays on

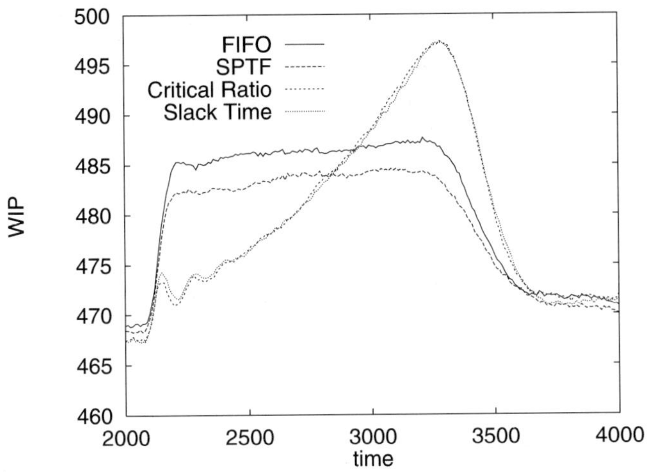

Fig. 6. WIP evolution for shifted Erlang delay.

approximately the same level for about 1250 hours and goes down to the steady-state level, whereas for CR and ST the WIP is constantly increasing during that period. The WIP decrease is considerably slower than for the constant delay fab. The oscillations disappear since the lots are no longer synchronized because they experience random delays due to the shifted Exponential delay unit.

The two modeling approaches of constant delays and shifted Exponential delays are the two extremal cases of delay time variability that were taken into consideration. In the following, we apply a delay model that is closer to real fab behavior: an Erlang-5 distributed delay time with a mean of 75 hours shifted by 50 hours.

Figure 6 shows the WIP evolution for shifted Erlang delay.

The observed behavior is a mixture of the constant delay scenario and the shifted Exponential scenario. Apart from a few oscillations right after the repair of the bottleneck machines, the oscillation disappear due to the randomness of the delay time. Again, FIFO and SPTF dispatching lead to approximately constant WIP levels, and CR and ST to constantly increasing WIP levels.

With respect to cycle times, the shifted Erlang system shows the behavior presented in Figure 7.

As for the constant case, the FIFO, CR, and ST cycle time sequences are approximately directly proportional to the respective WIP levels. In contrast to the constant case SPTF cycle time graph, the shifted Erlang one has no peaks higher than the FIFO curve. Due to the randomness in the delay times, the order of the lots is somewhat rearranged during the passage of the delay

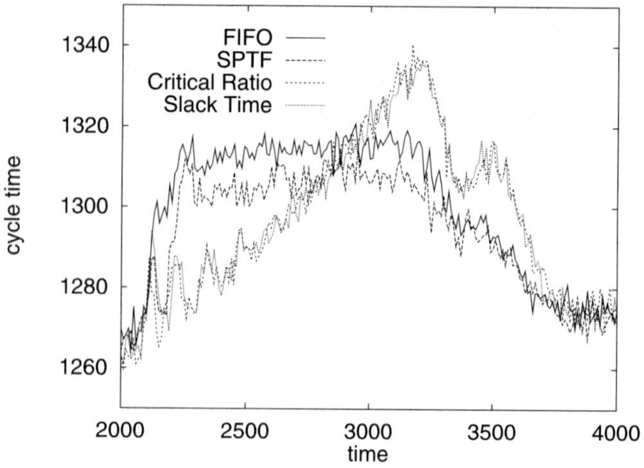

Fig. 7. Cycle time evolution for shifted Erlang delay.

unit. Thus, there are not always the same lots competing for service at the bottleneck work center.

Table 1 shows the average and variance of the cycle times observed in the time interval from 2000 hours to 4000 hours. With respect to the average cycle times, all four dispatch rules provide the same results. The variances of cycle times, however, are considerably different. The due-date based dispatch rules lead to a variance in cycle times that is about twice as large as the SPTF variance. The FIFO variance lies between that of SPTF and Slack Time.

From the results of our experiments, we draw the following conclusions. After a catastrophic failure of the bottleneck work station, the reduced fab model introduced in Section 2.1 shows essentially the same behavior as reported from a real wafer fab. Under the regime of CR dispatch, WIP level and cycle times increase and reach their maxima several days to weeks after the end of repair. This behavior can be observed for delay unit models with different variability. In our experiments the variability ranged from none, i.e., constant delays, to shifted Exponential. For small amounts of variability the WIP level tends to oscillate considerably. With respect to the average WIP

Table 1. Cycle times from 2000 hours to 4000 hours

Dispatch rule	Average	Variance
FIFO	1300.0	284.6
SPTF	1294.2	210.0
Critical Ratio	1297.0	409.1
Slack Time	1296.8	388.0

level and cycle time, FIFO dispatch leads to about the same results as CR for the period from leaving the steady-state until returning to it. Under the FIFO rule, however, less variations in WIP and cycle time are observed and the maxima of both measures are smaller. If the delay time variability is not too small, SPTF provides even better results.

With respect to explaining the WIP increase even weeks after a catastrophic failure, we conclude that this fab behavior is caused by the combination of due-date oriented dispatch and the cyclic nature of the flow of lots through the fab. Only these two typical characteristics of wafer fabrication together, lead to the blocking of fresh lots at the bottleneck work center and induce the constant WIP increase.

Avoiding the WIP Increase

Since unnecessary WIP is a waste of money and cycle times that are longer and more variable than expected are causing trouble, fab managers try to avoid such situations.

For our catastrophic failure scenario, we will not be able to apply complex strategies due to the simplicity of the model and the lack of parameters to play with. In the following, we present two simple strategies: changing of the dispatch rule and stopping lot release during repair time.

The strategy of changing dispatch rules to avoid the WIP increase is based on the following observation. For the first period of time after the repair of the bottleneck tool, CR outperforms FIFO with respect to WIP level. During the second period, however, FIFO leads to smaller amounts of inventory than CR (cf. Figure 6). Unfortunately, as shown in Figure 8, this strategy does not work.

Even though the dispatch rule changes from CR to FIFO at time 2700 hours, WIP increase as if the rule would have been left unchanged. This indicates that the reordering of the lots that takes place after the repair determines the evolution of the WIP to a large extent. Later changes have only a minor influence.

As a second strategy, we tried an approach suggested by Goldratt [12]. As soon as the bottleneck work center of a fab that is needed for the processing of each product goes down the fab has zero capacity. Thus, it makes no sense to release new material into the fab since it will be blocked. The lot release should be restarted as soon as the bottleneck tool group is up again. Figure 9 shows the WIP graphs for a fab where lot release was stopped from 2000 hours to 2050 hours.

During the repair time of 50 hours, the WIP level drops considerably. Later on, the WIP evolution is identical to the conventional system shifted by the WIP drop. For FIFO, the maximum WIP level is only slightly higher than steady-state. For CR and ST, the maximum WIP level is higher than for FIFO but considerably lower than for the original system. With respect to the cycle times, the stopping of lot releases has only marginal effects. In

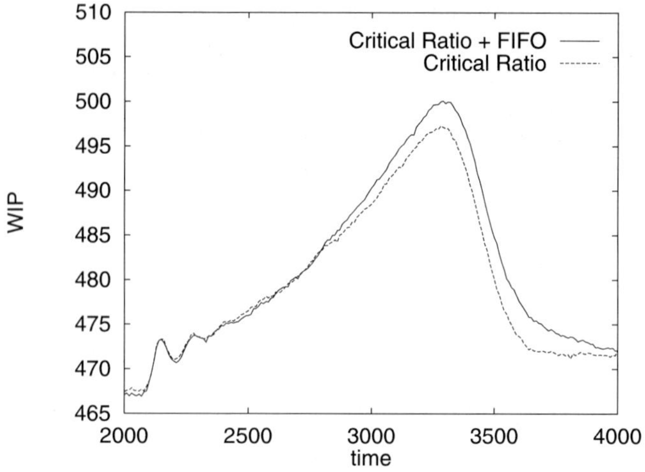

Fig. 8. WIP evolution for changing dispatch from CR to FIFO.

summary, the stop strategy provides a clear benefit with respect to WIP level but no effect with respect to cycle times. In addition, one has to take into account that there must be some spare capacity at the bottleneck because the lots that were not released to the fab during repair time will have to be released to it later on.

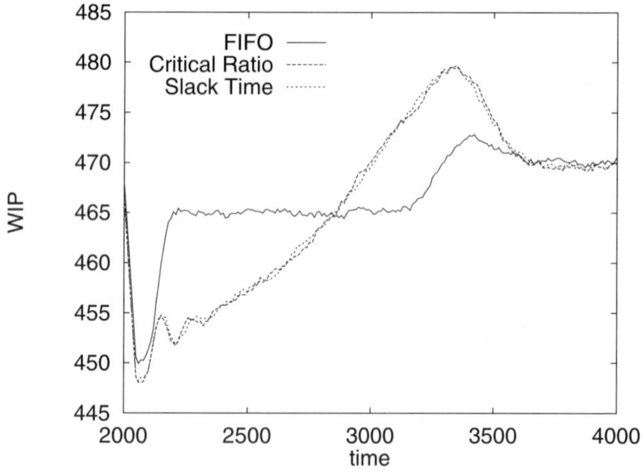

Fig. 9. WIP evolution for stopping lot releases.

2.2 Estimation of the Cycle Time Distribution

In the above study, we were not able to obtain real fab measurements to support the parameterization of our simple fab model. In particular, we had to assume that the delay time variation lies between the one of a constant and a shifted Exponential distribution.

In this study, we present a statistical analysis of the lot intervisit times of the bottleneck work center in a realistic semiconductor manufacturing facility. By means of this analysis, we are able to fit the parameters of an improved simple fab model, and to compare the cycle time distributions of the full fab model with those predicted by the simple model.

By means of simple fab models, we intend to foster our basic understanding of the behavior of wafer fabs under the regime of different lot release and dispatch rules. If the simple modeling approaches mimic accurately the full fabs, these models can be applied for the development of new control strategies for wafer fabs.

Full Factory Model

As test fab for our experiments, we choose the slightly modified MIMAC fab 6 testbed data set [13]. Fab 6 consists of 228 machines and 97 operators. It manufactures 9 types of wafers each of which has more than 10 layers and requires more than 250 process steps. The modified fab produces no scrapped wafers and has only 6 products because there are 3 products that do not require the bottleneck work center for being processed. With respect to dispatch rules it should be noted that setup avoidance is always used in our experiments. The time between lot starts of each product is constant.

We use the Factory Explorer simulation tool to collect the following datasets from the modified MIMAC 6 full fab model [15]. Given a bottleneck load and a dispatch rule for all work centers/tool groups, we record for each product the following delays. We consider the time from lot start until it first reaches the bottleneck, the *start delay*, for each cycle separately the time it takes to reenter the bottleneck after leaving it, the *cycle delay*, and the time from leaving the bottleneck for the last time until having left the fab, the *final delay*. For each considered time period several thousand measurements are taken. The measured intervals consist of processing times, setup times, and waiting times. Then, for each of the data sets a theoretical distribution is selected. The decision is based on Q-Q-plots and sums of squared differences of the measurements' histograms and several distribution candidates. In addition, the autocorrelation function of each dataset is computed for the first 30 lags.

We consider the following four scenarios: FIFO and CR with a bottleneck load of 80% and 95%, respectively. For a review of dispatch rules see [9] or [4].

For each scenario the simulation is run for 10 years of fab time. The first year's measurements are not considered to avoid an initialization bias. We

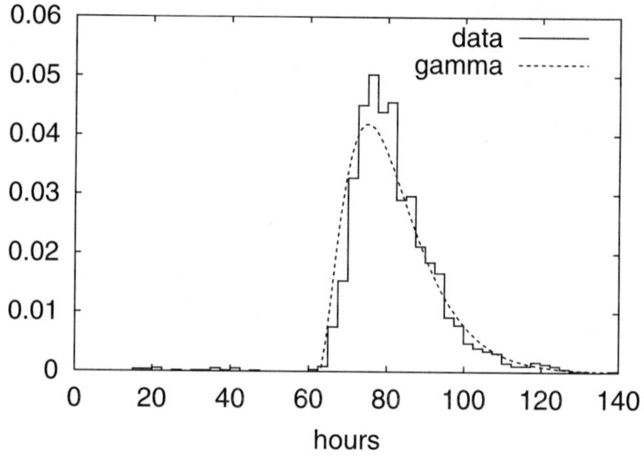

Fig. 10. Example distribution of a cycle delay.

obtain for each time period of interest at least 2000 measurements. For each scenario more than 100 distributions have to be fitted. To keep the model simple we intend to use only one class of distributions to model all delays. It turns out that the class of shifted Gamma distributions provides the best match among all tested candidates. For each shifted Gamma distribution $gamma(\alpha, \beta) + \Delta$ three parameters have to be estimated: the shape parameter α, the scale parameter β, and the shift parameter Δ. First, we determine the set of parameters that minimizes the squared distance of the empirical density of the measured data and the shifted Gamma density function. Then, while keeping the scale parameter, the two other parameters are recomputed in order to obtain the same mean and variance for empirical data and theoretical density function. This method offers the best result with respect to providing a good match in shape of the distributions of the delays while achieving the exact values for mean and variance. As an example, Figure 10 shows the empirical distribution of the first cycle delay of product 1 of the CR 95% scenario. The other fitted Gamma distributions show approximately the same level of accuracy.

For the 95% scenarios, almost all sequences of measurements show considerable correlation. In most cases, the lag-1 coefficients of correlation are larger than 0.5 and the decays of the autocorrelation functions are slower than exponential for at least the first ten lags. Figure 11 shows the empirical autocorrelation curve for the first 30 lags of the aforementioned sequence of measurements.

These correlations originate basically from the fact that subsequent lots of the same product and the same layer, i.e., the same bottleneck inter-visit cycle, see the fab and its machines in roughly the same state. Due to dispatch

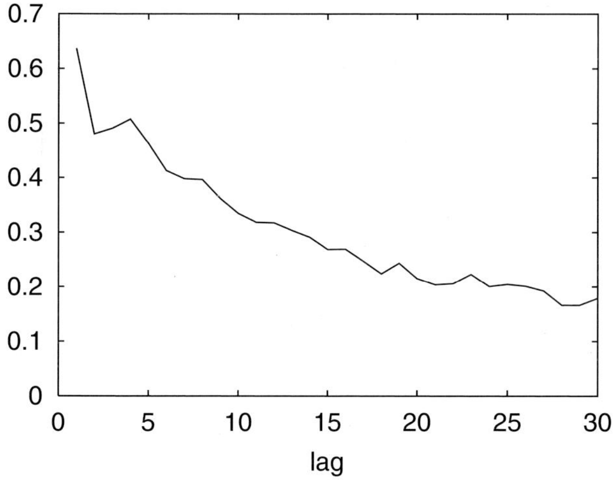

Fig. 11. Example correlations of a cycle delay.

rules such as FIFO or CR, overtaking of lots is avoided to a large extent. In addition, lots are grouped while waiting for batch completion at batch machines such as oxidation ovens.

Simple Factory Model

In order to predict the cycle time distributions correctly, the model used in Section 2.1 has to be modified. The general idea of a detailed model of the bottleneck work center and delay units representing all other work centers is kept. The bottleneck model includes the number of machines, the processing times of each product, and the application of the full fab dispatch rule. There is one delay unit for the time spent by a lot from release to first entering the bottleneck work center, one delay unit for the period of time that it takes from departing the bottleneck machines until reentering the bottleneck queue, and a delay unit for the time from leaving the bottleneck for the last time until finishing the final processing step. The model is depicted in Figure 12.

All delays are modeled by shifted Gamma distributions that are parameterized as mentioned in Section 2.2. For each product the delays are determined individually. The same holds for each product's cycle delays.

To model correlated delays, we choose the following approach. If we add two random variables that are Gamma distributed with a common shape parameter the result is Gamma distributed with the same shape parameter and a scale parameter that is the sum of the two single ones [6]. Given a sequence of random variables X_i that are $gamma(\frac{\alpha}{n}, \beta)$ distributed for a positive integer n, the sum $Z_j = \sum_{i=j-n+1}^{j} X_i$ is $gamma(\alpha, \beta)$ distributed for $j \geq n$. The coefficients of correlation result in $\rho_j = \frac{n-j}{n}$ for $0 \leq j \leq n$, and $\rho_j = 0$ for

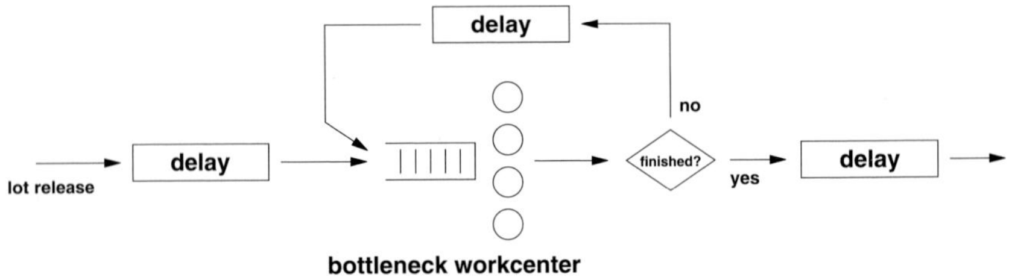

Fig. 12. Modified simple factory model.

$j > n$, because we keep $n-1$ values and replace just a single value to compute Z_j from Z_{j-1}.

This simple approach facilitates a delay model with Gamma distributed delays having a linearly decreasing correlation structure. The modeling of delays with other correlation structures, such as autoregressive processes, while still providing Gamma distributed values is considerably more complex than the above method [14].

Results

The simple factory model is implemented in ARENA 3.01 [16]. The duration of each run is 10 years of simulated time. Measurements taken during the first year are discarded.

To determine whether both the simple factory model and the full factory model exhibit the same behavior, our primary goal is to match cycle times for each product in mean, variance, and shape of distribution for both models. In the following experiments, lot release and, as a consequence, bottleneck loads are exactly the same for both models.

We first consider FIFO dispatching. In the 80% load scenario the correlations of the delays are low compared to the 95% case. Thus, we used uncorrelated delay units for the simple fab model. Figure 13 shows the cycle times of product 1. The histograms of the cycle times for the simple and the full factory models match well. The same holds for all other products.

In the FIFO 95% load scenario, the delays are considerably correlated with empirical lag-1 coefficients of correlations ranging from about 0.5 up to 0.9. To keep the model simple, we apply two correlation scenarios: moderate with a lag-1 correlation of 0.66 ($n = 3$) and strong with a lag-1 correlation of 0.9 ($n = 9$). Without modeling the correlations the histogram shapes look similar but the mean cycle times are too low. Introducing positive correlation has a clumping effect on the lots because consecutive lots of the same product have similar delays. It turns out, however, that this lot clumping does not result in higher cycle times as expected. Figure 14 depicts the product 1 cycle time histograms of the full fab and the simple fab with uncorrelated delays. The

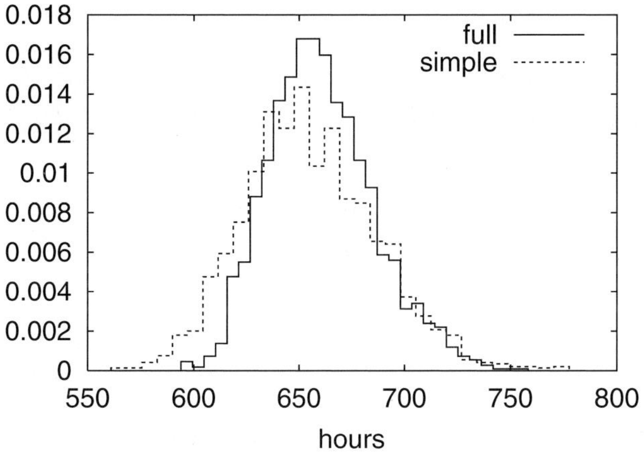

Fig. 13. Product 1 cycle times (FIFO 80%).

histograms for the correlated delays are not shown because they almost match the uncorrelated one.

In the following, CR dispatching is considered. The FIFO dispatching at 95% load leads to an average flow factor over all products of about 2.1, where the flow factor is defined as the ratio of average cycle time and raw processing time. The target flow factor for the CR 95% scenario is set to 2.1. This results in shorter cycle times for all but one product and a considerable reduction of

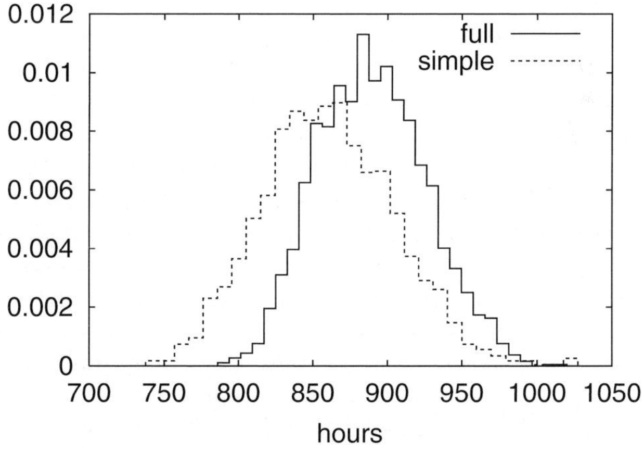

Fig. 14. Product 1 cycle times (FIFO 95%).

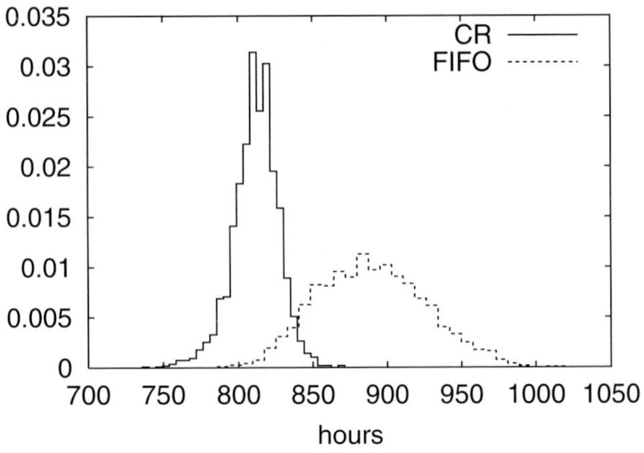

Fig. 15. Product 1 cycle times (FIFO/CR 95%).

variance of cycle times for all products. This is a typical result for switching from FIFO to CR in a wafer fab [11]. In Figure 15 cycle time histograms of product 1 under the regime of FIFO and CR are provided.

Figure 16 shows the cycle time histograms of the CR 95% scenario. In contrast to the FIFO scenarios, the shapes of the histogram curves do not match well for either strength of correlation.

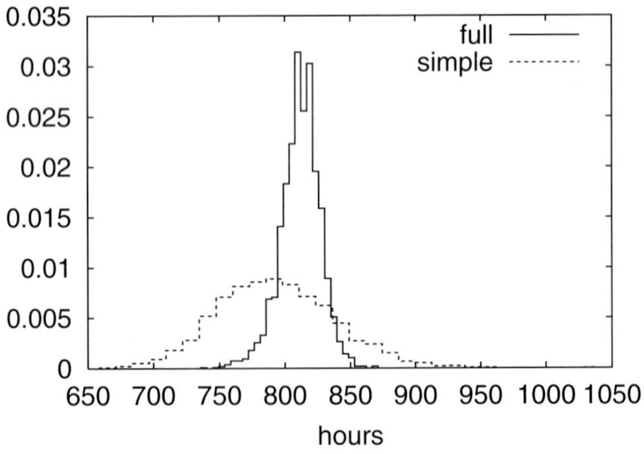

Fig. 16. Product 1 cycle times (CR 95%).

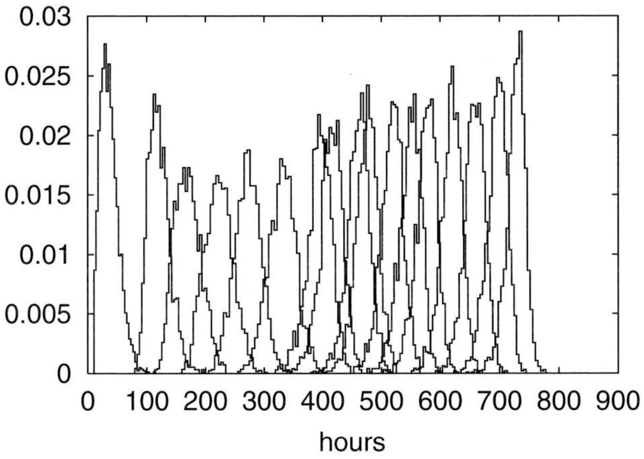

Fig. 17. Full factory model histograms of cycle completion times.

This result is caused by a special property of the CR dispatch rule. The application of CR not only completely avoids overtaking of lots of the same product and cycle, such as FIFO does, but also to a large extent overtaking of lots of the same product that are in different cycles, and of lots of different products. Here, lot overtaking is defined as lots being processed earlier at the bottleneck than lots that are closer to their due date. In the full factory model this kind of overtaking only rarely happens because at each machine the lots are processed according to their due dates. In the simple model, however, this reordering of lots does only take place at the bottleneck work center. Only the effect of overtaking of lots of the same product and cycle is reduced by introducing correlation that leads to lot clumping.

The variance reduction effect of CR is visualized in Figure 17 and Figure 18. In both cases lots have the same distribution of cycle delays for each cycle, but the shapes of histograms of the periods of time taken from lot start to finishing a particular cycle are considerably different. In case of full factory with CR, most of the histograms of consecutive cycles are clearly separated (cf. Figure 17) whereas the histograms overlap in the case of simple factory with CR (cf. Figure 18). Due to CR, lots of one product being in the same cycle are grouped together with respect to cycle times in the full factory model.

Table 2 shows the mean and standard deviation values μ and σ of the cycle times of product 1 for the considered scenarios. In case of FIFO, the simple factory model values are close to those of the full model. For the CR scenario, however, the values provided by the simple model are considerably lower than for the full model.

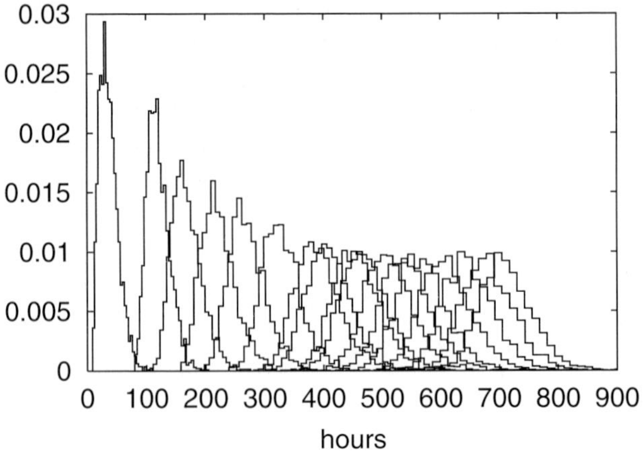

Fig. 18. Simple factory model histograms of cycle completion times.

Table 2. Mean and variance of product 1 cycle times

		FIFO		CR	
		full	simple	full	simple
95%	μ	889.60	860.56	811.85	793.26
	σ	36.344	44.355	15.262	44.801
80%	μ	660.65	655.37		
	σ	24.811	31.466		

3 Conclusion

In Section 2.1, we presented a reduced wafer fab model that exhibits essential features of a real wafer fab. It consists of a detailed model of the bottleneck work center and a delay unit that models the remaining machines of the fab. Lots released to the fab model have to cycle through bottleneck and delay unit repeatedly in order to model the layered nature of semiconductor manufacturing. For our experiments, four dispatch rules at the bottleneck work center are assumed. Two of them without due date dependence (First In First Out, Shortest Processing Time First), and two of them with due dates involved (Critical Ratio, Slack Time).

This fab model is used to assess the evolution of the WIP level and cycle time of the fab after recovering from a catastrophic failure, i.e., a complete failure of all bottleneck machines for a longer period of time.

Particular attention is devoted to the modeling of the delay time for the delay unit. It turns out, however, that the dispatch rules have a greater effect on the behavior of the fab than the choice of the delay time model.

Using the proposed model, we were able to reproduce fab behavior as observed in real semiconductor manufacturing facilities. It turns out that the phenomenon of increasing WIP is mainly caused by a combination of the due-date oriented dispatching and the cyclic nature of the lot flow. Using the simple strategy of stopping lot releases during repair time, we provide first results on how to partially avoid the tremendous increase in inventory after a bottleneck breakdown.

In Section 2.2, we presented a simple factory model that is intended to predict the cycle time distributions of the lots in a semiconductor fab and to facilitate the understanding of the basic fab behavior under the regime of different dispatching and lot start rules.

The model is well suited to predict the cycle times in the FIFO case. For CR, however, the dispatch rule avoids overtaking of lots with a later due date if other lots with a closer due date are already waiting for a resource to become available. This property of CR reduces both mean and variance of the cycle times. The current version of the simple factory model is not capable of avoiding lot overtaking. This results in cycle times that have a higher variance than those of the full factory model.

In summary, the application of simple models for complex production systems is a helpful tool to analyze and understand the principal system behavior. For detailed quantitative estimates of the system performance, there are scenarios where these models are too simple to provide accurate results.

References

1. Fowler, J.W., Fu, M.C., Schruben, L.W., Brown, S., Chance, F., Cunningham, S., Hilton, C., Janakiram, M., Stafford, R. and Hutchby, J.: Operational Modeling & Simulation in Semiconductor Manufacturing. In: Proceedings of the 1998 Winter Simulation Conference (1998) 1035–1040
2. Kempf, K.G.: Simulating Semiconductor Manufacturing Systems: Successes, Failures, and Deep Questions. In: Proceedings of the 1996 Winter Simulation Conference (1996) 3–11
3. van Zant, P.: Microchip Fabrication. McGraw-Hill, New York, 2nd Edition (1990)
4. Atherton, L.F. and Atherton, R.W.: Wafer Fabrication: Factory Performance and Analysis. Kluwer Academic Publishers, Boston (1995)
5. Fowler, J. W., Hogg, G. L., and Mason, S.J. Workload Control in the Semiconductor Industry. Production Planning and Control **13** (2002) 568–578
6. Law, A.W. and Kelton, W.D.: Simulation Modeling and Analysis McGraw-Hill, New York, 3rd edition (2000)
7. Rose, O.: WIP Evolution of a Semiconductor Factory After a Bottleneck Workcenter Breakdown. In: Proceedings of the 1998 Winter Simulation Conference (1998) 997–1003
8. Rose, O.: Estimation of the Cycle Time Distribution of a Wafer Fab by a Simple Simulation Model. In: Proceedings of the SMOMS '99 (1999 WMC) (1999)

9. Wein, L.M.: Scheduling semiconductor wafer fabrication. IEEE Transactions on Semiconductor Manufacturing **1** (1998) 115–130
10. Lovegrove, W., Hammond, J. and Tipper, D.: Simulation methods for studying nonstationary behavior of computer networks. IEEE Journal on Selected Areas in Communications **8** (1990) 1696–1708
11. Brown, S., Fowler, J., Gold, H. and Schömig, A.: Measurable improvements in cycle-time-constrained capacity. In: Proceedings of the 6th International Symposium on Semiconductor Manufacturing (1997)
12. Goldratt, W.D. and Cox, J.: The Goal: A Process of Ongoing Improvement. North River Press, New York, 2nd edition (1994)
13. Fowler, J. and Robinson, J.: Measurement and Improvement of Manufacturing Capacities (MIMAC): Final Report. SEMATECH Technical Report 95062861A-TR (1995)
14. Cario, M.C., and Nelson, B.L.: Autoregressive to anything: Time-series input processes for simulation. Operations Research Letters **19** (1996) 51–58
15. Chance, F.: Factory Explorer 2.8 User Manual Wright Williams & Kelly, Pleasanton (2002)
16. Kleton, W.D., Sadowski, R.P. and Sadowski, D.A. Simulation with Arena. McGraw-Hill, New York (1997)

Logistics Networks: Coping with Nonlinearity and Complexity

Karsten Peters[1], Thomas Seidel[2], Stefan Lämmer[2], Dirk Helbing[3,4]

[1] Institute for Logistics and Aviation, TU Dresden, 01062 Dresden
karsten.peters@tu-dresden.de
[2] Institute for Transport and Economics,TU Dresden, 01062 Dresden
seidel|laemmer@vwi.tu-dresden.de
[3] Chair of Sociology, in particular of Modeling & Simulation, ETH Zurich UNO D11, Universitätstrasse 41, 8092 Zurich, Switzerland dhelbing@ethz.ch
[4] Collegium Budapest – Institute for Advanced Study, Szentháromság u. 2, H-1014 Budapest, Hungary

1 Introduction

Nowadays the complexity of logistics is a buzzword spreading in business, media and everyday practice. However, the study of logistics networks from the point of view of complex dynamical systems theory has started only recently. In the past decade, physicists have been more and more interested in interdisciplinary fields such as biophysics, traffic physics, econophysics, or sociophysics [28, 7, 13]. Also, the study of production processes and logistics networks has become attractive [27, 37, 1], although the title of the book "Factory Physics" [23] suggests that there should be some connection. In fact, it is quite natural to study production and logistics from the point of view of material flows [14]. Therefore, many-particle approaches such as Monte-Carlo simulations and fluid-dynamic models [8, 15, 31, 2] should be applicable to logistics systems. As we will discuss in the following, this is really the case.

Our contribution is structured as follows: In the first part (Sec. 2) we highlight some sources of complex dynamical behavior in logistic systems and discuss the impact and implications of these mechanisms. In particular we address the nonlinearities in material flow processes (Sec. 2.1), caused by control algorithms. Moreover, the dynamics induced by connecting several logistic sub-systems in networks has itself important influence and may lead to complex dynamics and instabilities, even if a single subsystem behaves stable and regular (Sec. 2.2). Thirdly, we discuss phenomena like the "slower is faster effect" emerging due to nonlinear interactions in many particle systems as they are regularly found in transport systems in logistics (Sec.2.3).

The second part of this paper, i.e. Sec. 3, introduces two basic control principles, which may improve the controllability of complex logistic systems.

On the one hand side, system wide coordination by means of synchronization can be reached by a local coupling of neighboring control elements, while centralized control is not necessarily required (Sec. 3.1). But also by establishing suitable local interaction mechanisms local coordination may eventually spread all over the system (Sec. 3.2). Both principles benefit from the interesting features of self-organizing systems: based on nonlinear interactions, a locally emerging pattern may have global effects.

2 Sources of Complex Behavior in Logistic Systems

Modern logistics systems inside a factory as well as the corresponding logistics systems for supply and distribution are typically large networks of various production and storage facilities. Their dynamics is governed by complex, system immanent inter-dependencies involving both deterministic and stochastic elements as well as unforeseen external influences. In certain regards logistics and production networks may be described as coupled dynamical queuing systems. Considering discontinuities in the processes and a generally non-synchronous flow of material and information in network like systems, modeling an understanding of the dynamics of these complex systems is challenging. The traditional approaches to queuing systems focusing mostly on equilibrium solutions or computationally costly discrete event simulations are limited if it comes to larger systems. Moreover, usually a huge number of non-stationary system components, e.g. for transport tasks are included in logistic systems. Thus logistic systems, in particular transportation systems, fit into the class of multi particle systems which reveal a number of quite surprising dynamical properties [13].

In the following we shall discuss typical sources of complex behavior in logistic systems and recent approaches to modeling and understanding of the related complex dynamical behavior.

2.1 Discontinuities in Processes

Production and supply processes are usually not constant in time. Weekly and seasonal cycles, for example, generate oscillatory behavior. However, a closer look at logistics systems uncovers, that within a logistics process several material flows (i.e. different products and educts) will interfere at nodes of logistics networks. These nodes may be workstations in production or warehouses and terminals in supply networks where material flos compete for scarce resources (capacities, time, vehicles, ...). For the majority of all logistic systems a parallel service of several intersecting flows or conflicting tasks is impossible, unsafe or inefficient. Alternating exclusion of conflicting tasks is frequently required in the organization of production processes, road traffic or in communication networks. Instead of parallel processing a sequential switching between different tasks must be organized (see Fig.1).

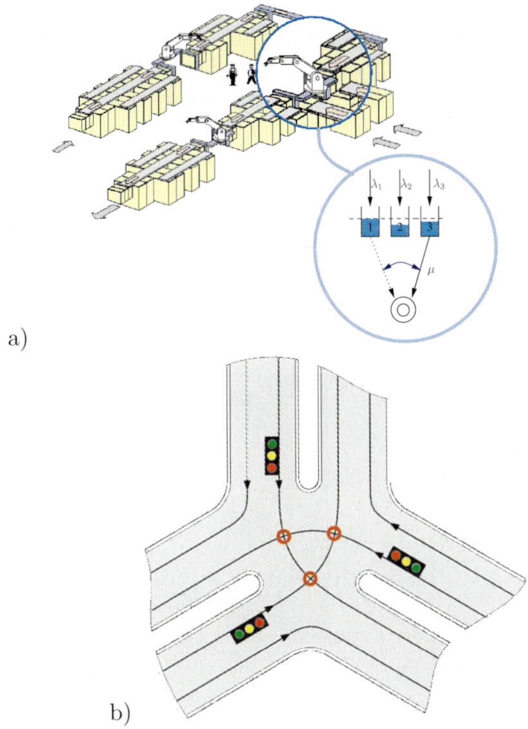

Fig. 1. Situations, where conflicts in the service of different material flows at intersection points must be resolved are common in logistics networks.
(**a**) Illustration of a production system in which a robot has to serve three different processes. Such switched server systems can be understood as hybrid dynamical systems: We have different buffer queues which are continuously filled with materials but only one of the buffers can be emptied at a time. Moreover, switching between containers takes a setup time τ^{setup}. Therefore, switching reduces service time and is discouraged (after [32]).
(**b**) The same situation occurs in traffic networks. At a single intersection the service, i.e. the green light of corresponding traffic lights is switched between the different lanes to allow all vehicles to cross the intersection without conflicts. While switching from one state to another, all traffic lights are set to red for a period of τ^{setup}

If we consider logistics systems as dynamic systems, control related switchings between different operation modes and parameters are essential nonlinearities leading to complex behavior in these systems. At a reasonable level of abstraction it is possible to neglect stochastic influences and to approximate the material flows as a continuous flows. This approach leads to models of dynamical systems consisting of piecewise defined continuous time evolution processes interfaced by some logical or decision making process. These systems have to be described by continuous as well as by discrete (logical) variables in the framework of hybrid dynamical systems.

Figure 2 illustrates a system in which a robot has to serve three different processes. Such systems can be mapped to so-called switched server systems: Let us assume we have different buffers which are continuously filled with material at certain rates. All buffers have finite storage capacity, and only one buffer can be served at a time. Moreover, switching between different buffers

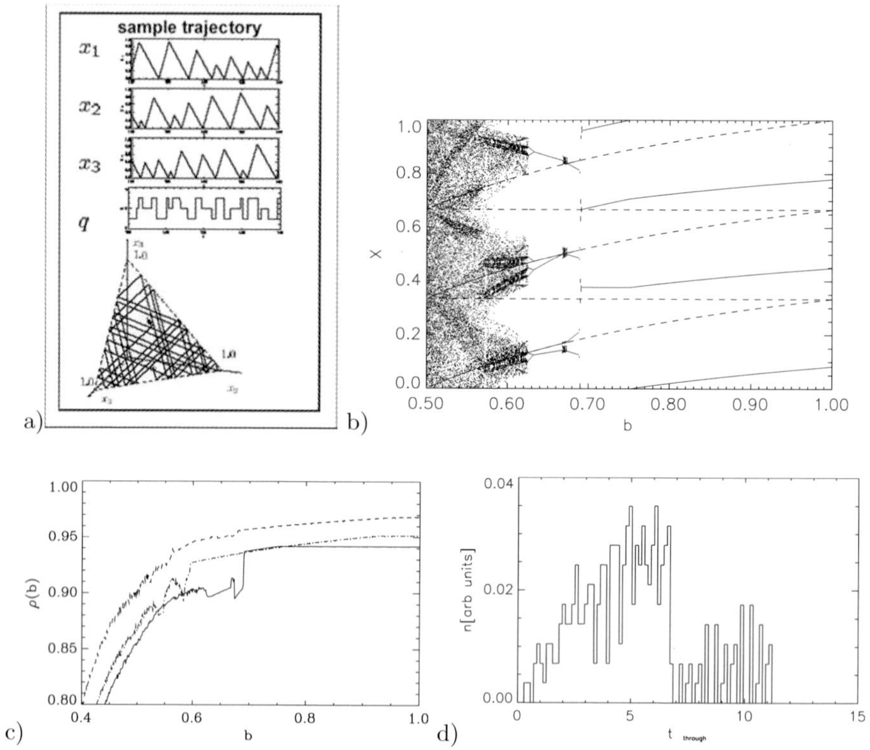

Fig. 2. Illustration of a switched server model . The graphics (**a**) on the right illustrates the strange billiard dynamics of the trajectories, which is often chaotic. By changing the capacity b of buffers, different dynamic behavior of the system is induced, as depicted in a bifurcation diagram (**b**). The dynamics directly determines the reached throughput rate $\rho(b)$ of the manufacturing system (**c**). Note, that a lossless production will result in $\rho = 1.0$ and the restricted capacity of buffers can dramatically decrease the throughput, and even induce surprising jumps in the production rate (The different curves correspond to different feeding rates of the input buffers.). The dynamics determines also the distribution of throughput-times, which can be multi-modal even for such relatively simple systems, as the histogram (**d**) depicts (after [31])

or different products, respectively, takes a setup time. Therefore, switching reduces service time and is discouraged. As a consequence, one may decide to empty the currently served container completely and then switch to the fullest container in order to avoid that its capacity is eventually exceeded. However, if the capacity of any buffer is reached before an other product buffer is served completely, the service of the latter must be interrupted to serve the filled buffer anyway. Such a control algorithm (which is usually implemented in form of priority policies in production systems) leads, if mapped to the state space of the system, to a strange billiards kind of dynamics (i.e. trajectories reminding of billiard balls reflecting at some non-rectangular boundaries). These dynamics can be investigated by considering the mapping of boundary points onto other boundary point under the dynamics of the system. Depending on the properties of this mapping it may easily happen that this dynamics tends to be chaotic [32, 31] (see Fig. 2). In particular, production speeds for different product and the capacities of buffers or material flow relations are crucial parameters, which play the rule of bifurcation parameters in the dynamic system. If critical parameters values are reached or crossed, the dynamic behavior of the system can, according to bifurcation theory fundamentally change, often unexpected.

Moreover, the complicated, even sometimes chaotic dynamics has implications for the relevant performance metrics of manufacturing and logistics. Depending on the dynamic regime, throughput times and their distributions, for instance, are changing. For this reason, many statistical distributions of arrival or departure rates may actually be a result of non-linear interactions in the system. The complex dynamics has important consequences for the system, making it difficult to predict the future behavior and process times. This complicates control a lot. Moreover, non-linear interactions often imply (phase) transitions from one dynamical behavior to another one at certain critical parameter thresholds. Such transitions are often very unexpected and mix up the schedules. We actually conjecture that the phase space (i.e. a representation of dynamic behaviors as a function of parameter combinations) is in many cases fractal, i.e. subdivided into small and often irregularly shaped areas. An understanding of such systems requires a solid knowledge of the theory of complex systems. For this reason, production and logistics are interesting areas for fundamental and applied research of theoretical physicists. We should particularly underline that, due to the many parameters in production systems (e.g. production speeds, minimum or maximum treatment times in time-critical processes, etc.), the sensitivity to (even small) parameter changes is mostly large. Therefore, the performance for a new combination of parameter values is only poorly estimated by interpolation between previous experimental measurements, which are usually available for a few parameter sets only. That is, the predictability and robustness of production processes is often low, while they are an important aspect for efficient and reliable production.

2.2 Network Effects

Several previous works uncovered a variety of oscillatory and even chaotic behavior in production systems and supply chains [25, 32, 17, 36]. Whereas traditional modeling approaches focused on stochastic queuing models recently the formulation of continuous flow models for the information and material flows in supply networks lead to new insights. By formulating the dynamics of warehouses or suppliers in a supply network in form of coupled differential equations by denoting the inventory of a node j in the supply chain as

$$\frac{dN_i}{dt} = Q_i^{\rm in}(t) - Q_i^{\rm out}(t) = \underbrace{\sum_{j=1}^{u} d_{ij} Q_j(t)}_{\text{supply}} - \underbrace{\left[\sum_{j=1}^{u} c_{ij} Q_j(t) + Y_i(t) \right]}_{\text{demand}} . \quad (1)$$

Here u production units or suppliers $j \in \{1, \ldots, u\}$ deliver d_{ij} products of kind $i \in \{1, \ldots, p\}$ per production cycle to other suppliers and consume c_{kj} goods of kind k per production cycle. The coefficients c_{kj} and d_{ij} are determined by the respective production process, and the number of production cycles per unit time (e.g. per day) is given by the production speed $Q_j(t)$. That is, supplier j requires an average time interval of $1/Q_j(t)$ to produce and deliver d_{ij} units of good i. The above formulation allows for an investigation of supply chains in terms of the stability theory for dynamical systems. In particular the influences of different supply network topologies, which enter into the model via the supply matrix d_{ij} are accessible for a mathematical analysis.

Helbing et al. [17, 21, 22] found both convective and absolute instabilities in continuous models of supply chains, which induce in certain parameter regions increasing oscillations along a supply chain. These observations may explain the so called bull-whip or Forrester-effect from first principles. The investigation reveals additionally, that sometimes even small changes in the supply network topology can have enormous impact on the dynamics and stability of a supply network.

The results of such models may therefore allow for a systematic approach to the design of robust supply networks under dynamically changing demands.

2.3 Dynamic Interactions and "Slower is Faster" Effects

One characteristic feature of driven multi-component or driven many-particle systems operated near capacity is the possibility of mutual obstructions due to competition for scarce resources (time, space, energy, etc.). From game theory it is known that the selfish, local optimization of the behavior of all elements can lead to system states far off the system optimum. That is, even if all parts of the system perform well and serve the local demands best, the overall result can be very bad. This is obviously the case, if the different processes are not well synchronized or coordinated otherwise. Therefore, it can be better to wait for some time rather than starting activities immediately upon availability. Such actionism could produce overcrowding, bottlenecks, and inefficiencies in other parts of the system. We will illustrate this, sometimes rather surprising effect for different examples of transport and production systems.

Slower is Faster Effects in AGV Systems

Nowadays material transport within nodes of logistic networks, such as warehouses, manufacturing systems or logistics terminals, is often done by automated guided vehicles (AGV), i.e. without any drivers. These vehicles move along virtual tracks. We have studied such a logistics system, namely a container terminal in a harbor. There, containers had to be moved from the ships to the container storage area and back (see Fig. 3). However, instead of moving most of the time, they often obstruct each other. This is, because each

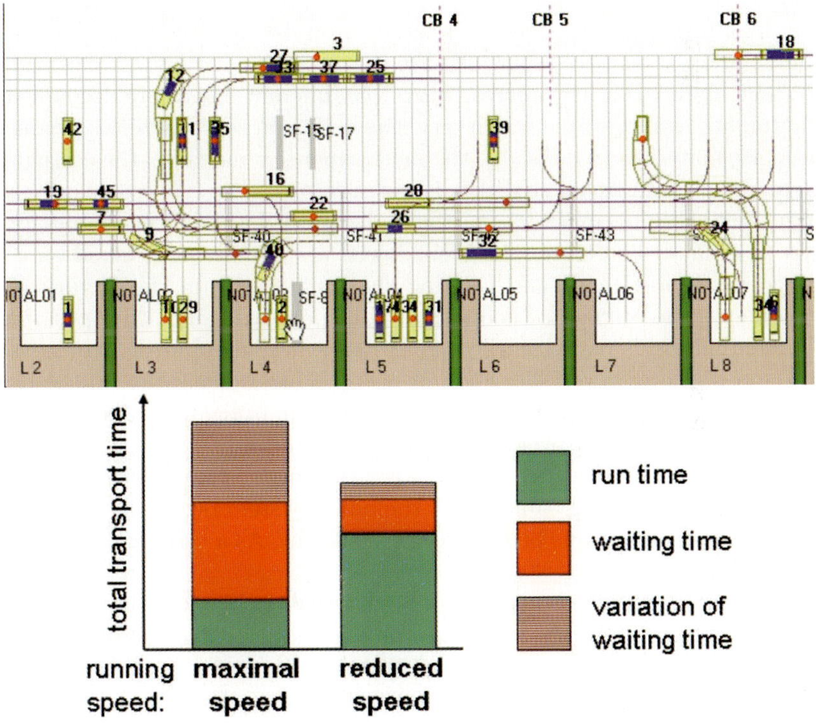

Fig. 3. Top: Illustration of a container terminal in a harbor. The containers are moved by automated guided vehicle along virtual tracks. **Bottom**: The total transport time is composed of the run time and the waiting time, which varies a lot. Furthermore, it strongly depends on the vehicle speed

vehicle is surrounded by a safety zone, which may not be entered in order to avoid accidents and damage of goods.

According to the "slower-is-faster effect" , the logistic processes may be more efficient, if the speed of the automated guided vehicles is reduced. In fact, there should be an optimal speed. Reducing the speed will, of course, increase the run times. However, it also allows one to reduce the safety zones, since these are essentially proportional to the speed. As a consequence, the vehicles will not obstruct each other so often. Thus, their waiting times will go down. Moreover, the predictability of the whole system is improved due to the reduced fluctuations, such that also the scheduling of transport jobs has to deal with smaller safety margins. These effects can overcompensate the increase in the run times. Therefore, reducing speed can sometimes make logistic processes more efficient.

Dynamic Interactions and the Optimization of Buffer Loads

A similar observation can also be made in storage units with automated material handling systems. Here we refer to a production layout which can be found in many industries as for instance the packaging industry. We consider a two stage manufacturing process, where the different stages are decoupled by an automated storage area which serves as buffer. In some factories producing packaging materials (see Fig. 4), the most expensive machine, the corrugator, brakes down quite frequently, if it is not new anymore. The corrugator produces the packaging material (corrugated paper), i.e. the basic input for all other machines. Therefore, the breakdowns are usually considered as causing the main bottleneck and limiting the profitability of the factory. As a consequence, the corrugator is often run full speed whenever it is operational.

Potentially, this causes earlier breakdowns, which could be avoided by a lower speed of operation. Moreover, it also produces congestion in the buffer system. Whenever the system exceeds a certain utilization ("work in process"), so-called cycling procedures are needed to find the stacks which were required for further processing (see Fig. 4). These cycling procedures take over-proportionally more time, when the buffer system becomes fuller. In this way, the buffer operations become quite inefficient. As a consequence, the real bottleneck is the buffer system.

In principle, there are two ways to solve this problem: Either to increase the buffer storage capacity or to stop the corrugator, when the buffer system reaches a certain utilization (see Fig. 5). While the corrugator is turned off, it can be cleaned and fixed. This proactive maintenance can reduce the number and duration of unplanned downtimes. Therefore, reducing the speed of the system increases its throughput and efficiency again. Our event-driven, calibrated simulations of the production system indicate that the expected increase of production efficiency would be of the order of 14%.

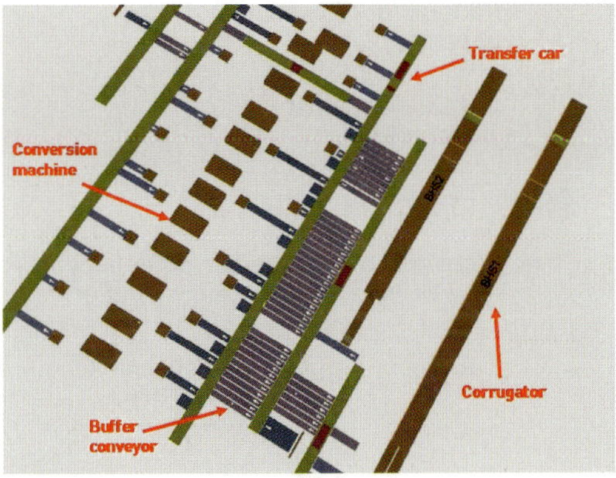

a)

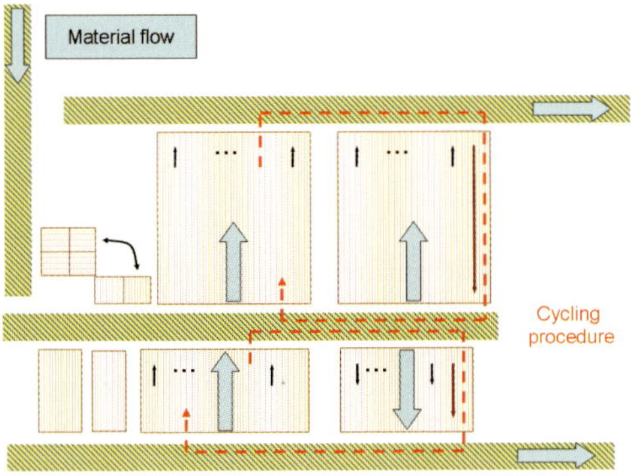

b)

Fig. 4. Top: Illustration of a plant in the packaging industry producing boxes. The corrugator produces corrugated paper board, which is processed (e.g. cut and printed) by conversion machines. Transfer cars transport the materials, and buffers allow for a temporary storage.
Bottom: Illustration of the cycling procedure, by which a specific stack is moved out of the buffer for further processing. This procedure can be quite time-consuming, if there is a lot of work-in-process (WIP) in the buffer

3 Self-organization in Complex Systems as a Possible Control Paradigm

Besides the dynamical phenomena discussed above, the optimization of production processes is often an NP-hard problem . For this reason, it is common to use precalculated schedules and designs determined off-line for certain assumed boundary conditions (e.g. given order flows). This is mostly done with methods from Operations Research (OR) or event-driven simulations. However, in reality the boundary conditions are varying in an unknown way, so that the outcome may be far from optimal, as the optimal solution is

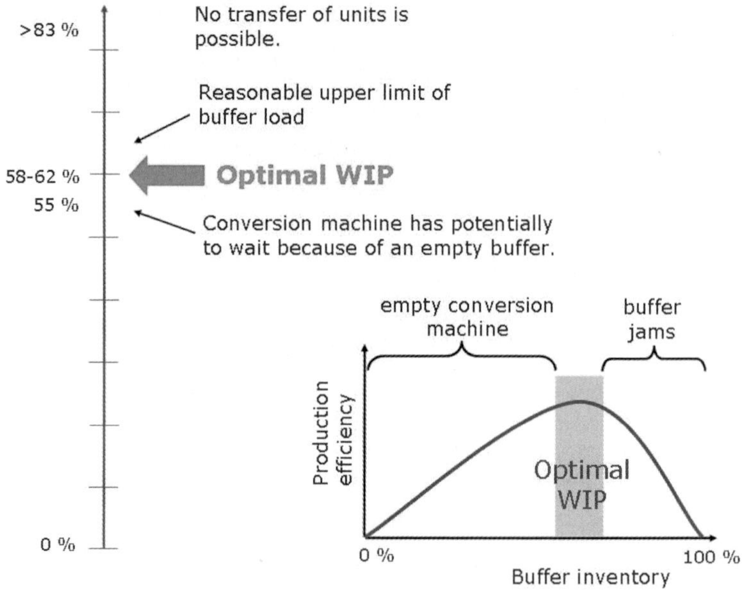

Fig. 5. Top and **middle**: The optimum buffer utilization is neither close to zero nor close to capacity. In the first case, the buffer may be emptied, so that further production steps are delayed. In the second case, buffer operation becomes inefficient or even almost impossible. In the illustrated practical example, the optimum buffer utilization is between 58% and 62% (only!). Stopping further transfer to the buffer at this level is reasonable and allows for proactive maintenance of the upstream production machine(s). According to computer simulations, this strategy is expected to reach a reduction in production times

sensitive to parameter changes (see Sec. 2). Therefore, adaptive on-line control strategies would be desirable. Although they cannot be expected to be system-optimal, the higher degree of flexibility promises a higher average performance, if the adaptation manages only to drive production *close* to the system optimum.

As the exact system optimum can mostly not be determined on-line, one needs to find suitable heuristics. Typical and powerful heuristic methods are, for example, genetic or evolutionary algorithms. However, their speed is also not sufficient for adaptive on-line control. Therefore, we are currently seeking for better approaches that make use of characteristic properties of material flow systems such as conservation laws. A typical example is the continuity equation for material flows. Such equations can also be formulated for merging, diverging, and intersecting flows, although this is sometimes quite demanding. Moreover, most of the logistic material flow networks are subject to continuous demand variations and unforeseen failures. Besides adaptivity and optimality, robustness and flexibility are important requirements for control concepts.

Can we learn from the stable, smooth, and efficient flow of nutrients and other chemical substances in circulatory systems of biological organisms? Synchronized dynamics of a population of cells often plays an important role for it [35, 41]. For example, our heart functions as an pump through the appropriate synchronized dynamics of a population of cardiac cells. This synchronization is realized through appropriate designs of cardiac cells and their network architecture of local interactions. Another interesting example is found in amoeboid organisms [29, 40], where the rhythmic contraction pattern produces streaming of protoplasm. Synchronization phenomena have been intensively studied for these biological systems during the last decades by means of mathematical models, in particular coupled phase-oscillator models [4, 12, 24, 35, 39].

3.1 Bundling Effects and the Synchronization of Switching Controls

Let us consider at first the problem of coordinating the switching between conflicting flow directions in a material transport network, which is a directed graph with of a set of nodes and links. Material can move between the nodes with a finite velocity. Thus any moving element experiences a delay t_{ij} between its departure at one node i and its arrival at the next node j. Whereas a distinct subset of nodes may act as a source or sink of moving material, we shall concentrate on those nodes where the flow of material is conserved, i.e.

$$\sum q^{\text{in}} = \sum q^{\text{out}}. \tag{2}$$

Here $\sum q^{\text{in}}$ and $\sum q^{\text{out}}$ denote the average rate of incoming and outgoing material, respectively. Each node has to organize the routing of materials arriving through incoming links towards its outgoing links. All allowed connections between incoming and outgoing links can be described through discrete states of the respective node. As long as such a state is 'active', material can flow from a subset of incoming links through the node and leave through outgoing links. All other flow relations are blocked. Usually the switching between different discrete states needs a certain time interval τ, called switching- or setup- time. Depending on the flow rates, the duration of these discrete states may vary.
A major simplification of the problem can be obtained if we consider a cyclic service sequence, we can assign a periodic motion to every switching node. Thus, a node can be modeled as a hybrid system consisting of a phase-oscillator and a piecewise constant function M that maps the continuous phase-angle $\varphi(t)$ to the discrete service state $s(t)$, e.g. $M : \varphi(t) \to s(t)$.
The switched service of different flows leads to convoy formation processes. This implies highly correlated arrivals at subsequent nodes, which requires to optimize M with respect to a minimal delay of the material. Whereas the map M can be optimized according to the actual local demand, the phase-angle φ is coupled to the oscillatory system of the neighboring nodes. Thus with a suitable synchronization mechanism we can achieve a coordination of the

switching states on a network wide level. In consequence we suggested in a previous paper [26] an adaptive decentralized control concept consisting of two parts:
(a) Phase-synchronization of all oscillators in the network, based on local coupling between intermediate neighbors.
(b) Mapping of phase-angles to the switching states based on local optimization.
For the sake of concreteness, we applied our method to the control of traffic lights at intersections of road networks. The objective of our decentralized control method is the network wide coordination of the individual switching sequences based on a local coupling between the intersections in the road network. By modeling each intersection i as an oscillator, characterized by its phase-angle φ_i and its effective frequency $\omega_i = \dot{\varphi}$, coordination is achieved by synchronizing the oscillator network. Hereby, for providing a common time scale and allowing the intersections to trigger the switching cycles right at the best time, we used a phase-locked state where the phase difference between neighboring oscillators is fixed [35].

Therefore we applied a coupling between any oscillator $i = 1, 2, \ldots N$ and its nearest neighbors $j \in \mathcal{N}_i$ with adjustments of phases and frequencies on two different timescales.

Figure 6 shows a simulation of the control concept developed in [26]. Adaptive traffic light synchronization is an example for a potentially self-organized coordination among locally coupled service stations. The fraction of green lights for the east-west direction in a Manhattan-like road network varies in an oscillatory way, as expected. However, in contrast to precalculated traffic light schedules, the oscillations are adaptive and provide a flexible response to the local, stochastically varying traffic situation. The travel times of vehicles traversing a regular lattice road network are reduced due to a network wide coordination of traffic lights.

3.2 Congestion and Adaptive Re-routing

However, many times, materials can be processed by multiple machines with different technical and performance specifications. In the simplest case, one has I identical machines for parallel processing. The question is, which stacks should be sent to what machine? Therefore, if different alternative production paths are available, adaptive routing is an issue (see Fig. 7). Due to the finite setup times, it is normally not reasonable to send different stacks of the same job to different machines. Moreover, depending on capacity utilization, it may be costly to use *all* available machines. Obviously, the optimum usage of parallel processing capacity must be load dependent.

Here, a biologically inspired approach can help. When the trails are wide enough, ants are known to establish one path between the nest and a single food source. After some time, this path corresponds approximately to the shortest path. This means a minimization of “transport costs”, i.e.

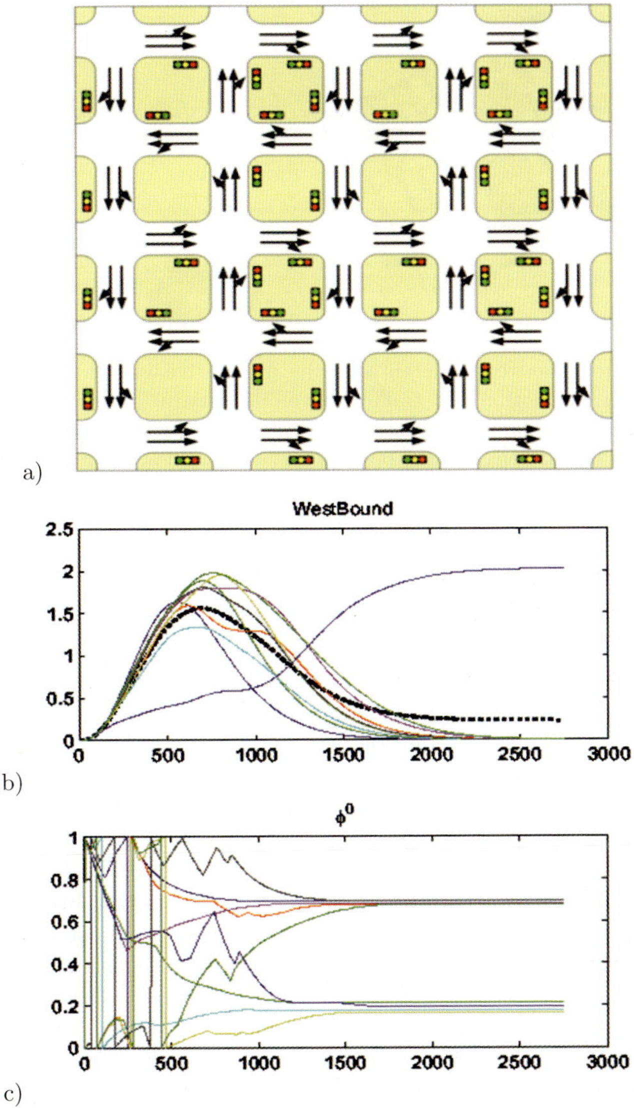

Fig. 6. The travel times of vehicles traversing a regular lattice road network (**a**) are reduced due to a network wide coordination of traffic lights. From a random initial state, the decentralized adaptive control concept let the intersections exponential convergence towards a globally synchronized state (**b,c**). Both, the synchronized cycle times as well as the locally adjusted switching states, allow the emergence of green waves and the reduction of overall travel times

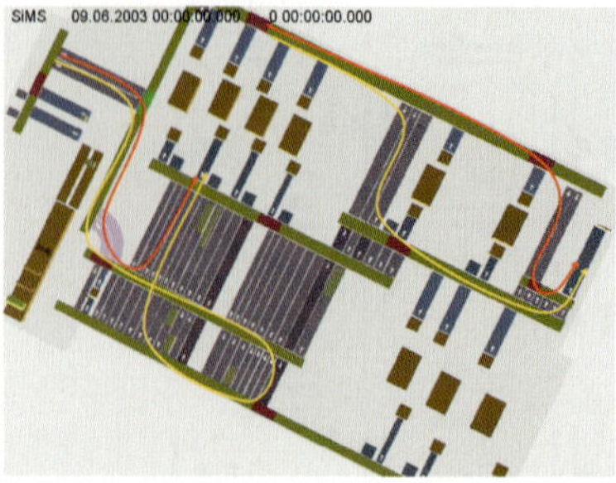

a)

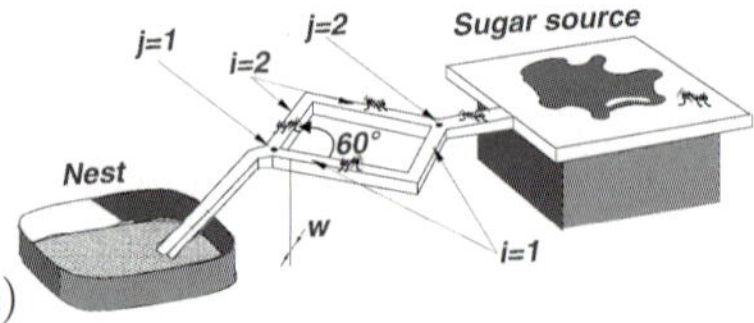

b)

Fig. 7. The choice between different available alternative ways in crowded situations occurs both in production and traffic, as well as in natural systems. Ants are able to adapt to this situation by a simple, local interaction mechanism.
a) Illustration of different routing options for material flows in a packaging factory.
b) Illustration of an experiment on the route choice behavior of ants. Neither of the two branches of the bridge from the nest to the food source provided enough transport capacity alone. Although the chemical attraction is in favor of one ant trail only, repulsive pushing interactions can lead to the establishment of additional trails, if the transport capacity of one trail is too low (after [11])

optimum usage of resources. The underlying mechanism is a random exploration behavior in combination with a pheromone-based attraction via particular chemicals, which leads to a re-enforcement of the most utilized trails. In the end, one of the trails survives, while the others fade away.

It is interesting to see what happens if the capacity of the ant trail is too small to keep up the desired level of material flow (food). This has been tested by connecting nest and food source by two narrow bridges only [11]. In such cases, ants are using additional trails to keep up the desired level of throughput. The underlying mechanism is a repulsive interaction among ants. In an encounter of two ants at a bifurcation point, an ant is likely to be pushed to the alternative bridge. This can lead to the establishment of additional and stable ant trails [11, 33, 34] (see Fig. 7).

The microscopic simulations support the conjecture that the discovered collision-based traffic optimization principle in ants generates optimized traffic for a wide range of conditions. It optimizes the utilization of available capacities and minimizes round-trip times. Simulation results for a symmetrical bridge with four branches is shown in Fig. 8. It can be seen that the majority of ants uses one of the large branches, and within both branches, the majority of ants uses one of the small branches. Which of the branches is preferred basically depends on random fluctuations or the initial conditions. When the overall ant flow is increased, the ant flows on the branches become more equally distributed. Above a certain overall flow, the usage of the bridge is completely symmetrical, as for a bridge with two branches. On a

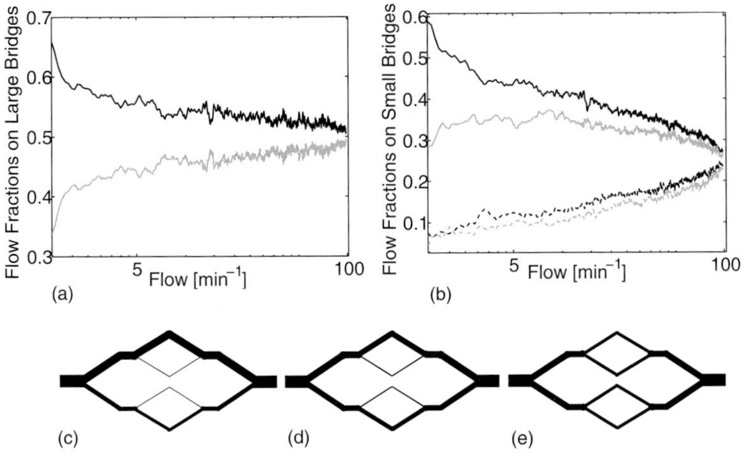

Fig. 8. Simulation results for a completely symmetrical bridge with two large branches that are again subdivided into two small branches each. The overall width of the bridge (i.e. all branches) is the same everywhere. (**a**) Fraction of ants on the two large branches as a function of the overall ant flow ϕ. (**b**) Fraction of ants on the four small branches. The lower figures give a schematic representation of the distribution of ant flows over the large and small branches (**c**) at very low flows, (**d**) at medium flows and (**e**) at large flows. (After [34])

logarithmic flow scale, it appears that we do not have a discontinuous transition from the use of one small branch to the use of two small branches and another transition to the use of three or four branches. However, the sharpness of the transition is certainly a matter of the choice of parameters.

Therefore, optimal machine utilization could be implemented via an ant algorithm [9, 3]. When the utilization (demand) is low, just one of the machines would be operated. Otherwise, based on suitably defined repulsive interactions, jobs would be distributed over more alternative machines or routes, depending on queue lengths and capacities.

4 Summary and Outlook

In this chapter we discussed sources of complexity in logistics networks. Considering the importance of a well functioning and effective logistics for our modern society and the wealth of their citizens the importance of new methods for understanding and optimizing of these systems can hardly be underestimated. As we have shown, the science of complex systems, including nonlinear dynamics, the study of network properties and many particle or traffic physics can help in this regard.

In the light of the complexity related point of view, optimization and control of logistics networks tends to be a challenging task. Therefore, new heuristics are needed to reach an adaptive, but highly performable operation. In this respect, we favor a *self-organization* approach. It requires a suitable design of the interactions in the system, otherwise it easily ends up in a local optimum or becomes unstable, such that breakdowns and slower is faster effects occur.

Biologically inspired methods are a promising approach, here. We have mentioned ant algorithms and synchronization principles, but learning from biology may also include the study of the production of artifacts in biological systems like cells and tissues and the transfer of biological organization principles to production plants and supply networks. In fact, this field, recently referred to as *biologistics* in Ref.[14] can foster a significant advancement in our ability to cope with the increasing complexity of logistics networks.

Acknowledgments

The authors would like to thank Dieter Armbruster, Audrey Dussutour, Anders Johansson and Hiroshi Kori for the great collaboration and interesting discussions. Furthermore, they are grateful for partial financial support by SCA, the EU project MMCOMNET, and the German Research Foundation (DFG).

References

1. Armbruster, D., Mikhailov, A.S. and Kaneko, K. (eds.) Networks of Interacting Machines: Production Organization in Complex Industrial Systems and Biological Cells World Scientific, Singapore (2005)
2. Armbruster, D. Marthaler, D., Ringhofer, C. SIAM J. Multiscale Modeling and Simulation **2** (2004) 43
3. Armbruster, D., de Beer, C., Freitag, M., Jagalski, T. and Ringhofer, C.: Physica A **363** (2006) 104
4. Blekhman, I.: Synchronization in science and technology Asme Press, New York (1988)
5. Camazine, S., Deneubourg, J.–L., Franks, N.R., Sneyd, J., Theraulaz, G. and Bonabeau, E.: Self-Organization in Biological Systems. Princeton University Press, New Jersey (2003)
6. Chen, H, Yao, D.D. Fundamentals of queueing networks Springer, New York (2001)
7. Chowdhury, D., Santen, L. and Schadschneider, A.: Statistical physics of vehicular traffic and some related systems. Phys. Rep. **329** (2000) 199–329
8. Daganzo, C.: A Theory of Supply Chains. Springer, New York (2003)
9. Dorigo, M., Maniezzo V. and Colorni, A.: The ant system: Optimization by a colony of cooperating agents. *IEEE Trans. Syst. Man, Cybernetics, B* **26**, (1996), pp. 1
10. Diakaki, C., Dinopoulou, V., Aboudolas, K., Papageorgiou, M., Ben-Shabat, E., Seider, E., Leibov, A.: Transport Res. Board **1856** (2003) 202
11. Dussutour, A., Fourcassie, V., Helbing, D. and Deneubourg, J.-L.: Optimal traffic organization in ants under crowded conditions. Nature **428** (2004) 70

12. Ermentrout, B.: J. Math. Biol. **15** (1991) 339
13. Helbing, D.: Rev. Mod. Phys. **73** (2001) 1067
14. Helbing, D., Armbruster, D., Mikhailov, A. and Lefeber, E. (eds.) Special Issue: Information and Material Flows in Complex Networks. Physica A, **363** (2006)
15. Helbing, D.: New Journal of Physics **5** (2003) 90
16. Helbing, D., Lämmer, S. and Lebacque, P. in: Deissenberg, C., Hartl, R.F. (eds.) Optimal Control and Dynamic Games. Springer, Dortrecht (2005)
17. Helbing, D. : in: Radons, G. Neugebauer, R. (eds.) Nonlinear Dynamics of Production Systems Wiley, New York (2004)
18. Helbing, D.: A section-based queueing-theoretical traffic model for congestion and travel time analysis in networks. Journal of Physics A: Mathematical and General **36** (2003) L593
19. Helbing, D. and Lämmer, S.: Verfahren zur Koordination konkurrierender Prozesse oder zur Steuerung des Transports von mobilen Einheiten innerhalb eines Netzwerkes [Method to Coordinate Competing Processes or to Control the Transport of Mobile Units within a Network] pending patent DE 10 2005 023 742.8.
20. Helbing, D., Johansson, A., Mathiesen, J., Jensen, M.H. and Hansen, A.: Analytical approach to continuous and intermittent bottleneck flows. Physical Review Letters **97** (2006) 168001
21. Helbing, D., Lämmer, S., Brenner, T. and Witt, U.: Physical Review E **70** (2004) 056118
22. Helbing, D., Lämmer, S., Seidel, T., Seba, P. and Płatkowski, T.: Physical Review E **70** (2004) 066116
23. Hopp, W.J. and Spearman, M.L. Factory Physics. McGraw-Hill, Boston (2000)
24. Kuramoto, Y.: Chemical Oscillations, Waves, and Turbulence Springer, New York (1984)
25. Larsen, E.R., Morecroft, J.D.W. and Thomsen, J.S: urop. J. Op. Res. **119** (1999) 61
26. Lämmer, S., Kori, H., Peters, K. and Helbing, D.: Physica A **363** (2006) 36
27. Mikhailov, A.S. and Calenbur, V. From Cells to Societies: Models of Complex Coherent Action Springer, Berlin (2002)
28. Nagatani, T.: Rep. Prog. Phys. **65** (2002) 1331
29. Nakagaki, T., Yamada, H. and Ueda, T.: Biophys. Chem. **84** (2000) 195
30. Papageorgiou, M.: Concise Encyclopedia of Traffic and Transportation Systems Pergamon Press, Oxford (1991)
31. Peters, K., Worbs, J., Parlitz, U. and Wiendahl, H.P.: In: Radons, G. and Neugebauer, R.: (eds.) Nonlinear Dynamics of Production Systems. Wiley, New York (2004) pp. 39–54 .
32. Peters, K., Parlitz, U.: Int. J. Bifurcation and Chaos **13** (2003) 2575
33. Peters, K., Johansson, A. and Helbing, D.: Swarm intelligence beyond stigmergy: Traffic optimization in ants. Kuenstliche Intelligenz **4** (2005) 11
34. Peters, K., Johansson, A., Dussutour, A. and Helbing, D.: Analytical and numerical investigation of ant behaviour under crowded conditions. Advances in Complex Systems **9** (2006) 337–352
35. Pikovsky, A., Rosenblum, M. and Kurths, J.: Synchronization: a universal concept in nonlinear sciences Cambridge University Press (2001)
36. Ponzi, A., Yasutomi, A. and Kaneko, K.: A non-linear model of economic production processes. Physica A **324** (2003) 372

37. Radons, G. and Neugebauer R. (eds.) Nonlinear Dynamics of Production Systems Wiley, New York (2004)
38. Rem, B. and Armbruster, D.: Chaos **13** (2003) 128
39. Strogatz, S.H.: Nature **410** (2001) 268
40. Tero, A., Kobayashi, R. and Nakagaki, T.: Physica D **205** (2005) 125
41. Winfree, A.T.: The Geometry of Biological Time. Springer, New York (1980)

Repeated Auction Games and Learning Dynamics in Electronic Logistics Marketplaces: Complexity, Bounded Rationality, and Regulation through Information

Miguel A. Figliozzi[1], Hani S. Mahmassani[2], Patrick Jaillet[3]

[1] The University of Sydney, The Institute of Transport and Logistics Studies, Faculty of Economics & Business, Sydney, NSW 2006, Australia miguel@itls.usyd.edu.au

[2] University of Maryland, College Park, Department of Civil & Environmental Engineering, Martin Hall, College Park, MD 20742, USA masmah@umd.edu

[3] Massachusetts Institute of Technology, Department of Civil & Environmental Engineering, Cambridge, MA 02139-4307, USA jaillet@mit.edu

1 Introduction

Online markets for transportation services, in the form of Internet sites that dynamically match shipments (shippers' demand) and transportation capacity (carriers offer) through auction mechanisms are changing the traditional structure of transportation markets. Beyond changes in market structure, Internet auctions have emerged as an effective catalyst to sell/buy through electronic marketplaces. Transaction time, cost, and effort could be dramatically reduced, creating new markets and connecting buyers and sellers in ways that were not previously possible [22].

McAffee and McMillan [27] define auctions as market institutions with an explicit set of rules determining resource allocation and prices on the basis of bids from the market participants. Two types of resources could be traded in transportation marketplaces: (a) loads, or demands of shippers, being "sold" to the lowest bidder– this would be the case of extra supply looking for scarce demands; and (b) capacity, i.e. the capacity to move goods, by a given mode from location A to location B, being "sold" to the highest bidder. The buyer of such capacity could be a shipper wishing to move a load, a carrier needing the extra capacity to move contracted loads, or a third party hoping to make a profit by reselling this capacity.

The focus of this chapter is the study of transportation marketplaces (TM) that enable the sale of cargo capacity based mainly on price (case a), yet can still satisfy the customers level of service demands. Specifically, this chapter considers the reverse auction format (also known as procurement auctions), where shippers post loads, triggering carrier bids. The market is comprised of shippers that independently call for shipment procurement auctions and the carriers that participate in them (we assume that the probability of two auctions being called at the same time is zero). Auctions are performed one at a time as shipments arrive to the auction market. The market generates a sequence of auctions (procurement, bidding, and auction resolution) that take place in real time, thereby precluding the option of bidding on two auctions simultaneously. The behavioral aspects of auction market behavior are more readily articulated without the added complexity of the combinatorial aspect. However, other market settings are possible. Markets where carriers bid on configurable bundles of loads give rise to combinatorial auctions. Nandiraju and Regan [28] present a comprehensive survey of freight transportation electronic marketplaces.

In auction markets, prices are not negotiated; they are generated as the outcome of carrier bids and a predefined set of rules. These rules precisely define a strategic environment, therefore allowing the study and analysis of carriers' behavior (expressed through bids). As such, auctions provide a useful laboratory to gain insight into carriers' behavior in a freight market. Auction-based electronic marketplaces give rise to new dimensions in the behavior of the principal freight transportation decision agents, especially with regard to learning in a competitive bidding environment. While the area of freight demand, and the underlying behavioral dimensions, have received limited attention in the travel behavior research community, behavioral considerations play a critical role in determining the performance of auction-based electronic freight markets, and the policy implications of different marketplace rules and regulatory requirements. Furthermore, the behavioral dimensions at play in electronic freight markets are examples of more general behavior mechanisms in competitive decision situations that extend beyond the realm of freight transportation (e.g. airline schedule and pricing decisions).

This chapter has nine sections. Next section introduces mathematical notation and describes the marketplace framework and operation. Section 3 articulates a framework to study carrier behavior. Section 4 identifies the characteristics of transportation auctions as well as associated sources of complexity and bounded rational behavior. Two sources of bounded rational behavior, knowledge acquisition and problem solving capabilities, are analyzed in Sections 5 and 6. Section 7 discusses learning in a TM setting. Reinforcement learning and fictitious play are analyzed and adapted to the particularities of a transportation marketplace. Section 8 presents different computational experiments aimed at studying the properties of different auction settings and learning methodologies. Section 9 ends with a chapter summary and conclusions.

2 Description of Transportation Marketplaces

The TM enables the sale of cargo capacity based mainly on price, while still satisfying customer level of service demands. The specific focus of the study is the reverse auction format, where shippers post loads and carriers compete over them (bidding). The auctions operate in real time and transaction volumes and prices reflect the relative status of demand and supply. A framework to study transportation marketplaces is presented by Figliozzi et al. [10] . The market is comprised of shippers that independently call for shipment procurement auctions , and carriers, that participate in them (we assume that the probability of two auctions being called at the same instant is zero). Auctions are performed one at a time as shipments arrive to the auction market. Shippers generate a stream of shipments, with corresponding attributes, according to predetermined probability distribution functions. Shipment attributes include origin and destination, time windows, and reservation price. Reservation price is the maximum amount that the shipper is willing to pay for the transportation service. It is assumed that an auction announcement, bidding, and resolution take place in real time, thereby precluding the option of bidding on two auctions simultaneously.

Consider a TM in which n carriers are competing; a carrier is denoted by $i \in \Im$ where $\Im = \{1, 2..., n\}$ is the set of all carriers. Let the shipment/auction arrival/announcement epochs be $\{t_1, t_2, ..., t_N\}$ such that $t_i < t_{i+1}$. Let $\{s_1, s_2, ..., s_N\}$ be the set of arriving shipments. Let t_j represent the time when shipment s_j arrives and is auctioned. There is a one to one correspondence between each t_j and s_j (i.e. for each t_j there is just one s_j). Arrival times and shipments are not known in advance. The arrival instants $\{t_1, t_2, ..., t_N\}$ follow some general arrival process. Furthermore, arrival times and shipments are assumed to come from a probability space (Ω, F, P), with outcomes $\{w_1, w_2, ...w_N\}$ Any arriving shipment s_j represents a realization at time t_j from the aforementioned probability space, therefore $w_j = \{t_j, s_j\}$.

In an auction for shipment s_j, each carrier $i \in \Im$ simultaneously bids a monetary amount $b_j^i \in R$ (every carrier must participate in each auction, i.e. submit a bid). A set of bids $b_j^{\Im} = \{b_j^1, ..., b_j^n\}$ generates publicly observed information y_j. Under maximum information disclosure, all bids are revealed after the auction, i.e. $y_j = b_j^{\Im}$. Under minimum information disclosure, no bids are revealed after the auction, i.e. $y_j = \{\}$. Each carrier is informed only about his bidding outcome: successful or unsuccessful. The fleet status of carrier i when shipment s_j arrives is denoted as z_j^i, which comprises two different sets: S_j^i (set of shipments acquired up to time t_j by carrier $i \in \Im$) and V_j^i (set of vehicles in the fleet of carrier i, vehicle status updated to time t_j). The estimated cost of serving shipment s_j by carrier $i \in \Im$ of type z_j^i is denoted $c^i(s_j, z_j^i)$. Let I_j^i be the indicator variable for carrier i for shipment s_j, such that $I_j^i = 1$ if carrier i secured the auction for shipment s_j and $I_j^i = 0$ otherwise. The set of indicator variables is denoted $I_j^{\Im} = \{I_j^1, ..., I_j^n\}$ and $\sum_{i \in \Im} I_j^i \leq 1$. Let π_j^i be the profit obtained by carrier i for shipment s_j, then $\pi_j^i = (b_j^i - c^i[s_j, \theta_j^i])I_j^i$. Bidders have private costs when each bidder knows the cost of the object at the time of bidding. This cost is the disutility that the bidder himself obtains from the consumption, use, possession or service of the auctioned item. Let $\Im = \{1, 2, ..., n\}$ be the set of bidders and θ^i denote the private information that buyer (seller) i possesses about the value (cost) of the item being auctioned. Private values are assumed in this chapter, therefore, $\theta^i_j = \{z^i_j, a^i, c^i\}$ is the private information of any carrier $i \in \Im$ at time t_j. Carrier $i \in \Im$ is uncertain about $\theta^i_j = \{z^{-i}_j, a^{-i}, c^{-j}\}$ at time t_j, the proprietary private information regarding competitors' fleet status, assignment, and cost functions respectively. The superscript $-i$ is used to indicate the set of competitors of carrier i.

3 Determinants of Carrier Behavior

In a TM, carrier behavior is defined as a sequence of bids taken by a carrier. This section looks into the elements or factors that determine carrier behavior. These factors are: carrier technology, bounded rationality, information availability, and strategic setting. Though all the factors are somewhat related, the first two are prominently intrinsic to the carriers own characteristics, while the last two are predominantly linked to environmental or somewhat extrinsic factors. In this section, the discussion is limited to highlight the link between them and carrier behavior.

3.1 Carrier Technology

Carrier technology or the sophistication of the dynamic vehicle routing problem (DVRP) solution has an important role in bidding. In the bidding decision making process the carrier technology determines the number of feasible schedules to be evaluated. Therefore, unsophisticated DVRP technologies seriously limit the quality and quantity of alternatives that could be evaluated [8].

3.2 Auction Rules - Information Revelation

Different auction payment rules lead to different bidding functions. Information revelation rules can also play a significant role [11]. The information that is revealed (before bidding begins or after each auction) can influence how, how much, and how fast carriers can learn or acquire knowledge about the strategic setting and competitors behaviors. The information that could be available after auctions are resolved includes: bids placed, number of carriers participating, links (names) between carriers and bids, and payoffs. The information that could be available before bidding begins includes: some carriers individual characteristics (e.g. fleet size or previous performance/profits from public financial reports), information about who knows what, information asymmetries, or common knowledge about previous items. Private information (as defined in Section 2) is not included since it involves proprietary information that usually is to the best interest of the carrier to keep private.

Two extreme information scenarios can be defined: maximum and minimum. A maximum information environment is defined as an environment where all the information, mentioned in the previous paragraph, is revealed. On the other hand, an environment where no information is revealed is called a minimum information environment. These two extreme scenarios can approximate two realistic situations: maximum information would correspond to a real time internet auction where all auction information is equally accessed by participants; minimum information would correspond to a shipper telephoning carriers for a quote. The shipper calls back just the selected carrier (if any is selected).

3.3 Strategic Setting

In this chapter, it assumed that a carrier operates in an environment determined by the other carriers behaviors; a carrier uses a model of the behavior of the other carriers as an input to his decision problem. Under this interpretation a carriers bidding function suits a carriers best interest, assuming that competitors bidding functions pursues competitors best interests. This is defined as a competitive strategic environment.

A diametrically different environment is a collusive or collaborative environment. One danger of auctions is the possibility that buyers/sellers who repeatedly participate in the same auctions could engage in collusive behavior. This topic is of primordial importance in the field of Industrial Organization a general references to this area include the work of Tirole and Martin [38, 25]. As a general rule, the more information is revealed, the easier collusion becomes. Even in minimum information settings collusion is possible. Blume and Heidhues [3] study collusion in repeated first-price auctions under the condition of minimal information release by the auctioneer. In each auction a bidder only learns whether or not he won the object. Bidders do not observe other bidders bid, who participates or who wins in cases in which they are not the winner. Even under these restrictive assumptions, for large enough discount factors, collusion can nevertheless be supported in the infinitely repeated game. Nevertheless, it may entail complicated inferences and full monitoring among them. Marshal and Marx [24] analyze bidder collusion in first and second price auctions and Symmetric Independent Private Value assumptions (SIPV) assumptions. The SIPV assumptions are strong but simplify the bidding problem significantly. In general, SIVP models can be studied analytically [20]. As detailed in Section 4, the TM characteristics render the bidding problem intractable.

The two environments, competitive and collusive, are nonetheless connected since underlying every negotiation or agreement there is a game-like component [32]. From each carrier's individual perspective, the incentives (and legal or market risks) of collaborating with competitors has to prevail over the profits that can be obtained when each party acts separately (competitive environment). The auction rules, e.g. first price, second price, open, closed, etc., do affect carriers strategies. For a general introduction to auction types and bidding strategies, the reader is referred to the comprehensive book by Krishna [20].

3.4 Bounded Rationality

Bounded rationality limitations affect a) the knowledge that a carrier is able to acquire, and b) the bidding problem that the carrier can solve. Given the carrier's rational limitations, fleet technology, information available, and a competitive strategic setting the carrier ends up solving a bidding problem that it is constrained by his/her rational or computational constraints. Bounded rationality in a TM is studied in Section 4.

3.5 Framework for Carrier Behavior

Figure 1 presents a schematic overview of the process that brings about carriers' behavior in a TM. A shipper's decision to post a shipment in the auction

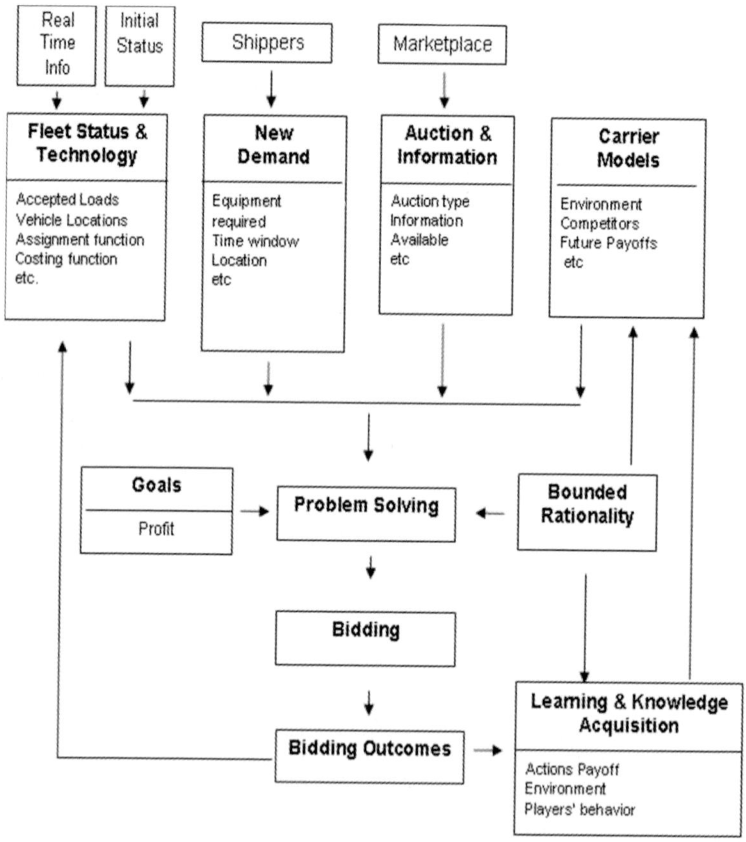

Fig. 1. Carrier behavior in a sequential auction transportation marketplace (TM)

market initiates an auction. Carriers respond to auctions postings. Carriers attempt to maximize profits by adjusting their behaviors in response to interactions with other carriers and their environment. Bounded rationality limitations are pervasive and affect how a carrier models, evaluates, and optimizes his action as indicated by the arrows in . Carriers also must abide by the constraints and the physical feasibility specified by their assignment strategies and pool of awarded shipments.

In this framework, carriers' learning and knowledge about other competitors' behavior types evolve jointly over time and their strategies at a given moment are contingent on interactions that have occurred or will occur in a path-dependent time line. Past decisions are binding and limit the future actions of carriers, therefore behavioral rules are state-conditioned and the carriers co-adapt their behavior as the marketplace evolves over time. Carriers' internal events are the assignment, pickup, and delivery of loads, mostly operational decisions. Carriers repeatedly engage in bidding interactions modeled as non-cooperative games. However these repeated bidding interactions may also be the only means of communication for a carrier to "identify" or "manipulate" other competitors.

4 Bounded Rational Behavior in the Transportation Marketplace

From the carrier point of view, the cost and value of transportation services are hard to quantify. The value of a traded item (shipment) may be strongly dependent upon the acquisition of other items (e.g. nearby shipments). In addition, the value of a shipment is related to the current spatial and temporal deployment of the fleet. The geographic dispersion of both demand and supply, uncertain demand arrival rates, and realizations over time and space, contribute to a dynamic and stochastic environment. These factors further increase the uncertainty of a shipment's cost and value.

Auctions, as a device to match supply and demand, provide a powerful mechanism to allocate resources, especially when the latter have uncertain or non-standard value. Auction analysis can be quite challenging, especially in a stochastic setting such as transportation. In this kind of setting, carriers face a complicated decision problem, which stems partly from the strategic inter-dependency among competitors' decisions, costs, and profits. Auctions have been widely studied by economists, leading to recent advances in the theoretical understanding of different auction types and designs. These models have been mainly focused on one-time auctions with symmetric risk-neutral agents that bid competitively for a single or multiple units. Optimal bidding strategies have been found in many auction environments, however the case of sequential auctions, with bidders with multiunit demand/supply curves, remains intractable [20]. Another source of complexity arises from the need to solve fleet management problems (vehicle routing problems with time windows, penalties, etc) to obtain the cost information for a shipment. These are NP hard problems, which cannot generally be solved optimally for realistic fleet sizes in a dynamic and stochastic environment.

Competition in a TM is an ongoing and sequential process, and thus naturally represented as an extensive-form game. The standard definition of rationality (for economists at least) requires that agents automatically solve problems that are in fact beyond the capabilities of any agent [7]. The problem is intractable and well beyond the conceptual and computational abilities of ordinary humans or decision support systems. In addition, response time limitations or the framing effects and cognitive limitations of the human mind prevent bidders from behaving rationally. The framing and cognitive limitations of human thinking have been widely studied and reported [6, 18], mainly in the psychology and economics literature. Therefore, the basic motivation for studying models of bounded rationality in TM environments comes about from the implausibility of perfect rationality models.

When the complexity of the auction problem precludes bidders from implementing a full game theoretic approach, computational agents (or human beings with the help of decision support systems) need to simplify or alter the original choice or decision problem. Bounded rational behavior, as studied in this research, is born out of these simplifications or alterations to the original insurmountable problem. This chapter attempts to provide a behavioral framework to understand how carriers can tackle the overwhelming complexity of the problems they face in a TM (complex detailed stories, numerous current options, future infinite contingent options, and the potential consequences).

Bounded rational bidders solve a less complex problem than fully rational bidders. The type of problem they solve is directly influenced by available response time, existing computational/material resources, and their own cognitive/decision-making process. Although the result of bounded rational deliberation would not necessarily be an equilibrium solution, the bounded rational response would have more bearing on how ordinary carriers or human decision makers would act in sequential auction TM. The introduction of bounded rational decision makers radically alters the notion of equilibrium and decision making. Game theory assumes that players know the prevailing equilibrium and act consequently. For bounded rational agents, the equilibrium, if any, is not known beforehand, it is built.

Bounded rational behavior is born out of simplifying a (complex) problem or the cognitive/material limitations of the decision maker (or decision support system). Therefore, bounded rationality is always associated with the notion of deficiency or insufficiency of a positive quality (of a rational player). Although bounded rationality as a research topic is not new, it was first proposed by Simon [35], many modeling issues surrounding bounded rational decision making have not yet been fully addressed. Bounded rationality and learning in games are currently very active areas of research; however general and comprehensive models that integrate how agents (or humans) acquire, process, evaluate, search for information, and make decisions are still mostly open. As expressed by Aumman, "here is no unified theory of bounded rationality, and probably never will be"[1].

Rationality assumptions are very convenient from a modeling point of view. The self-referential nature of rationality (coupled with common knowledge in games) imposes astringent limitations on how a rational agent (player in a game) foresees his competitors' behavior and how the competitors foresee other players' behavior. Bounded rationality come with an embarrassment of riches in terms of the number of possible deviations from a fully "rational" model. When bounded rational behavior appears, it may take on many different forms. Bounded rational decision makers do not necessarily choose equally, even when having the same knowledge or information. Furthermore, there may be many "plausible" bounded rational models that can explain a given social or economic phenomenon. Correspondingly, the many possible ways each bounded rational bidder can model his competition adds a class of uncertainty not found where all players are perfectly rational.

Determining the bounded rationality of a carrier is crucial since it is equivalent to determining how the carrier bids (i.e. his bidding function) in a TM. Similarly, determining that all carriers are rational is equivalent to determining how the carriers bid (i.e. their bidding function) in a SIPV setting. A bidding function, as understood in this research, is a process, whose inputs are a carrier's private information and his knowledge about the auction and competitors, and whose output is a bid. Given the plethora of games and decision problems, bounded rational behavior is hard to define, classify, and model in general terms. When the restrictions of rationality are lifted, any general assumption about the behavior of the bidders that is not properly justified, introduces a strong sense of arbitrariness. In order to avoid this kind of arbitrariness, the discussion of bounded rationality is limited to the TM context. Any departure from the rationality model is connected to carriers' cognitive and problem solving processes.

Bounded rationality can stem from different cognitive and computational/physical limitations, in the TM context, the following classification of sources is proposed:

- Bounded Recall and Memory: a carrier has limited memory (physical capacity) to:
  - record and keep past data/information
  - simulate and record data of all future possible paths in the decision tree
- Processing Speed: time is valuable in a dynamic setting. Most practical problems have a limited response time that may limit the solution quality or decrease the effectiveness of delayed response.

- Data Acquisition and Transmission: data acquisition and processing is usually costly. Furthermore, the transmission of data among agents can be noisy. In a world with bounded resources (budget/memory/attention), deciding how, how much, and what type of information should be acquired, kept, transmitted, or analyzed can lead to complex decision problems.
- Knowledge Acquisition: in a dynamic strategic situation, as data is being revealed or obtained, carriers have the potential to acquire knowledge (truths about competitors or the environment) from logical and sound inferences. In particular, the decision maker may have limited ability to discover competitors behavior, which may involve modeling and solving complex logical and econometrics problems.
- Problem Solving: as a carrier participates in a TM market, it is required to make decisions (bidding or fleet management decisions). These decisions may lead the carrier to formulate and solve complex optimization problems. In particular, the decision maker may have limited ability to predict or model the impact of his own actions on future fleet operational costs or on his competitors behavior.

Although the five aspects of bounded rationality are somewhat interrelated, this research focuses on the knowledge acquisition and problem solving aspects. Memory and processing speed are physical limitations. It is assumed carriers have enough material resources and response time/speed to implement bidding and fleet management strategies with different degrees of sophistication. Carriers have limitations to formulate and elucidate knowledge acquisition problems. Similarly, carriers have limitations to formulate and solve complex optimization problems. The data available to carriers is only limited to data publicly and freely disclosed after each auction, which renders the data acquisition problem trivial. No transmission losses or alterations are considered.

The focus of this research is on the knowledge acquisition and problem solving aspects, as they capture how carriers can frame and solve TM problems. Therefore, the emphasis is on the more "mental" processes that determine behavior rather than on the physical limitations. Knowledge acquisition and problem solving in a TM are analyzed in the next two sections respectively.

5 Knowledge Acquisition in a Transportation Marketplace

In a TM, each carrier is aware that his actions have significant impact upon his rival's profits, and vice-versa. In the perfect rational model, common knowledge and logical inferences allow the estimation of the impact of a carriers actions on competitors' profits and vice-versa. It is implicit that a rational bidder bids as a rational bidder. In a bounded rational model, a carrier faces two basic types of uncertainties regarding the competition: (a) an uncertainty relative to the private information of his opponents, and (b) a strategic uncertainty relative to bounded rationality type of the others players.

The first type of uncertainty is about $\theta_j^{-i} = \{z_j^{-i}, a^{-i}, c^{-i}\}$ for a carrier $i \in \Im$ at time t_j, the private information regarding competitors' fleet status, assignment, and cost functions respectively. This type of uncertainty is also present in most game theoretic auction models (games of incomplete information). The second type of uncertainty is about the bidding strategies that the competitors use, $b^{-i} = \{b^1, ..., b^{i-1}, b^{i+1}, ..., b^n\}$ the set of bidding functions of all carriers but carrier i. It is implicit that a bounded rational bidder bids accordingly, i.e. as a bounded rational bidder. However, it is not evident for the competition to determine what "type" of bounded rationality a carrier has. This type of uncertainty is not present in game theoretic auction models. Depending on a carrier's ability to elucidate uncertainties (a) and (b), two extreme cases may take place:

1. No knowledge acquisition. The carrier cannot form a useful model of competitors' be-havior that links their private information and their bids. In this situation, the "best" a carrier can do is to observe market prices and estimate them as the result of a random process. This is similar to assuming that competitors are playing $b^{-i}(\xi)$ =f(ξ) or simply $b^{-i}(\xi) = \xi$, where ξ is a random process that is not linked in any way to carrier i bidding, capacity/deployment, and history of play or to the competitors' private information $\theta_j^{-i} = \{z_j^{-i}, a^{-i}, c^{-i}\}$
2. Full knowledge acquisition. The carrier knows $\theta_j^{-i} = \{z_j^{-i}, a^{-i}, c^{-i}\}$ and also $b^{-i} = \{b^1, ..., b^{i-1}, b^{i+1}, ..., b^n\}$ therefore carrier i is able to precisely foresee what the competition is going to bid for shipment s_j. However, carrier i still has uncertainties about the future bids, simply because carrier i does not know the future realizations of the demand. Nevertheless, carrier i can estimate future prices not just as a stationary random process but as a function of shipment arrival distribution, shipment characteristics distribution, competitors' behavior, and competitors' private information. This is ξ =f($\Omega, \theta_j^{-i}, b^{-i}$).

In game theoretic terms the former case is not possible since there is no "strategic" game if players cannot speculate about the competitors' actions. The latter case corresponds to a game of perfect and complete information *if all the players are rational* and the private information is common knowledge. Knowledge states in between the two extreme cases correspond to games of imperfect information,*if all the players are rational* and there is uncertainty about the players' private information.

The uncertainty about the players' private information can be expressed as p($\theta_j^{-i} \mid \theta_j^i, h_{j-1}$). In a game of incomplete information each player (bidder) has expectations (beliefs) about the competitors' private values. Following Harsanyi's [17] modeling of games of incomplete information, players' types $\theta_j^{\Im} = \{\theta_j^i\}_{i=1}^n$ are drawn from some probability density function p($\theta_j^1, ..., \theta_j^n$)

where types θ_j^i belong to a space Θ^i. The conditional probability about his opponents' types $\theta_j^{-i} = \{\theta_j^1, ..., \theta_j^{-i}, \theta_j^{+i}, ..., \theta_j^n\}$ given his own type θ_j^i is denoted $p(\theta_j^{-i} \mid \theta_j^i, h_{j-1})$. This is what characterizes and complicates the solution of a dynamic game of incomplete in-formation. Since the players do not know the competitors' types at the start of each auction, they have to update these conditional probabilities (beliefs about the competitors' status) as public information is revealed and the game evolves.

Acquiring knowledge about the competitors' private information and bounded rationality type poses a potentially highly complicated econometric/logical problem. A carrier's behavior is likely to be affected by his own history and how the carrier perceives and models the stra-tegic situation. From the public information (revealed after each auction) and its own private information a carrier needs to build a model of the private information and bounded rationality type of his competitors.

Even in simple auctions, the econometric models can quickly become extremely complex and data are usually not rich enough to successfully estimate those structurally complex models [21]. Furthermore, the complexity of the underlying DVRP adds hurdles to the problem. However, the most challenging obstacle may come from the competitors, which may be "sophisticated" enough to realize that they are bidding against other bidders who are also learning and may adjust their behavior accordingly, in order to obstruct the process of knowledge acquisition. This type of sophistication is particularly important when the fact that the same carriers interact repeatedly is common knowledge.

In most game theoretic models, a simple private value probability distribution, symmetry, rationality, and common knowledge assumptions permit a closed analytical solution. In equilibrium bidders know the competitors' bidding function, however, they do not know the reali-zation of the competitors' private value, therefore they do not know the competitors' actual bid. Conversely, in a TM, private values are not random but correlated, the status of a carrier at time t_j provide useful information to estimate the status of the carrier at time t_{j+1}. A bidder may potentially obtain information about competitors' private values and bidding functions if the bidder invests resources to infer them. Market settings, such as auction data disclosed and number of competitors, strongly affect the difficulty of the inference process.

Summarizing, repeated interaction can lead to learning and knowledge acquisition. This research distinguishes among the two. Learning takes place in the no-knowledge case; the carrier does not get to know the competitors' behavioral processes just the price function as a random process. Learning is superficial, it is merely phenomenological. In the full knowledge case, the carrier acquires knowledge about the competitors' behavioral processes. Knowledge acquisition is deeper; it is causal. Learning in a TM environment is discussed in Section 7.

6 Problem Solving in a Transportation Marketplace

The previous section focused on "what can be learnt or known" about the competition. This section specifically contemplates "how carriers come up" with a bid or decision given what has been learnt or what knowledge has been acquired about a problem. Usually, models in which decision makers are assumed rational do not explain the procedures by which decisions are taken, rational procedures are implicitly embedded in the answer or approach. Further-more, economic models pay no or little attention to how hard it is to make decisions. Conversely, bounded rational decision maker models detail the procedural aspects of decision making. Those detail procedures are the essence of a bounded rational decision making model. The degree of intricacy of the decision making procedure is used in the last part of this chapter to classify bounded rational behaviors. As a carrier participates in a TM market, it is required to make decisions, to choose among alternative future paths. Each decision poses a problem that the carrier has to solve (not necessarily optimally). The rest of this section analyzes, in this order, the type of decision a carrier faces in a TM and how bounded rationality can appear in the steps of a decision making process.

From the carriers' point of view, the choice problems that take place in a TM are either bidding or operational (fleet management) decisions. Bidding decisions may carry a strategic value since they directly affect competitors' profits. Bidding decisions are also the result of a bounded rational decision process, a carrier's choice and therefore can reveal or transmit in-formation about a carrier's decision making process or intentions. Operational (fleet management) decisions mostly affect a carriers' own fleet status (private information). Therefore, operational decisions are considered non-strategic and take place as new information arrives: auctions are won or shipments are served. This type of decision, for example, includes the estimation of a shipment value or service cost, the rerouting of the fleet after a successful bid, the reaction to unexpected increase in travel times, etc. A detailed formulation of the value or service cost problem is found in Figliozzi et al. [14, 15].

There are several factors that contribute to the complexity of biding in a TM. These factors are: competitors bounded rationality, knowledge about the competitors, look-ahead depth, and the type of auction utilized. This section analyzes the first three factors. The auction characteristics that significantly affect bidding complexity, from the bidder's perspective, are: (a) the use of incentive compatible mechanisms and (b) the number of item being auctioned, e.g. combinatorial auctions are more demanding computationally than single item auctions. Incentive compatible mechanism simplify considerable the bidder's problem because the optimal bid is the cost or reservation value, regardless of the actions of the competitors [9].

Sophisticated bounded rational players have a "model" of the other players. For example, in the work of Stahl and Wilson [36] and Vidal and Durfee [39], players model other play-ers' cognitive process and decision rules up to a finite number of steps of iterated thinking. The number of iterations that a player can perform is a measure of the sophistication of a player. A zero level player does not model his opponents, it simply ignores the fact that other agents exit. Reinforcement learning is an example of this type of agent sophistication. A one level agent models only the frequency or another statistic that represents other players' ac-tions. Fictitious play is an example of this type of agent sophistication. A two level agent can simulate the other agents' internal reasoning process (i.e. a model of level zero or level one agents) and take an action by taking into account how the other players (of level zero or one) are going to play. A level three agent can build models, simulate them, and act in response to the behavior of players up to level two. Recursively, a level four agent can model the actions of level three agents and so on. Perfectly rational agents can follow the recursion to an infinite level. Then, if the level of rationality of a player is denoted by L^i, then that player can model the most sophisticated of his competitors up to a level $L^{-i} = L^i - 1$.

Section 5 dealt with the level of knowledge about the competition. A player with no-knowledge about the competition can only implement a level zero or level one type of player since it cannot link his actions (bids) to the consequences that his actions have. A player with full knowledge could possibly foresee (if it could only solve the corresponding problems) the behavior of any player type. However, the complexity increases as the level type to be implemented increases, i.e. as the competitors bounded rationality sophistication increases. The carrier with full-knowledge knows $\theta_j^{-i} = \{z_j^{-i}, a^{-i}, c^{-i}\}$ about the competition and also $b^{-i} = \{b^1, ..., b^{i-1}, b^{i+1}, ..., b^n\}$. Therefore, carrier i can compute precisely what the competition is going to bid for shipment s_j. However, carrier i still has uncertainties about the future bids, simply because carrier i does not know the future realizations of the demand. Nevertheless, carrier i can estimate future prices, not just as a random process but as a function of shipment arrival and characteristics distribution, competitors' behavior and competitors' private information. When the knowledge is imperfect, complexity further increases since there is a probability distribution over the competitors' private information space. Furthermore, the probability distribution is a function of the history of play and the competitors' fleet management strategies. In mathematical notation, the probability distribution of competitors' future private information is $\mathrm{p}(\theta_N^{-i} \mid h_N)$.

The third factor is the look-ahead depth. In a sequential auction setting like a TM, bids affect future auctions profits. The look-ahead depth is the number of future auctions that are taken into account when estimating how a bid may affect future auctions profits. A zero step look-ahead (or myopic) analysis does not consider future auction profits, just the profit for the cur-rent auction. A one-step look-ahead analysis considers one future auction, current plus the following auction profits. Similarly, a m-step look-ahead analysis considers m future auctions, current plus the following m auction profits. When the analysis is myopic, shipment s_j is known and the uncertainties are reduced to a minimum. Projecting one step into the future, the arrival time

(t_{j+1}) and characteristics of shipment s_{j+1} are uncertain. Furthermore, if the link between bidding and future prices ξ_{j+1} is incorporated, the optimal bid for shipment s_j takes into account its impact on competitors' bids (prices) in the next auction. Then, for ship-ment s_{j+1} the price function at time t_{j+1} is a function of the previous bids and the unknown previous arrival $\xi_{j+1}(s_j, b_j^{*i})$. In the one-step problem, the arrival and characteristics of s_{j+1} are uncertain, but the future history h_{j+1} is a function of the already known s_j. Projecting two steps into the future, the estimation of the future price function ξ_{j+2} becomes more complex. The price function ξ_{j+2} for shipment s_{j+2} is a function of the yet unknown s_{j+1} and the two previous bids $\{b_j^{*i}, b_{j+1}^{*i} \mid h_{j+1}\}$. Moving one extra step into the future increases the problem complexity significantly. For shipment s_{j+2} the price function at time t_{j+2} is a function of the previous bids and the unknown previous arrival $\xi_{j+2}(s_j, s_{j+1}(t_j, \Omega), b_j^{*i}, b_{j+1}^{*i} \mid h_{j+1})$. Calculation of future price functions is increasingly difficult as uncertainties and dependencies on earlier (but not yet realized) bids and shipments accumulate. When the look ahead is up to shipment s_N, the number of decision variables $B^{*i} = \{b_j^{*1}, ..., b_N^{*i} \mid h_N\}$ to be estimated is: $\sum_{k=0}^{N-j} p^k$.

When the number of players (bidders) is n after each auction there are n possible outcomes and future histories. If backward induction is used, for each possible history it is necessary to estimate an optimal bid, the total number of decision variables increases exponentially with the number of future look-ahead steps. Let denote by $\Sigma = \{s_j, s_{j+1}(t_j, \Omega), ..., s_N(t_{N-1}, \Omega)\}$ the set of shipments to be analyzed. Then, the future price function when earlier bids affect future prices and the carrier has imperfect information is a function of ξ_N =f$(b_j^{*i}, ..., b_{N-1}^{*i}, \Sigma, p(\theta_N^{-i} \mid h_N))$.

Table 1 puts the three factors together. The table is set up in such a way that the complexity of the price function ξ increases, moving downward or rightward. With higher levels of competitors' bounded rationality, the complexity of the problem increases exponentially with the number of iterations and players to be simulated.

The symbol $\langle\cdot\rangle^{nL-i}$ is used to denote the number of iterations as a function of the number of players and the highest level of iterations that the competition can sustain. A "fully rational" equilibrium, is a special case of the imperfect knowledge case when all players are rational and $L^{-i} \to \infty$. In the game theoretic case, it is common knowledge that all the bidders are simultaneously foreseeing and simulating each other's bids and decisions at infinitum. Each cell of Table 2 is a different decision theory problem that can potentially be expressed as a mathematical program or algorithm. It was mentioned that the complexity increases moving downward or rightward.

The problem solving capabilities of the carrier determines the type of problem the carrier solves. For example, a carrier may have imperfect information about the competitors; however, problem solving limitation may force him to solve a myopic problem assuming no-knowledge about the competition. When cost or time limitations are added to the problems, carriers can choose to ignore part of his knowledge in order to get a reasonable answer in a

Table 1: Bidding Complexity as a function of price function (ξ) complexity

			Look-Ahead Depth		
Knowledge Level	**Own Type** L^i	**Comp. Type** L^{-i}	**Myopic** $\Sigma = \{s_j\}$ $B^{*i} = \{b_j^{*i}\}$	**1-step** $\Sigma = \{s_j, s_{j+1}(t_j, \Omega)\}$ $B^{*i} = \{b_j^{*i}, b_{j+1}^{*i} \mid h_{j+1}\}$	**Multi-step** $\Sigma = \{s_j, \ldots, s_N(t_{N-1}, \Omega)\}$ $B^{*i} = \{b_j^{*i}, \ldots, b_N^{*i} \mid h_N\}$
NO	$L^i = 0$	-	Reinforc. Learning	-	-
	$L^i = 1$	-	Fictitious Play Stationary ξ	Fictitious Play Stationary ξ	Fictitious Play Stationary ξ
FULL	$L^i = 1$	-	Acceptance Rejection	Acceptance Rejection Stationary ξ	Acceptance Rejection Stationary ξ
	$L^i \geq 2$	$L^{-i} \leq 1$	Acceptance Rejection	Optimal Pricing Non-stationary $\xi_{j+1} = \mathrm{f}(b_j^{*i}, \Sigma)$	Optimal Pricing Non-stationary $\xi_N = \mathrm{f}(b_j^{*i}, \ldots, b_{N-1}^{*i}, \Sigma)$
	$L^i = m$	$2 \leq L^{-i} < m$	Iterated Acceptance Rejection	Iterated Optimal Pricing Non-stationary $\xi_{j+1} = \left\langle \mathrm{f}(b_j^{*i}, \Sigma) \right\rangle^{nL^{-i}}$	Iterated Optimal Pricing Non-stationary $\xi_N = \left\langle \mathrm{f}(b_j^{*i}, \ldots, b_{N-1}^{*i}, \Sigma) \right\rangle^{nL^{-i}}$
IMPERFECT	$L^i = 1$	$L^{-i} \leq 1$	Fictitious Play $\xi_j = \mathrm{f}(\mathrm{p}(\theta_j^{-i} \mid h_j))$	Acceptance Rejection Stationary $\xi_{j+1} = \xi_j$	Acceptance Rejection Stationary $\xi_N = \ldots = \xi_{j+1} = \xi_j$
	$L^i \geq 2$	$L^{-i} \leq 1$	Fictitious Play $\xi_j = \mathrm{f}(\mathrm{p}(\theta_j^{-i} \mid h_j))$	Optimal Pricing Non-stationary $\xi_{j+1} = \mathrm{f}(b_j^{*i}, \Sigma, \mathrm{p}(\theta_{j+1}^{-i} \mid h_{j+1})$	Optimal Pricing Non-stationary $\xi_N = \mathrm{f}(b_j^{*i}, \ldots, b_{N-1}^{*i}, \Sigma, \mathrm{p}(\theta_N^{-i} \mid h_N))$
	$L^i = m$	$2 \leq L^{-i} < m$	Iterated Fictitious Play $\xi_j = \left\langle \mathrm{f}(\mathrm{p}(\theta_j^{-i} \mid h_j)) \right\rangle^{nL^{-i}}$	Iterated Optimal Pricing Non-stationary $\xi_{j+1} = \left\langle \mathrm{f}(b_j^{*i}, \Sigma, \mathrm{p}(\theta_{j+1}^{-i} \mid h_{j+1}) \right\rangle^{nL^{-i}}$	Iterated Optimal Pricing Non-stationary $\xi_N = \left\langle b_j^{*i}, \ldots, b_{N-1}^{*i}, \Sigma, \mathrm{p}(\theta_N^{-i} \mid h_N) \right\rangle^{nL^{-i}}$

reasonable time, in the spirit of the "satisfying" rule as proposed by Simon [34]. According to Simon, economic agents do not always optimize fully, they optimize up to a satisfying level. Level that depends on personal characteristics and circumstances.

Simplifying (downgrading complexity) the problem due to bounded rational limitations is always possible. It can be interpreted that each problem type (each cell) of table 1 is a different way of measuring how desirable each possible bid is, for a given DVRP technology. If the value of knowledge can be defined as the profit difference that a carrier can obtain going from the no to full knowledge assumption, likewise, the value of computational power is the profit difference that a carrier can obtain from solving a more complex problem due to the increased performance of his computational resources. Summarizing, based on their knowledge level and problem solving capabilities, agents differ in the type of problem they can solve.

7 Learning in a Transportation Marketplace

The reminder of this chapter studies the bidding behavior of carriers in a no-knowledge and no-strategic environments of Table 1. Henceforth, it is assumed that carriers bid trying to maximize their profits but limited by their bounded rational limitations. In this competitive setting, three different auction formats are compared using computational experiments. These auction formats are second price auctions, first price auction with minimum information disclosure, and first price auctions with maximum information disclosure.

The high complexity of acquiring and using knowledge about competitors' behaviors was discussed in Section 5, even in TM market/model that has been streamlined to the indispensable elements. Knowledge acquisition and its use can be considerably more complex in a more complete model where other critical constraints and variables are added (for example, getting drivers home, variation in travel times, delays incurred while unloading the truck, etc). Furthermore, noisy information transmission, as reported by Powell et al. [31], even among agents that respond to the same carrier (i.e. drivers, dispatchers, decisions support systems), seem to sustain the notion that perfect knowledge about competitors' private information and behavior could only be possible in a flight of the imagination. Imperfect knowledge is possible, but at the cost of even higher modeling complexity. Given the high level of complexity of full or imperfect knowledge assumptions, it is methodologically sensible to first focus on behaviors and settings which are more plausible for implementation in real-life TM marketplaces. The first tool that bounded-rational agents use to cope with insurmountable complexity is simplification. Henceforward, it is assumed that carriers can acquire a limited knowledge of the competitors' behavior. Carriers' knowledge is limited to learn about the distribution of past market prices or the relationships between realized profits and bids.

In an auction context, learning methods seek good bidding strategies by approximating the behavior of competitors. Most learning methods assume that competitors' bidding behavior is stable. This assumed bidding stability is akin to believing that all competitors are in a strategic equilibrium. Walliser [40] distinguishes four distinct dynamic processes to play games. In a decreasing order of cognitive capacities they are: eductive processes, epistematic learning (fictitious play), behavioral learning (reinforcement learning), and evolutionary processes. An eductive process requires knowledge about competitors' behavior (agents simulate competitors' behavior). Epistemic and behavioral learning are similar to fictitious play and reinforcement learning respectively (fully described in the next section). In the evolutionary process, a player has (is born with) a given strategy; after playing that strategy the player dies and reproduces in proportion to the utilities obtained (usually in a game where it has been randomly matched to another player).

This reminder of this chapter studies the two intermediate types of learning. Eductive-like type of play requires carriers to have almost unbounded computational power and expertise. On the other hand, evolutionary model players seem too simplistic: they have no memory, and simply react in response to the last result. Furthermore, the notion that a company is born, dies, and reproduces with each auction does not fit well behaviorally in the defined TM. Ultimately, neither extreme approach is practically or theoretically compelling in the TM context. Carriers that survive competition in a competitive market like truck-load (TL) procurement cannot be inefficient or unskilled. They are merely limited in the strategies they can implement. It is assumed that carriers would like to implement the strategy (regardless of its complexity) that ensured higher profits, but they are restricted by their cognitive and informational (which give rise to bounded rationality).

In practical and theoretical applications, the process of setting initial beliefs has always been a thorny issue. Implemented learning models must specify what agents initially know. Ideally, how or why these initial assumptions were built should always be reasonable justified or explained. In this respect, restricting the research to the TM context has clear advantages. Normal operating ratios in the TL industry range from 0.90 to 0.95 [37]. It is expected that operating ratios in a TM would not radically differ from that range. If prices are too high shippers can always opt out, abandon the marketplace and find an external carrier. Prices cannot be substantially lower because carriers would run continuously in the red, which does not lead to a self-sustainable marketplace. Obviously, operating ratios fluctuations in a competitive market are expected, in response to natural changes in demand and supply. However, these fluctuations should be in the neighborhood of historical long term operating ratios unless the market structure is substantially changed. Another practical consideration is the usage of ratios or factors in the trucking industry. Traditionally, the trucking industry has used numerous factors and indicators to analyze a carrier's performance, costs, and profits. It seems natural that some carriers would obtain a bid after multiplying the estimated cost by a bidding coefficient or factor. Actually, experimental data show that the use of multiplicative bidding factors is quite common [30].

7.1 Learning Mechanisms

In reinforcement learning the required knowledge about the game payoff structure and competitors behavior is extremely limited or null. From a single carrier's perspective the situation is modeled as a game against nature; each action (bid) has some random payoff about which the carrier has no prior knowledge. Learning in this situation is the process of moving (in the action space) in a direction of higher profit. Experimentation (trial and error) is necessary to identify good and bad directions.

Let M be the ordered set of real numbers that are multiplicative coefficients $M = \{mc_0, ..., mc_K\}$, such that if $mc_k \in M$ and $mc_{k+1} \in M$, then $mc_k < mc_k + 1$. Using multiplicative coefficients the profit obtained for any shipment s_j, when using the multiplicative coefficient mc_k is equal to:

$$\pi_j^i(mc_k) = (mc_k c_j^i - c_j^i)I_j^i = c_j^i I_j^i(mc_k - 1) \tag{1}$$

$$\pi_j^i(mc_k) = (b_j^{(2)} - c_j^i)I_j^i \tag{2}$$

The first equation applies to first price auctions while the second equation applies to second price auctions. In the second price auction the payment depends on the value of the second best bid which is represent by the term $b_j^{(2)}$. Adapting the reinforcement model to TM bidding, the probability $\varphi_j^i(mc_k)$ of carrier i using a multiplicative coefficient mc_k in the auction for shipment s_j is equal to:

$$\varphi_j^i(mc_k) = (1 - \lambda\pi_{j-1}^i(mc_k))\varphi_{j-1}^i(mc_k) + I_{j-1}^i(mc_k)\lambda\pi_{j-1}^i(mc_k) \tag{3}$$

Narenda and Tatcher showed that a players' time average utility, when confronting an opponent playing a random but stationary strategy, converges to the maximum payoff level obtainable against the distribution of opponents' play. The convergence is obtained as the reinforcement parameter λ goes to zero. To use equation (3), each bidder only needs information about his bids and the result of the auction. To use this model the profits $\pi_{j-1}^i(mc_k)$ must be normalized to lie between zero and one so that they may be interpreted as probabilities. The indicator variable $I_j^i(mc_k)$ is equal to one if carrier i used the multiplicative coefficient mc_k when bidding for shipment s_j, the indicator is equal to zero otherwise. The parameter λ is called the reinforcement learning parameter, it usually varies between $0 < \lambda < 1$.

The reinforcement is proportional to the realized payoff, which is always positive by assumption. Any action played with these assumptions, even those with low performance, receives positive reinforcement as long as it is played. Therefore, a mediocre action can be reinforced while at the same time "better" actions are negatively reinforced. Furthermore, in an auction context there is no learning when the auction is lost since $\pi_{j-1}^i(mc_k) = 0 \; \forall mc_k \in M$ if $I_{j-1}^i = 0$. Borgers and Sarin [4] propose a model that deals with the aforementioned problems. In this model the stimulus can be positive or negative depending on whether the realized profit is greater or less than the agent's "aspiration level". If the agent's aspiration level for shipment s_j is denoted ϱ_j^i and the effective profit is denoted:

$$\tilde{\pi}_{j-1}^i(mc_k) = \pi_{j-1}^i(mc_k) - \varrho_j^i, \tag{4}$$

and the probability becomes:

$$\varphi_j^i(mc_k) = (1 - \lambda\tilde{\pi}_{j-1}^i(mc_k))\varphi_{j-1}^i(mc_k) + I_{j-1}^i(mc_k)\lambda\tilde{\pi}_{j-1}^i(mc_k) \quad (5)$$

When $\varrho_j^i = 0$, the equation (5) provides the same probability updating equation as (3). Borgers and Sarin explore the implications of different policies to set the level of the aspiration level. These implications are clearly game dependent. A general observation applies for aspiration levels that are unreachable. In this case equation (4) is always negative; therefore the learning algorithm can never settle on a given strategy, even if the opponent plays a stationary strategy.

These learning mechanisms were originally designed for games with a finite number of actions and without private values (or at a minimum for players with a constant private value). In the TM context, the cost of serving shipments may vary significantly. Furthermore, even the best or optimal multiplier coefficient can get a negative reinforcement when an auction is lost simply because the cost of serving a shipment is too high. This negative reinforcement for the "good" coefficient creates instability and tends to equalize the attractiveness of the different multiplicative coefficients. This problem worsens as the number of competitors is increased causing a higher proportion of lost auctions, i.e. negative reinforcement. This chapter utilizes a modified version of the stimulus response model with reinforcement learning that better adapts to TM bidding [8, 13]. Each multiplicative coefficient m_k has an associated average profit value $\bar{\pi}_j^i(m_k)$ that is equal to:

$$\bar{\pi}_j^i(m_k) = \frac{\sum_{t\in\{1,\ldots,j\}} \pi_t^i(s_t) I_t^i(m_k)}{\sum_{t\in\{1,\ldots,j\}} I_t^i(m_k)}$$

The aspiration level is defined as the average profit over all past auctions: $\bar{\varrho}_j^i = \frac{\sum_{t\in\{1,\ldots,j\}} \pi_t^i(s_t) I_t^i}{j}$

Therefore the average effective profit is defined as $\bar{\pi}_{j-1}^i(mc_k) = \bar{\pi}_{j-1}^i(mc_k) - \bar{\varrho}_j^i$. Probabilities are therefore updated using equation (6).

$$\varphi_j^i(mc_k) = (1 - \lambda\bar{\pi}_{j-1}^i(mc_k))\varphi_{j-1}^i(mc_k) + I_{j-1}^i(mc_k)\lambda\bar{\pi}_{j-1}^i(mc_k) \quad (6)$$

With the latter formulation (6), a "good" multiplicative coefficient does not get a negative reinforcement unless its average profit falls below the general profit average. At the same time, there is learning even if the auction is lost. The learning mechanism that uses equation (6) is named as Average Reinforcement Learning (ARL) henceforth.

Stimulus-response learning requires the least information and can be applied to both first and second price auctions. The probability updating equations (3) and (6) are the same for first and second price auctions. Therefore the application of the reinforcement learning model does not change with the auction format that is being utilized in the TM. Using this learning method, a carrier does not need to model neither the behavior nor the actions of competitors. The learning method is essentially myopic since it does not attempt to measure the effect of the current auction on future auctions. The method clearly fits in the category of no-knowledge/myopic carrier bounded rationality. Since the method is myopic, for the first price auction the multiplicative coefficients must be equal or bigger than one, i.e. $mc_0 \geq 1$. A coefficient smaller than one, generates only zero or negative profits. In a second price auction the multiplicative coefficients can be smaller than one and still generate positive profits since the payment is dependent on the competitors' bids. In both types of auctions it is necessary to specify not just the set of multiplicative coefficients but the initial probabilities. If equation (5) is used it is also necessary to set the aspiration level. If equation (6) is used it is necessary to set the level of the initial profits but not the aspiration level. A uniform probability distribution is the classical assumption and indicates a complete lack of knowledge about the competitive environment.

Summarizing, in reinforcement learning, the agent does not consider strategic interaction. The agent is unable to model an agent play or behavior but his own. This agent is informed only by his past experiences and is content with observing the sequence of their own past actions and the corresponding payoffs. Using only his action-reward experience, he reinforces strategies that succeeded and inhibit strategies which failed. He does not maximize but moves in a utility-increasing direction, by choosing a strategy or by switching to a strategy with a probability positively related to the utility index. Reinforcement learning (and its variants) is a strategy that is designed to operate in an environment where the player (carrier) is unable to see the competitors' actions. Therefore, it is able to strongly reinforce (positively or negatively) only one action: the last action played. Unlike reinforcement learning, fictitious play requires the observation of competitors' actions. A good introduction to types of learning employed in this chapter (reinforcement learning and fictitious play) can be found in the work of Fudenberg and Levine [16].

Fictitious play came about as an algorithm to look for Nash equilibrium in finite games of complete information [5]. It is assumed that the carrier observes the whole sequence of competitors' actions and draws a probabilistic behavioral model of the opponents' actions. The agent does not try to reveal his or her opponents' bounded rationality from their actions although the agent may eventually know that opponents learn and modified their strategies too. The agent models not behavior but simply a distribution of opponents' actions. Players do not try to influence the future play of their opponents. Players behave as if they think they are facing a stationary, but unknown, distribution of the opponents' strategies. Players ignore any dynamic links between their play today and their opponents' play tomorrow. A player that uses a generalized fictitious play learning scheme assumes that his opponents' next bid vector is distributed according to a weighted empirical distribution of their past bid vectors. The method cannot be straightforwardly adapted to games with an infinite set of strategies (for example the real numbers in an auction). Two ways of tackling this problem are: a) the player divides the set of real numbers into a finite number of subsets, which are then associated with a strategy or b) the player uses a probability distribution, defined over the set of real number to approximate the probabilities of competitors play. In either case, the carrier must come up with a estimated stationary price function ξ (in our experiments carriers estimate a normal distribution using on competitors' past bids). If a second price auction format is used in the TM, the carrier bids using:

$$b_j^{*i} \in \underset{b \in R}{argmax} E_{(\xi)}\{[\xi - c^i(s_j, z_j^i)]I_j^i\} \tag{7}$$

If a first price auction format is used in the TM, the carriers bid using:

$$b_j^{*i} \in \underset{b \in R}{argmax} E_{(\xi)}\{[b - c^i(s_j, z_j^i)]I_j^i\} \tag{8}$$

In the second price auction (equation 7) the best price is simply the corresponding cost $c^i(s_j, z_j^i)$ due to the special properties of one-item second price auctions (8) (independence between the winners bid and the corresponding payment). Equation 8 has to be solved numerically or analytically.

7.2 Automaton Interpretation

The previous sections have described reinforcement learning and fictitious play models of learning. Reinforcement learning and fictitious play were originally conceived as human methods of learning. However, they can also be used by machines or computerized systems. This section tries to link both views. An automaton is a self operating machine or mechanism. In a game context, an automaton is meant to be an abstraction of the process by which a player *implements* a given bounded rationality behavior. Rubenstein [33] replaces the notion of a strategy with the notion of a machine called finite automaton. In Rubenstein's model a finite automaton that represents player i, is a four-tuple (Z^i, z_0^i, b^i, a^i), where Z^i is a finite set of machine states (from this constraint the adjective "finite"), z_0^i is the initial state for carrier i,$b^i : Z^i \rightarrow A$ is an output function that produces an action (given the state of the automaton), and $a^i : Z^i \times A^{-i} \rightarrow Z^i$ is a transition function that updates the state of the automaton (given the actions taken by the competitors in the previous period). The set of possible actions is denoted by A. Adapting these concepts to this research, a TM automaton can be defined as an abstraction of the process by which a carrier implements a given bounded rational behavior in a TM. A TM automaton can be defined by the eight-tuple $(Z^i, z_0^i, \Xi, \xi_0^i, S, b^i, u^i, a^i)$ comprised by:

Z^i the set of possible states (private information states);
z_0^i the initial state for carrier i;
Ξ the set of possible price functions;
ξ_0^i the initial price function for carrier i;
$s_j \in S$ the stimulus sent by marketplace;
$b^i : Z^i \times \Xi S \to R$ the bidding (output) function;
$u^i : h \times \Xi \to \Xi$ the update function (updates the price function $\xi \in \Xi$); and
$a^i : Z^i \times S \to Z^i$ the assignment function (assignment if an auction is won).

A TM automaton would work in the following way: the initial state and price function are z_0^i and ξ_0^i respectively, the automaton chooses a bid $b^i(z_0^i, \xi_0^i, s_1)$ when the first shipment arrives. If carrier wins, the assignment function updates the carrier's status $a^i(z_0^i, s_1)$. The price function is updated based on the information revealed after the auction $u^i(h_1, \xi_0^i)$. When the second shipment arrives the same process is repeated but starting with the new state and price function z_1^i and ξ_1^i respectively. Once the initial conditions are set, the transitions, bidding, and updating are set by the arrival of shipments. A TM automata game takes place when a player cannot change the working of his machine during the course of the game. The two learning approaches described in this section, reinforcement learning and fictitious play, can be interpreted as the work of an automaton (which is valid in general for any learning strategy that seeks or uses no knowledge about the competitors' behavior). Therefore, the simulation results presented in the next sections can also be interpreted as the interaction or competition of TM automata (which may represent the behavior of human, computerized, or hybrid dispatchers). It is assumed in this research that for a given status, price function, and stimulus, an action has the same probability of being played; as if the decision process is *wired-up* and cannot change (data and information can change over time, but not the decision-making process). This is consistent (in the short-medium term) with the industry experience [31].

8 Experimental Results

Closed analytical solutions for the complex carriers' decision problem in a TM setting would require many simplifications that could compromise the validity of the results. Therefore, computational experiments and simulation are used as needed to enhance and extend simpler theoretical models. Furthermore, simulation is used to study the dynamics of carriers' behaviors and interactions in controlled and replicable experiments.

8.1 Simulation Framework

The following sections study truck-load (TL) carriers that compete over a square area; the sides' lengths are equal to 1 unit of distance. For convenience, trucks travel at constant speed equal to one unit of distance per unit of time. Demands for truckload pickup-and-delivery arise over this area and over time. Origins and destinations of demands are uniformly distributed over the square area, so the average loaded distance for a request is 0.52 units of distance. All the arrivals are random; the arrival process follows a time Poisson process. The expected inter-arrival time is $E[T] = 1/(K\lambda)$, where λ is the demand request rate per vehicle and K is the total market fleet size. The total market fleet size that was used in the results is 4 (though similar trends were obtained with larger fleets -8 vehicles- as long as the same arrival rate/fleet size ratio is used). Roughly, the average service time for a shipment is 0.77 units of time (approximately $\lambda = 1.3$). The service time is broken down into 0.52 units of time corresponding to the average loaded distance, plus 0.25 units of time that approximate the average empty distance (average empty distance vary with arrival rates and time windows considered). Different Poisson arrival rates per truck per unit of time are simulated (ranging from 0.5 to 1.5). As a general guideline, these values correspond to situations where the carriers are:

- $\lambda = 0.5$ (uncongested)
- $\lambda = 1.0$ (congested)
- $\lambda = 1.5$ (extremely congested)

The shipments have hard time windows. In all cases, it is assumed that the earliest pickup time is the arriving time of the demand to the marketplace. The latest delivery time (LDT) is assumed to be:
LDT = arrival time + 2 x (shipment loaded distance + 0.25) + 2 x uniform (0.0, 1.0). All the shipments have a reservation price distributed as uniform (1.42, 1.52). In all cases, reservation prices exceed the maximum marginal cost possible in the simulated area ($\approx$1.41 units of distance). It is also assumed that all the vehicles and loads are compatible; no special equipment is required for specific loads. In all the simulations, two carriers are competing for the demands. In all cases there is an initial warm up or learning period of 250 auctions.

Multiple performance measures are used. The first is total profits, which equal the sum of all payments received by won auctions minus the empty distance incurred to serve all won shipments (it was already mentioned that shipment loaded distances are not included in the bids, loaded distances cancel out when computing profits). The profit for a particular shipment is defined as the difference between the payment received and the increment of the empty distance cost necessary to serve this shipment. The second performance measure is number of auctions won or number of shipments served, an indicator of market share. The third is shippers' consumer surplus, which is the accumulated difference between reservation prices and prices paid. The fourth is total wealth generated that is equal to the accumulated difference between reservation price (of served shipments) and empty distance traveled.

The second price auction used in the TM operates as follows: (a) each carrier submits a single bid, (b) the winner is the carrier with the lowest bid (which must be below the reservation price; otherwise the auction is declared vacant), (c) the item (shipment) is awarded to the winner, (d) the winner is paid either the value of the second lowest bid or the reservation price, whichever is the lowest, and (e) the other carriers (not winners) do not win, pay, or receive anything. The same procedure applies to first price auctions but the winner is paid the value of the winning bid, only point (d) changes.

In real time situations, this is an increasingly difficult task when optimal decision-making involves the solution of larger NP hard problems and the necessity of taking into account the stochastic nature of future demands. Three levels of DVRP technologies were simulated. These technologies are presented in an order that shows an increasing level of sophistication.

1. Base or Naïve Technology: this type of carrier simply serves shipments in the order they arrive. If the carrier has just one truck, it estimates the marginal cost of an arriving shipment s_j simply as the additional empty distance incurred when appending s_j to the end of the current route. If the carrier has more than one truck, the marginal cost is the cost of the truck with the lowest appending cost. This technology does not take into account the stochastic or combinatorial aspect of the cost estimation problem and is considered one of the simplest possible. Nonetheless, it provides a useful benchmark against which to compare the performance of more complex and computationally demanding technologies.
2. Static Fleet Optimal (SFO): this carrier optimizes the static vehicle routing problem at the *fleet* level. If the carrier has just one truck, the technology is equivalent to the previous case. If the carrier has more than one truck, the marginal cost is the increment in empty distance that results from *adding* s_j to the *total pool of trucks and loads* yet to be serviced. If the problem where static, this technology would provide the optimal cost. Again, like the two previous technology, it does not take into account the stochastic nature of the problem. This technology roughly stands for the best a myopic (as ignoring the future but with real time information) fleet dispatcher can achieve. Carriers fleet assignment and cost estimation is based on the static optimization based approach proposed by Yang et al.[42].
3. One step Look ahead Fleet Optimal (1SLA): as the previous carrier, this carrier optimizes the static vehicle routing problem the fleet level. This provides the static marginal cost for adding s_j. However, this carrier also knows the distribution of load arrivals over time and their spatial distribution. Hence, the carrier can simulate whether and how much winning s_j affects the marginal cost of serving the next arriving load. This technology roughly stands for what a fleet dispatcher with real time information and knowledge of future (yet unrealized probabilistic demands) can do. However, 1SLA is not an “optimal” technology, rather it is a heuristic that tries to estimate how serving s_j affects the cost of serving the next shipment.

8.2 Analysis of Experimental Results

A significant characteristic of one-item second price auction is also cost bidding, i.e. one-item second price auctions are incentive compatible mechanisms. That characteristic cannot be necessarily maintained in multiunit sequential auctions setting such as the TM marketplace. Of the two learning methods proposed, only reinforcement learning can be applied to second price auctions since fictitious play in a single-item second price auction coincides with marginal cost bidding. Regardless of the price distribution, the expected profit is always optimized with marginal cost bidding. In the TM context, the objective of reinforcement learning is to "learn" what the best bidding coefficient is; the bidding coefficient that maximizes a carrier's profits. Which raises the question: in a TM second price auction environment can carriers be better off by using bidding factors? This question is answered using computational experiments. Two carriers using the same type of DVRP technology compete against each other. However, while one carrier bids the marginal cost (called MC carrier) the other bids the marginal cost multiplied by a bidding factor (called BF carrier). Eleven different bidding factors are utilized, ranging from 0.5 to 1.5. The impact of these factors on carrier BF's profits are depicted in Figure 2. The profit levels of a BF carrier when the bidding factor is equal to 1.0 are used as the reference or base level they correspond to 100% level. Both carriers are using the SFO technology.

8.3 Performance of Marginal Cost Bidding

The results depicted in Figure 2 show that for low arrival rates the best bidding factor is 1.0, corresponding to simply bidding the marginal cost. For

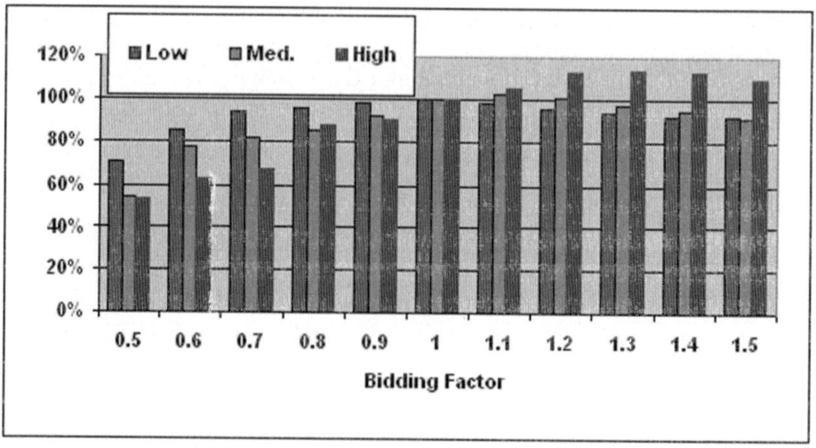

Fig. 2. Profit Level for a BF Carrier

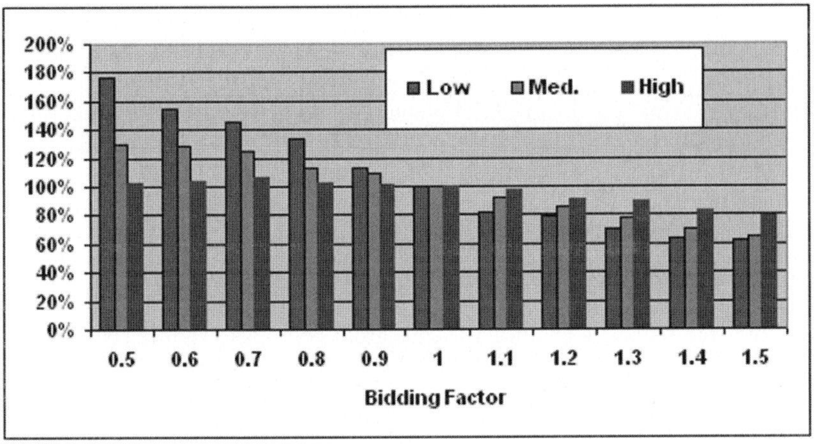

Fig. 3. Shipments Served by BF carrier

medium arrival rates the best bidding factor is 1.1. For high arrival rates the best bidding factor is 1.3. Regardless of the arrival rate level, the curve is quite flat around the "optimal". Furthermore, if the profits are connected the resulting curve is concave-shaped. A possible explanation to the results of Figure 2 may be obtained by analyzing how profits are generated. Total profits can be expressed as the average profit obtained per shipment multiplied by the number of shipments served. Figure 3 and Figure 4 show the impact of bidding factors on number of shipments served and average shipment served

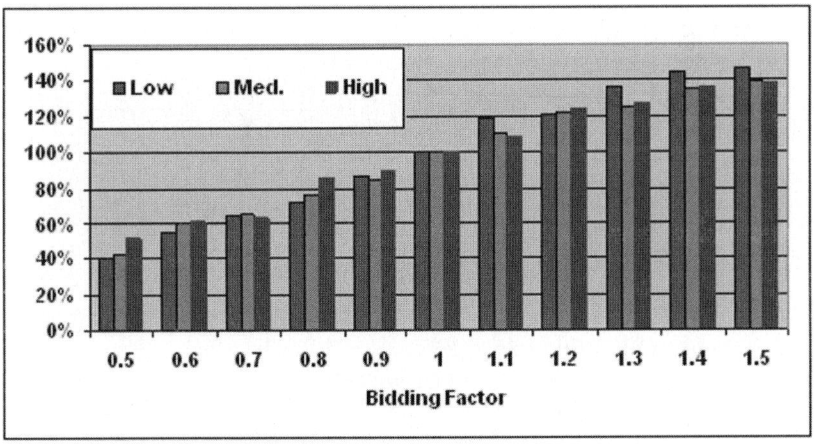

Fig. 4. Average Profit per Shipment Won for a BF Carrier

profit respectively. Again, the number of shipments served and average profit used as reference are those of a BF carrier when the bidding factor is equal to 1.0.

It is clear from Figure 3 and Figure 4 that, as expected, higher bidding factors increase the average profit per shipment won but decreases the number of shipments won. Vice versa, lower bidding factors decrease the average profit per shipment won but increases the number of shipments won. There are clearly two opposing forces at work when the bidding factor changes; this fact helps to explain the concave shape of the profit curve in Figure 2.

At this point, it has not yet been explained why the low arrival rate "optimal" bidding factor is around 1.0 (marginal cost case), while the "optimal" bidding factors are shifted to the right for higher arrival rates. The answer to this matter lies in the relation between profit elasticity and shipment served volume elasticity. To understand why profit elasticity and shipment served volume elasticity changes with the arrival rate is necessary to introduce Figure 5 and . They illustrate the different fleet utilization rates of carriers MC and BF respectively. Fleet utilization rate is defined as the average vehicle utilization. Vehicle utilization is defined as the percentage of the time a vehicle is moving (i.e. not idle).

With low arrival rates the utilization of the MC carrier is low (around 35% if the BF carrier uses a bidding factor equal to 1.0 - see Figure 5). Therefore when carrier BF increases his prices (utilizing higher bidding factors) carrier MC gains a significant percentage of the demand. This explains why in there is such an abrupt drop in demand (from 100 to 80%) when carrier BF moves from a bidding factor of 1.0 to 1.1. With higher arrival rates the fleet utilization of carrier MC is higher (at or over 70% - see Figure 5) and at very high

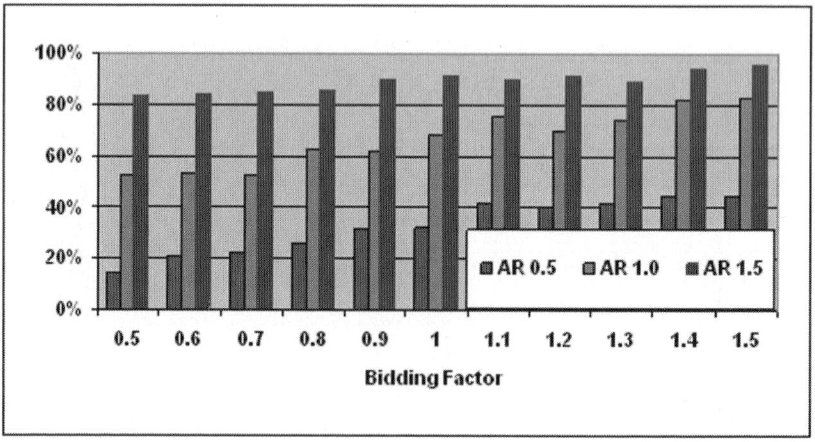

Fig. 5. Fleet Utilization (MC Carrier)

utilization rates it is more difficult to accommodate or to inexpensively add new shipments. As fleet utilization grows the capacity to serve new shipments decreases, therefore on average the opportunity costs of serving additional shipments starts to be significant. is the reverse mirror image of Figure 5. With high arrival rates carrier BF can rise prices substantially and still have a high fleet utilization; the increase in profits prevails over the decrease in shipments served. The explanation provided is plausible but not definitive. However, similar phenomena as the ones observed in Figure 2-6 have been widely recognized in the economics-industrial organization literature.

The incentives to increase prices as remaining market capacity decreases are contemplated in price-capacity oligopoly models. For example, in the Edgeworth-Bertrand model of competition, pricing is at marginal cost levels when demand is low, however prices increase after a critical capacity utilization threshold is surpassed. Similar intuition is obtained from Benoit and Krishna model of capacity constrained auctions, with limited capacity it is advantageous to speculate. Even in fleet management, the idea of filtering out shipments or similarly increasing the "admission" price of shipments under very high arrival rate conditions has been previously used (though not in a competitive environment). The Kim et al. study indicates that a fleet dispatcher under very high arrival rates (over capacity) is better off filtering out some demands (not being too close to capacity). Similar results are also found when carriers use other technologies such as the naïve or 1SLA. Figure 7 shows the profit changes when both carriers use naive technologies. Even when carriers have different technologies, similar results can be expected. Figure 8 show the profit changes for the BF carrier using naïve technology against a MC carrier using SFO technology.

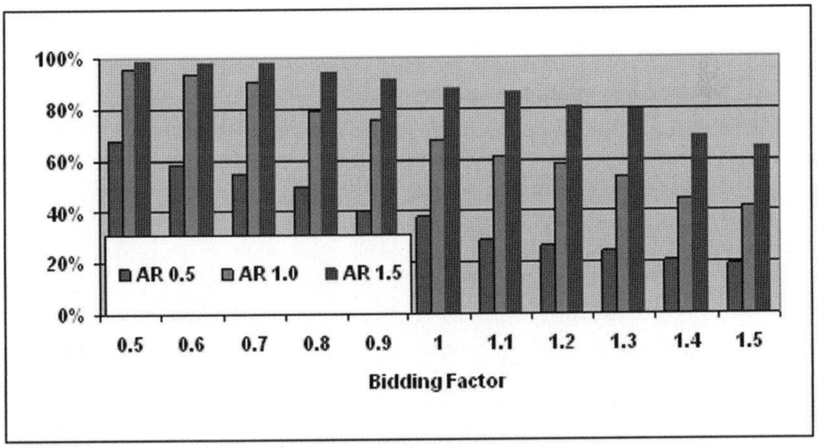

Fig. 6. Fleet Utilization (BF Carrier)

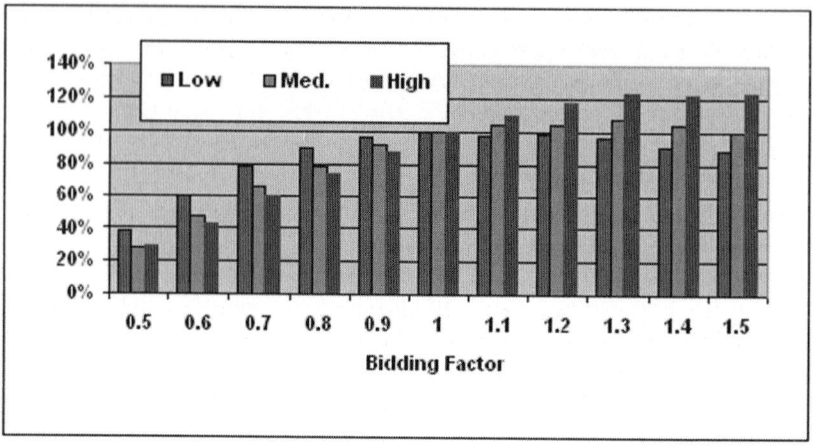

Fig. 7. Profit Level for a BF Carrier (both carrier use naïve technology)

The question that motivated these simulations was: in a TM second price auction environment can carriers be better off by using bidding factors? The answer is yes, but only at high arrival rates. This answer provides additional insights into the applicability of auction analysis to online algorithms/technologies. The results confirm the notion that DVRP technological leadership can be better exploited under low to moderate arrival rate conditions, where there is no incentive to adopt bidding factors that are not one. If there is an incentive to adopt bidding factors that are higher than one, there is an incentive to restrain capacity or to increase prices (profits are increased

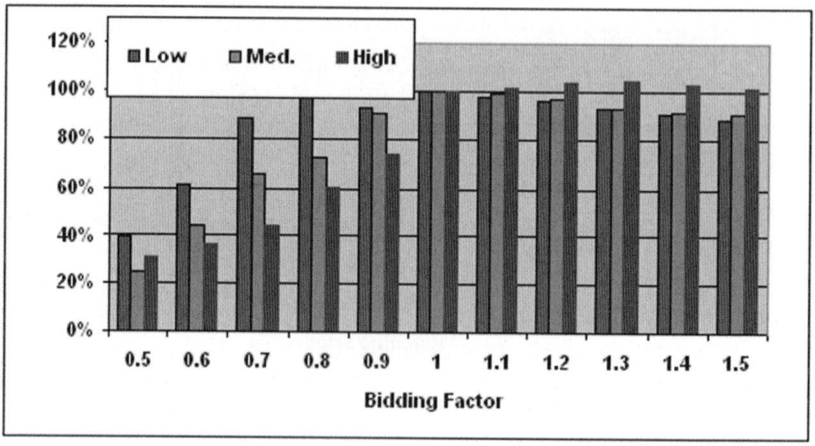

Fig. 8. Profit Level for a BF Carrier (SFO vs. naïve technology)

without increasing fleet management efficiency). As the arrival rate grows the advantage of being more efficient decreases; in general, scarcity exposes the incompetent while abundance hides inefficiencies.

8.4 Learning Methods Performance

The following results address the issue of learning performance of the two learning methods presented in this chapter. The previous results show that bidding factors can be used to increase carriers' profits in TM second price auctions with high arrival rates. Reinforcement learning could be used to "learn" which bidding factors produce a higher profits on average; as the auction results accumulates the most profitable bidding factors continuously increase their probability of being used. With low arrival rates, there is nothing to learn but the fact that marginal cost bidding (bidding factor 1.0) is the best alternative. Learning can be expensive though. For example, in a second price auction the longer it takes a bidder to learn that underbidding (bidding below his marginal costs) is not a good strategy, the more the bidder loses potential profits. The importance of the right learning coefficient then becomes evident. If the learning coefficient λ is too small learning is too slow; if λ s too big it may lock the learning algorithm in an undesirable bidding factor too quickly. Another important element is the number of alternatives that the learning algorithm must choose from; as a general rule, the more the alternatives the smaller the λ.

The speed of reinforcement learning can be quite slow in an auction setting like TM. The optimal bidding factor can be used and there is still roughly a 50% chance of losing (assuming two bidders with equal fleets and technologies). If the optimal bidding factor loses two or three times its chances of being played again may reduce considerably which hinders convergence to the optimal or even convergence at all. As discussed previously in this section, this issue can be avoided using "averages" (ARL method). Figure 9 illustrates the relative performance of Average Reinforcement Learning (ARL) and Reinforcement Learning (RL) in a first price auction. Both learning methods select a bidding factor among 11 different possibilities, ranging from 1.0 to 2.0 in intervals of 0.1. The learning factor is λ=0.10. Figure 9 shows the relative performance of ARL and RL after 500 auctions. It is clear that RLA obtains higher profits as the arrival rate increases. RL has a poorer performance because it cannot converge steadily to the optimal coefficient due to the reasons mentioned in the previous paragraph. The carrier RL tends to price lower (it keeps probing low bidding coefficients longer) and therefore serves a higher number of shipments. As shown in the previous section, as arrival rates increase after a critical point, a carrier can charge higher prices regardless of what the competitor is doing.

In first price auctions reinforcement learning and fictitious play can be used. The latter uses more information than the former. Therefore, it is expected that a carrier using fictitious play must outperform a carrier using

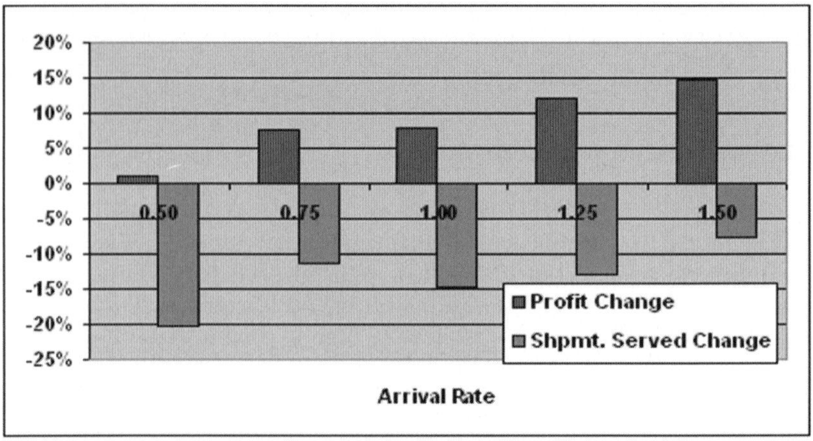

Fig. 9. ARL vs. RL (RL performance base of comparison)

reinforcement learning. Figure 10 shows the relative performance of Fictitious Play (FP) and ARL after 500 auctions. The ARL player is the same as in Figure 9. The FP carrier divides the possible competitors' bids in 15 intervals (from 0.0 to 1.5 in intervals of width 0.1) and start with a uniform probability distribution over them.

Clearly the FP carrier obtains higher profits across the board. The usage of a competitor past bidding data to obtain the bid that maximizes expected profits clearly pays off. In this case carrier ARL tends to bid less and serve

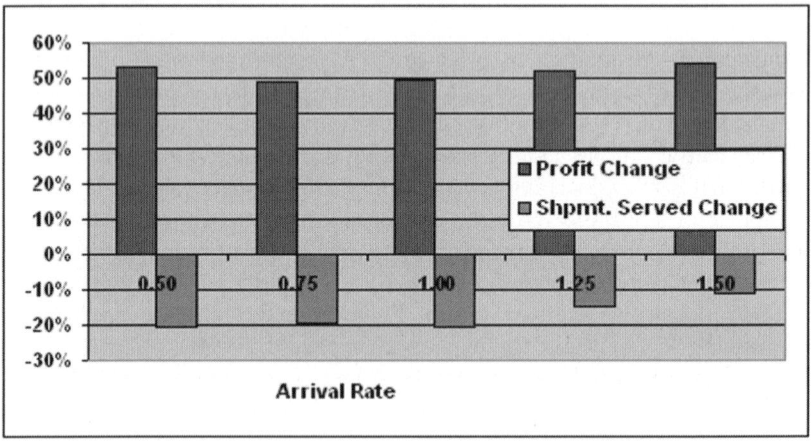

Fig. 10. ARL vs. FP (RL performance base of comparison)

more shipments, again, the difference diminished as the arrival rate increase. In the TM context even a simple static optimization provides better results than a search based on reinforcement learning. Not surprisingly, more information and optimization lead to better results. Therefore, if there is maximum information disclosure, carriers will choose to play fictitious play or a similar bidding strategy, especially since the complexity of FP (myopic) and ARL are not too different.

8.5 Comparing Auction Settings

The following results describe the outcomes of TM competition with different sequential auction settings. Within the competitive no-knowledge assumptions, three basic auction settings are compared: second price auction with marginal cost bidding, first price auction with reinforcement learning, and first price auction with fictitious play. Four different measures are used to compare the auction environments: carriers' profits, consumer surplus, number of shipments served, and total wealth generated.

To facilitate comparisons in all the four graphs that are presented subsequently, second price auctions with marginal cost bidding are used as the standard to measure up the two types of first price auction. All two carriers use SFO technologies0. Figure 11 illustrates the profits obtained by carriers. After the results of the previous section, it is not surprising that FP carriers obtain higher profits than ARL carriers. FP carriers use the obtained price information to their advantage. The highest carrier profit levels takes place with the second price auctions. These results do not alter or contradict theoretical results. With asymmetric cost distribution functions, Maskin and Riley show that there is not revenue ordering between independent value first and second price auctions. Figure 12 illustrates

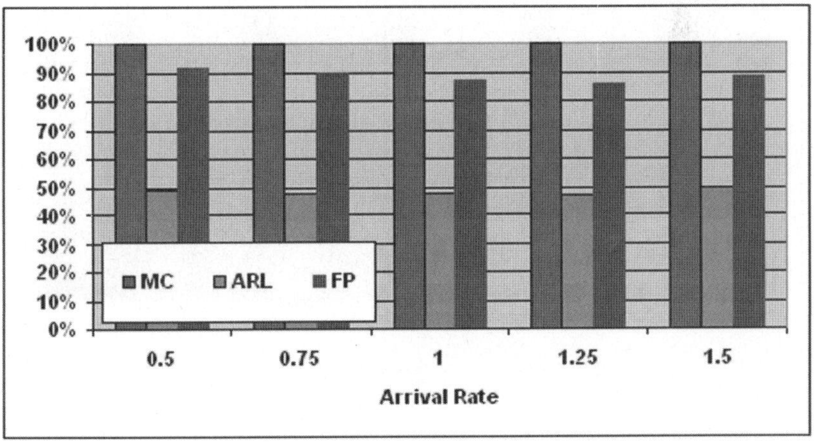

Fig. 11. Carriers' Profit level (Second Price Auction MC as base)

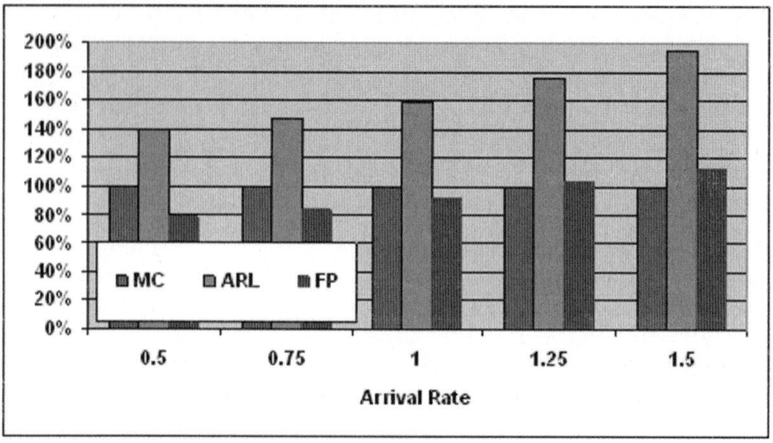

Fig. 12. Consumer Surplus level (Second Price Auction MC as base)

the consumer surplus obtained with the three auction types. Clearly, first price auction with reinforcement learning (minimum information disclosed) benefit shippers. Unsurprisingly, Figure 12 is almost the reverse image of Figure 11.

Figure 13 shows the number of shipments served with each auction setting. As expected, with second price auctions more shipments get served. Even in asymmetric auctions, it is still a weakly dominant strategy for a bidder to bid his value in a second price auction recall that this property of one-item second price auction is independent of the competitors' valuations. Therefore, in the

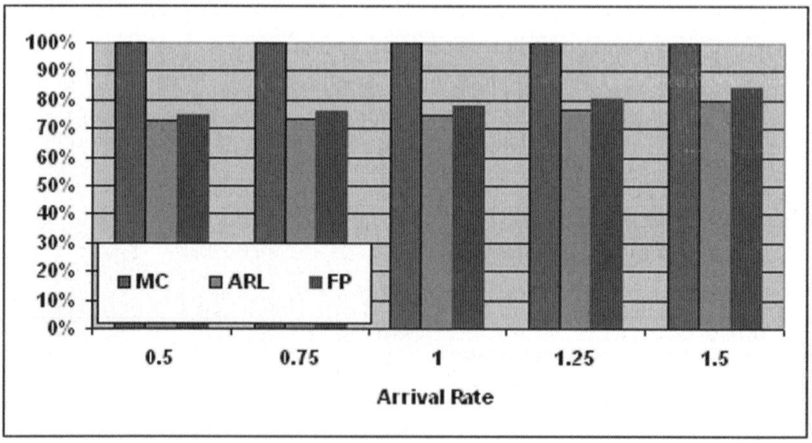

Fig. 13. Number of Shipments Served (Second Price Auction MC as base)

second price auction the shipment goes to the carrier with the lowest cost. In contrast, with ARL there is a positive probability that there are inefficient assignments since a higher cost competitor can use a bidding coefficient that results in a lower bid. Similarly with FP carriers, if the price functions are different (which is very likely since each carrier models the competitors' prices), a lower cost carrier can be underbid by a higher cost carrier with a positive probability. The results of Figure 12 and Figure 13 are similar to the insights provided by the reverse auction model with elastic demand, where introducing higher price uncertainty decreases prices (carriers' profits) but also decreases the probability of completing a potentially feasible transaction (number of shipments served). Figure 14 shows the wealth generated with each auction setting. Predictably, with second price auctions more wealth is generated. This is not surprising since marginal cost bidding is a "price efficient" mechanism. As the arrival rate increases the gap in total wealth generated tends to close up (Figure 14). Consistently, the lowest wealth generated corresponds to the case with FP bidders.

Summarizing, under the current TM setting, carriers, shippers, and a social planner would each select a different auction setting. Carriers would like to choose a second price auction. If first price auction are used, carriers would like to have maximum information disclosure. More information allows players to maximize profits, though total wealth generated is the lowest. Shippers would like to choose a first price auction with minimum information disclosure; more uncertainty about winning leads carriers to offer lower prices. However, the uncertainty leads to a reduction in the number of shipments served. Finally, from society viewpoint the most efficient system is the second price auction. More shipments are served and more wealth is generated.

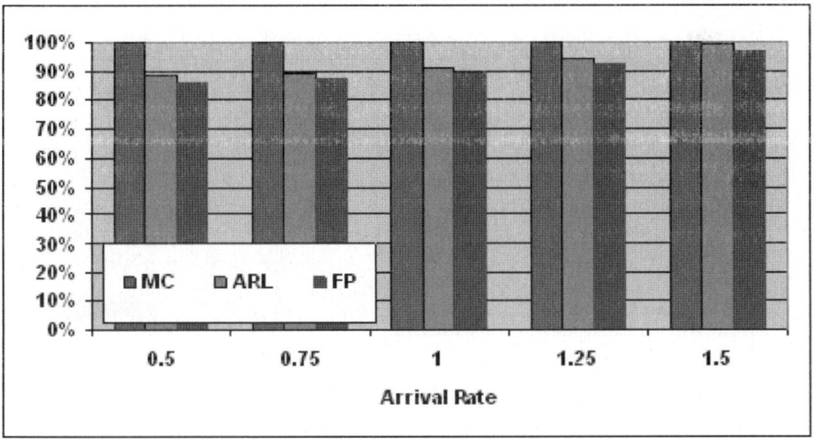

Fig. 14. Total Wealth Generated (Second Price Auction MC as base)

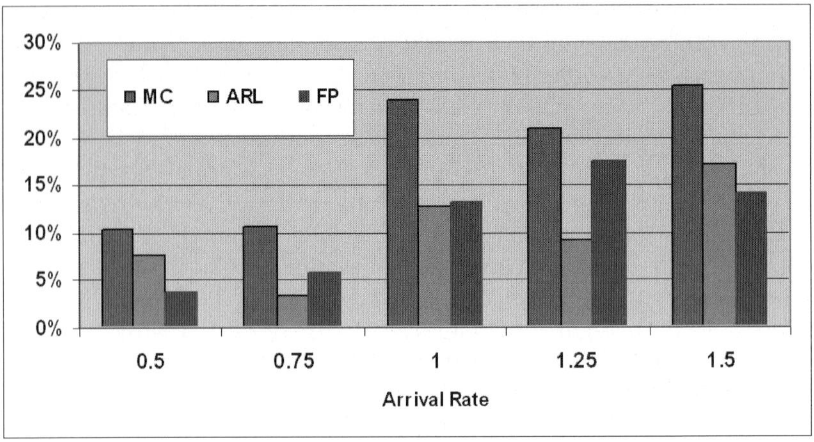

Fig. 15. Impact of Auction Type and Technology upgrading on Profits

8.6 Auction Settings and DVRP Technology Benefits

The final set of experiments looks at how auction settings impact the competitive edge that a more sophisticated DVRP can provide. Figure 15 illustrates the profit improvement of a carrier using a SFO technology over a carrier using the naïve technology. As expected, the second price auction better rewards a lower cost carrier. Again, this can be attributed to the lack of speculation about prices, which removes unnecessary speculation about competitors.

9 Conclusion

A competitive TM setting was analyzed to determine the likely sources of bounded rationality and the context of carriers' decision making process. Given the complexity of the bidding/fleet management problem, carriers can tackle it with different levels of sophistication. The complexity of the different bidding problems that a bounded rational carrier can be faced with was analyzed and classified. In the framework presented, sequential auctions can be used to model an ongoing transportation market, where the effect of carrier competition, knowledge and information availability, dynamic vehicle routing technologies, computational power, and decision making processes can be studied. This is an alternative framework to traditional models of behavior, equilibrium, decision-making, and analysis for transportation carriers. Decision making and behavior is defined as an expression of the goals and bounded rationality of the carrier as the type of pricing/bidding/fleet management problem that the carrier is able to tackle. Table 1 coupled with the appropriate learning mechanisms (for example reinforcement learning and fictitious play when they suit) embody the approach to carrier behavior proposed in this research.

Reinforcement learning and fictitious play, two learning methodologies for this type auction setting and assumptions are introduced and analyzed, as well as carrier learning and behavioral assumptions. Carrier's behavior is compared with the behavior of a machine. Computational experiments indicate that auction setting and information disclosure matters. Maximum information disclosure allows carriers to maximize profits at the expense of shippers' consumer surplus; minimum information disclosure allows shippers to maximize consumer surplus but at the expense of lowering the number of shipments served. Marginal bidding in second price auctions remains the most efficient incentive compatible auction mechanism, producing more wealth and more shipments served than first price auctions. It is demonstrated that under critical arrival rate there is no incentive to use bidding factors (no deviations from static marginal cost bidding). Furthermore, second price auction TM is the mechanism that provides the highest reward to carriers with more sophisticated DVRP technology.

References

1. Aumann R.: Rationality and Bounded Rationality. Games and Economic Behavior **21** (1997) 2–14.
2. Benoit, J. and Krishna, V.: Multiple-Object Auctions with Budget Constrained Bidders. Review of Economic Studies **68** (2001) 155–179.
3. Blume, A. and Heidhues, P.: Private monitoring in auctions. Journal Of Economic Theory **131** (2006) 179–211.
4. Borges, T. and Sarin, T.: Naïve reinforcement learning with endogenous aspirations. University College of London, Mimeo (1996)
5. Brown, G.: Iterative Solutions of games by fictitious play. In Activity Analysis of Production and Allocation In Koopmans, T. (eds.): Activity Analysis of Production and Allocation Wiley, New York (1951)
6. Camerer, C.: Individual Decision Making In Kagel J.H., Alvin, A.E. (eds.): The Handbook of Experimental Economics Princeton NJ, Princeton University Press (1995)
7. Colinsk J.: Why Bounded Rationality. Journal Economic Literature **34** (1996) 669–700.
8. Figliozzi, M.: Performance and Analysis of Spot Truck-Load Procurement Markets Using Sequential Auctions Ph. D. Thesis, School of Engineering, University of Maryland College Park (2004)
9. Figliozzi, M.: Analysis and Evaluation of Incentive Compatible Dynamic Mechanisms for Carrier Collaboration. Transportation Research Record **1966** (2006) 34–40.
10. Figliozzi, M., Mahmassani, H. and Jaillet, P.: Framework for study of carrier strategies in auction-based transportation marketplace. Transportation Research Record **1854** (2003) 162–170.
11. Figliozzi, M., Mahmassani, H. and Jaillet, P.: Modeling Carrier Behavior in Sequential Auction Transportation Markets. 10th International Conference on Travel Behaviour Research (IATBR) (2003)

12. Figliozzi, M., Mahmassani, H. and Jaillet, P.: Competitive Performance Assessment of Dynamic Vehicle Routing Technologies Using Sequential Auctions. Transportation Research Record **1882** (2004) 10–18.
13. Figliozzi, M., Mahmassani, H. and Jaillet, P.: Auction Settings and Performance of Electronic Marketplaces for Truckload Transportation Services. Transportation Research Record **1906** (2005) 89–97.
14. Figliozzi, M., Mahmassani, H. and Jaillet, P.: Quantifying Opportunity Costs in Sequential Transportation Auctions for Truckload Acquisition. Transportation Research Record **1964** (2006) 247–252.
15. Figliozzi, M., Mahmassani, H. and Jaillet, P.: Pricing in Dynamic Vehicle Routing Problems. Transportation Science (2007) *in press*
16. Fudenberg, D. and Levine, D. The Theory of Learning in Games In Kagel J.H., Alvin, A.E. eds.: The Handbook of Experimental Economics. MIT Press, Cambridge Massachusetts (1998)
17. Harsanyi, J.: Games with incomplete information played by Bayesian players. Management Science **14** (1967) 159–182 and 320–334.
18. Kagel, J.R.A.: Handbook of Experimental Economics. Princeton NJ, Princeton University Press (1995)
19. Kim Y., Mahmassani, H.S. and Jaillet, P.: Dynamic truckload truck routing and scheduling in oversaturated demand situations. In Transportation Network Modeling 2002 Washington, Transportation Research Board Natl Research Council
20. Krishna V.: Auction Theory. San Diego, Academic Press (2002)
21. Laffont, J.J.: Game theory and empirical economics: The case of auction data. European Economic Review **41** (1997) 1–35.
22. Lucking-Reiley, D. and Spulber, D.: Business-to-Business Electronic Commerce. Journal of Economic Perspectives **15** (2001) 55–68.
23. Mahmassani, H.: Freight and Commercial Vehicle Applications. In Hensher, D. (eds.) Travel Behaviour Research Pergamon, Elsevier Science (2001)
24. Marshall, R. and Marx, L.: Bidder Collusion. Working Paper, Penn State University, Duke University. (2002)
25. Martin, S.: Advanced Industrial Economics. Cambridge, Blackwell Publishers (1993)
26. Maskin, E. and Riley, J.: Asymmetric Auctions. Review of Economic Studies **67** (2000) 413–438.
27. McAffee, R. and McMillan, J.: Auctions and Biddings. Journal of Economic Literatur **25** (1987) 669–738.
28. Nandiraju, S. and Regan, A.: Freight Transportation Electronic Marketplaces: A Survey of Market Clearing Mechanisms and Exploration of Important Research Issue. Proceedings 84th Annual Meeting of the Transportation Research Board, Washington D.C (2005)
29. Narendra, K. and Thatcher, M.: Learning Automata: a survey. IEEE Transactions on Systems, Man, and Cybernetics **4** (1974) 889–899.
30. Paarsh, H.: Deciding between Common and Private Value Paradigms in Empirical Models of Auctions. Journal of Econometric **15** (1991) 191–215.
31. Powell, W., Marar, A., Gelfand, J. and Bowers, S.: Implementing Real-Time Optimization Models: A Case Application form the Motor Carrier Industry. Operations Research **50** (2002) 571–581.
32. Raiffa, H., Richardson, J. and Metcalfe, D.: Negotiation Analysis. Harvard, The Belknap Press of Hardvard University Press. (2002)

33. Rubinstein, A.: Modeling Bounded Rationality. Cambridge, MIT Press. (1998)
34. Simon, H.: A behavioral model of rational choice. Quarterly Journal of Economics **69** (1955) 99–118.
35. Simon, H.: Rational choice and the structure of the environment. Psychological Review **63** (1956) 129–138.
36. Stahl, D. and Wilson, P.: On Players Models of Other Players: theory and experimental evidence. Games and Economic Behavior **10** (1995) 213–254.
37. TCA: Truckload Carrier Association Website. Accessed December 16th, 2003
38. Tirole, J.: The Theory of Industrial Organization. Cambridge, MIT Press. (1989)
39. Vidal, J. and Durfee, E.: Recursive agent modeling using limited rationality. In Proceedings of the First International Conference on Multi-Agent Systems (ICMAS-95) Menlo Park, AAAI Press. (1995) 376–383.
40. Walliser, B.: A spectrum of equilibration processes in game theory. Journal of Evolutionary Economics **8** (1998) 67–87.
41. Wolfstetter, E.: Topics in microeconomics: industrial organization, auctions, and incentives. Cambridge, Cambridge University Press. (1999)
42. Yang, J., Jaillet, P. and Mahmassani, H.: Real-time multivehicle truckload pickup and delivery problems. Transportation Science **38** (2004) 135–148.

Part III

Traffic

Decentralized Approaches to Adaptive Traffic Control

Arne Kesting[1], Martin Schönhof[1], Stefan Lämmer[1], Martin Treiber[1], Dirk Helbing[2,3]

[1] Institute for Transport & Economics, TU Dresden, Andreas-Schubert-Str. 23, 01062 Dresden, Germany
kesting|schoenhof|laemmer|treiber@vwi.tu-dresden.de
[2] Chair of Sociology, in particular of Modeling & Simulation, ETH Zurich, UNO D11, Universitätstrasse 41, 8092 Zurich, Switzerland dhelbing@ethz.ch
[3] Collegium Budapest – Institute for Advanced Study
Szentháromság utca 2, H-1014 Budapest, Hungary

1 Motivation and Solution Approaches

Traffic congestion is a severe problem on freeways in many countries. According to a study of the European Commission [1], its impact amounts to 0.5% of the gross national product and will increase even up to 1% in the year 2010. Since in most countries, building new transport infrastructure is no longer an appropriate option, there are many approaches towards a more effective road usage and a more 'intelligent' way of increasing the capacity of the road network and thus of decreasing congestion. Due to the potential benefits and the expected technological progress, there is considerable research in the area of intelligent transport systems (ITS) [2, 3]. Examples of advanced traffic control systems are, e.g., ramp metering, adaptive speed limits, or dynamic and individual route guidance. The latter examples are based on a *centralized traffic management*, which controls the operation and the system's response to a given traffic situation.

However, traffic systems are highly complex multi-component systems suffering from instabilities and non-linear dynamics, including chaos. This is caused by the non-linearity of interactions, delays, and fluctuations, which can trigger phenomena such as stop-and-go waves, noise-induced breakdowns, or slower-is-faster effects. The recently upcoming information and communication technologies (ICT), including cheap optical, radar, video, or infrared sensors and mobile communication technologies promise new solutions leading from the classical, centralized control to decentralized approaches in the sense of collective (swarm) intelligence and ad hoc networks. Such concepts reduce the problem of data flooding by restricting to the locally relevant information only and reach more adaptiveness, flexibility, resilience and robustness with respect to local requirements and temporary failures.

Our main focus is on future adaptive cruise control (ACC) systems, which will not only increase the comfort and safety of car passengers, but also enhance the stability of traffic flows and the capacity of the road. We call this 'traffic assistance' [4, 5]. We present an automated driving strategy that adapts the operation mode of an ACC system to the autonomously detected, local traffic situation, see Sec. 2. The impact on the traffic dynamics is investigated by means of a multi-lane microscopic traffic simulation in Sec. 3. Moreover, vehicles will become automatic traffic state detection, data management, and communication centers when forming ad hoc networks through inter-vehicle communication (IVC) [6, 7, 8, 9]. In Sec. 4, we discuss the mechanisms and applicability of short-range inter-vehicle communication for detection dynamic congestion fronts on freeways. Adaptive, self-organized traffic control in urban road networks is another interesting application field for decentralized strategies. In Sec. 5, we present control principles that allow one to reach a self-organized synchronization of traffic lights. We conclude with a summary and outlook in Sec. 5.

2 From Adaptive Cruise Control to Traffic Assistance Systems

The recent development and availability of adaptive cruise control (ACC) systems extends earlier cruise control systems, which were designed to maintain a selected speed. An ACC system is able to detect and to track the leading vehicle, measuring the actual distance and speed difference to the vehicle ahead by a radar sensor. Together with the own vehicle speed, these input data allow the system to calculate the required acceleration or deceleration to maintain a selected safe time gap. The update time for the data is typically 0.1 s, i.e., much shorter than the reaction time of human drivers of about 1 s [10]. In commercially available ACC systems, the time gap can be adjusted by the user typically in the range between 1 s and 2 s. Additionally, the user is able to select the desired velocity. These systems offer a gain in comfort in applicable driving situations on freeways, but they are still restricted to free traffic and high speed regimes due to their limited speed and acceleration range. The next generation of ACC systems is constructed to operate in all speed ranges and most traffic situations on freeways including stop-and-go traffic. Furthermore, ACC systems have the potential to prevent actively a rear-end collision and, thus, to achieve also a gain in safety.

However, what is the effect of ACC systems on the overall traffic situation? Do they necessarily increase the instability of traffic flows in favor of more driving comfort? Or is it possible to increase the capacity and stability by suitable modification of vehicle interactions? If yes, how could such a traffic assistance system look like? In the following, we will give a short description of a new

ACC system that we have developed [4, 11, 5]: The proposed traffic-assistance system consists of several system components: The main operational layer is still the ACC system calculating the vehicle's acceleration (on a typical time scale of 0.1 s). The new feature of the proposed system is the strategic layer, which implements the changes in the driving style in response to the *local* traffic situation by changing some *parameters* of the ACC system. To this end, a detection algorithm determines, which of the five traffic situations mentioned above applies best to the actual traffic situation. The ACC parameter settings related to the detected traffic state changes typically on time scales of minutes and in a range of typically a few hundred meters. This is analogous to manual changes of the desired velocity or the time gap in conventional ACC systems by the driver, which, of course, is possible in the proposed system as well.

2.1 Implementation of the Adaptive ACC Driving Strategy

The design of an ACC-based traffic assistance system is subject to several, partly contradicting, objectives. On the one hand, the resulting driving behavior has to be safe and comfortable to the driver. This implies comparatively large gaps and low accelerations. On the other hand, the performance of traffic flow is enhanced by lower time gaps T and higher accelerations, which can be seen when considering the main aspects of traffic performance: The *static road capacity* C, defined as maximum number of vehicles per time unit and lane, is strictly limited from above by the inverse of the vehicle time gap, $C < 1/T$. Moreover, simulations show that higher accelerations increase both the *traffic stability* and the outflow from congested traffic, which is typically lower than the free-flow capacity [5]. Our approach to resolve these conflicting goals is based on following observations:

- Most traffic breakdowns are initiated at some sort of road inhomogeneities or infrastructure-based 'bottlenecks' such as on-ramps, off-ramps, or sections of road works [12, 13, 14].
- An effective measure to avoid or delay traffic breakdowns is to homogenize the traffic flow.
- Once a traffic breakdown has occurred, the further dynamics of the resulting congestion is uniquely determined by the traffic demand (which is outside the scope of this investigation), and by the traffic flow in the immediate neighborhood of the downstream boundary of congestion [15]. According to empirical investigations [12], the downstream boundary is mostly fixed and located near a bottleneck.
- Traffic safety is increased by reducing the spatial velocity gradient at the upstream front of traffic congestion, i.e., by reducing the risk of rear-end collisions.

In the context of the ACC-based traffic assistance system, we make use of these observations by *only temporarily changing the comfortable settings of the*

ACC system in specific traffic situations. The selected situations have to be determined autonomously by the equipped vehicles and they have to allow for specific actions to improve the traffic performance. To this end, we propose the following discrete set of five traffic situations and the corresponding actions:

1. **Free traffic**. This is the default situation. The ACC settings are determined solely by the goal of maximum driving comfort. Since modern ACC systems allow each driver to set the parameters for the time gap and the desired velocity individually, this may lead to different parameter settings.
2. **Upstream jam front**. Here, the objective is to increase safety by decreasing velocity gradients. Compared to the default situation, this implies earlier braking when approaching slower vehicles. Notice that the operational ACC layer always assures a safe approaching process independently from the detected traffic state.
3. **Congested traffic**. Since drivers cannot influence the development of traffic congestion in the bulk of a traffic jam, the ACC settings are reverted to their default values.
4. **Downstream jam front**. To increase the dynamic bottleneck capacity, accelerations are increased and time gaps are temporarily decreased.
5. **Bottleneck sections**. Here, the objective is to locally increase the capacity, i.e., to *dynamically compensate for the capacity drop*, which is the defining property of bottlenecks. This implies a temporal reduction of the time gap.

Notice that drivers typically experience the sequence of these five traffic states when travelling through congested traffic. In the following, we will discuss the implementation of the above ACC concept by adjusting three relevant driving parameters: The *acceleration parameter* a gives an upper limit for the acceleration $\dot{v}(t)$ of the ACC-controlled vehicle. Consequently, this parameter is increased when leaving congestion, i.e., when the state 'downstream front' has been detected. The *comfortable deceleration parameter* b characterizes the deceleration when approaching slower or standing vehicles. Obviously, in order to be able to brake with lower decelerations, one has to initiate the braking maneuver earlier, which corresponds to higher levels of anticipation. Since this smoothes upstream fronts of congestion, the parameter b is decreased when the state 'upstream front' has been detected. Notice that, irrespective of the value of b, the ACC vehicle brakes stronger than b if this is necessary to avoid collisions. Finally, the *safe time gap parameter* T is decreased if one of the states 'bottleneck' or 'downstream front' is detected.

In order to be acceptable for the drivers, the system parameters need to be changed in a way that preserves the individual settings and preferences of the different drivers and also the driving characteristics of different vehicle categories such as cars and trucks. Particularly, the preferred time gap T can be changed both by the driver, and by the event-driven ACC adaptation. This can be fulfilled by implementing *relative* changes by means of *multiplication factors* λ_a, λ_b, and λ_T defined by the relations

$$a^{(s)} = \lambda_a^{(s)}\, a, \quad b^{(s)} = \lambda_b^{(s)}\, b, \quad T^{(s)} = \lambda_T^{(s)}\, T. \tag{1}$$

Here, the superscript (s) reflects the traffic state, to which the respective value is applicable. Furthermore, a, b, and T denote the default parameters (e.g., $a = 1.4\,\mathrm{m/s^2}$, $b = 2\,\mathrm{m/s^2}$, and $T = 1.5\,\mathrm{s}$). In summary, this implementation can be represented in terms of a *strategy matrix* as depicted in Table 1. Of course, all changes are subject to restrictions by legislation (lower limit of T) or by properties of the vehicle (upper limit of a and b).

2.2 Autonomous Traffic State Detection

Let us now present a detection model for an automated, vehicle-based identification of the local traffic situation as required for the proposed driving strategy. Our detection model uses time-series data measured on-board of a vehicle. The Controller Area Network (CAN) of the vehicle provides the own speed, whereas the distance to the leader is measured by the radar sensor of the ACC system. Due to short term fluctuations, the time series data require a smoothing in time to reduce fluctuations. In our traffic simulator (cf. Fig. 1), we have used an exponential moving average (EMA) for a measured quantity $x(t)$,

$$x_{\mathrm{EMA}}(t) = \frac{1}{\tau} \int_{-\infty}^{t} dt' e^{-(t-t')/\tau}\, x(t'), \tag{2}$$

with a relaxation time of $\tau = 5\,\mathrm{s}$. The EMA allows for an efficient real-time update by using an explicit integration scheme for the corresponding ordinary differential equation

$$\frac{d}{dt} x_{\mathrm{EMA}} = \frac{x - x_{\mathrm{EMA}}}{\tau}. \tag{3}$$

Our autonomous traffic state detection distinguishes the five above mentioned traffic states according to the following criteria: The *free traffic state* is characterized by a high average velocity,

$$v_{\mathrm{EMA}}(t) > v_{\mathrm{free}}, \tag{4}$$

with a typical value for the threshold of $v_{\mathrm{free}} = 60\,\mathrm{km/h}$. In contrast, the *congested traffic state* is characterized by a low average velocity,

$$v_{\mathrm{EMA}}(t) < v_{\mathrm{cong}}, \tag{5}$$

with a threshold of $v_{\mathrm{cong}} = 40\,\mathrm{km/h}$. The detection of an upstream or downstream jam front relies on a *change* in speed compared to the past. Approaching an *upstream jam front* is characterized by

$$v(t) - v_{\mathrm{EMA}}(t) < -\Delta v_{\mathrm{up}}, \tag{6}$$

whereas a *downstream front* is identified by an acceleration period,

Table 1. The *driving strategy matrix* represents the implementation of the ACC driving strategy in a nutshell. Each of the traffic situations corresponds to a different set of ACC parameters. We encode the ACC driving characteristics by adapting the *safe time gap* T, the *maximum acceleration* a, and the *comfortable deceleration* b. λ_T, λ_a, and λ_b are the corresponding multiplication factors, see Eq. (1). For example, $\lambda_T = 0.5$ denotes a reduction of the default time gap T by 50% in bottleneck situations

Traffic situation	λ_T	λ_a	λ_b	Driving behavior
Free traffic	1	1	1	Default/Comfort
Upstream front	1	1	0.7	Increased safety
Congested traffic	1	1	1	Default/Comfort
Bottleneck	0.5	1.5	1	Breakdown prevention
Downstream front	0.5	2	1	High dynamic capacity

Fig. 1. Screenshot of our traffic simulator, showing an on-ramp scenario. For matters of comparison, two simulation runs are displayed. In the upper simulation, 100% of the vehicles are equipped with the ACC-based traffic assistance system. The different vehicle colors distinguish the five locally detected traffic states. The reference case of vehicles without ACC systems displayed in the lower simulation run shows congested traffic at the bottleneck under otherwise equivalent traffic conditions. In both simulations, the same upstream boundary conditions have been used

$$v(t) - v_{\mathrm{EMA}}(t) > \Delta v_{\mathrm{down}}. \tag{7}$$

Both thresholds are of the order of $\Delta v_{\mathrm{up}} = \Delta v_{\mathrm{down}} = 10\,\mathrm{km/h}$.

The most important adaptation for an efficient driving style requires knowledge about bottlenecks. The identification of this state requires information about the infrastructure, because bottlenecks are typically associated with spatial inhomogeneities in the freeway design such as on-ramps, off-ramps, lane closures, or construction sites [12]. This information may be provided by a digital map database containing the coordinates $(x_{\mathrm{begin}}, x_{\mathrm{end}})$ of the bottleneck area in combination with a positioning device (GPS receiver), which provides the actual vehicle position $x(t)$. Then, a bottleneck state is identified under the conditions

$$x(t) > x_{\mathrm{begin}} \quad \text{and} \quad x(t) < x_{\mathrm{end}}. \tag{8}$$

In addition to local information, non-local information improves the detection of dynamic congestion fronts and dynamic bottleneck sections like a temporary lane-closure due to an accident. To this end, we consider inter-vehicle communication in Sec. 4) (see Fig. 2).

It can happen that none of the proposed criteria is fulfilled or several criteria are met simultaneously. Therefore, we need a *heuristics* for the discrete

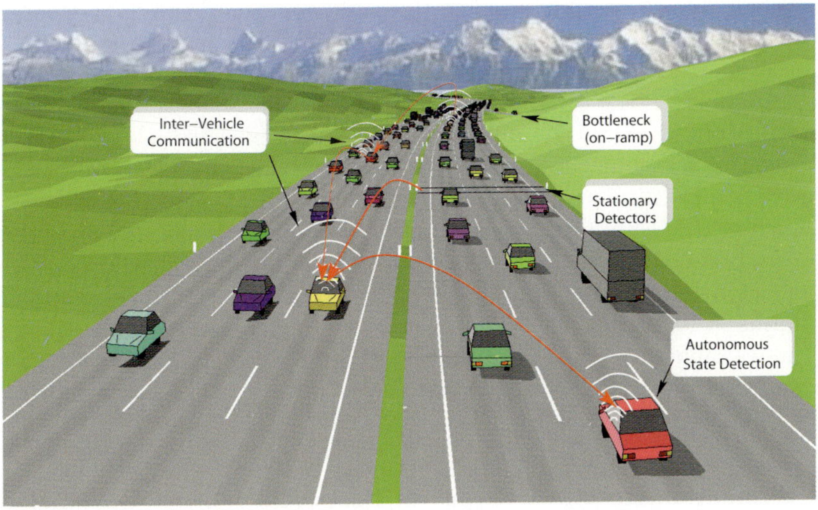

Fig. 2. Illustration of different information sources used for a vehicle-autonomous detection of the current traffic situation: (i) The radar sensor of the ACC system provides local floating-car data. (ii) A positioning device (GPS receiver) in combination with a digital map allows for a detection of stationary bottlenecks resulting from on-ramps, off-ramps, uphill gradients, etc. Additional information can be communicated by broadcast services (iii) either by stationary senders or detectors, or (iv) by inter-vehicle communication

choice problem. In our traffic simulations (cf. Fig. 1), we found that the following decision order is the most adequate one: *downstream front* → *bottleneck* → *traffic jam* → *upstream front* → *free traffic* → *no change*. This order also reflects the relevance of the driving strategy associated with these traffic states for an efficient traffic flow.

3 Simulating the Impact of ACC Systems on Traffic Flow

In order to evaluate the impact of the proposed traffic-adaptive driving strategy of our ACC vehicles (cf. Sec. 2), we have investigated a traffic scenario with an uphill gradient as typical representative for a stationary and flow-conserving bottleneck. We have carried out a simulation of a three-lane freeway section of total length 13 km (cf. Fig. 1). The inflow at its upstream boundary has been specified according to empirical detector data (time series of traffic flow and truck proportions) from the German freeway A8 East leading from Munich to Salzburg. As example for a flow-conserving bottleneck, we have modelled the uphill region with a gradient slope by locally increasing the safe time gap parameter by 30% for all vehicles in a range of 500 m around the bottleneck location at $x = 10\,\text{km}$ and smooth linear transitions around.

In our simulations, we have used the *Intelligent Driver Model* (IDM) [16], which has been successfully applied to describe real-world traffic phenomena [16]. Its input quantities such as distance, speed and relative speed to the predecessor are exactly those of an ACC system. The parameters for the simulations are given in Table 2. Lane changes have been simulated with the algorithm MOBIL [17], which is based on the expected advantage in the new lane in terms of the gain in possible acceleration or the avoidance of deceleration as calculated with the longitudinal driving model.

Each vehicle equipped with an ACC system has incorporated the driving strategy proposed in Sec. 2.1. While passing the uphill road section, the autonomous detection model (see Sec. 2.2) identifies the stationary bottleneck by means of its digital map. The relative parameter change of the time gap T, the maximum acceleration a, and the comfortable deceleration b are summarized by the 'driving strategy matrix' (see Table 1). For further details, we refer to Ref. [4, 18].

3.1 Simulation Results for Various Proportions of ACC Vehicles

Figure 3 illustrates the spatiotemporal dynamics of the traffic density for various proportions of ACC vehicles. The simulation scenario without ACC vehicles shows a traffic breakdown at $t \approx 16{:}20\,\text{h}$ at the uphill bottleneck at $x = 10\,\text{km}$ due to the increasing incoming traffic at the upstream boundary during the afternoon rush hour. The other three diagrams of Fig. 3 show the simulation results with ACC proportions of 10%, 20%, and 30%, respectively.

Table 2. Model parameters of the *Intelligent Driver Model* (IDM) for cars and trucks. The vehicle length has been set to 4 m for cars and 12 m for trucks. Vehicles equipped with the ACC system adapt their parameters T, a, and b to the detected traffic situation summarized in the 'driving strategy matrix' in Table 1. The website `http://www.traffic-simulation.de` provides an interactive simulation of the IDM in combination with the used lane-changing model MOBIL

IDM Parameter	Car	Truck
Desired velocity v_0	120 km/h	85 km/h
Safe time gap T	1.5 s	2.0 s
Maximum acceleration a	1.4 m/s^2	0.7 m/s^2
Desired deceleration b	2.0 m/s^2	2.0 m/s^2
Jam distance s_0	2 m	2 m

Increasing the proportion of vehicles applying the traffic-adaptive ACC driving strategy reduces the traffic congestion significantly. An equipment level of 30% ACC vehicles avoids the traffic breakdown almost completely. Already a proportion of 10% 'intelligent' ACC vehicles improves the traffic flow,

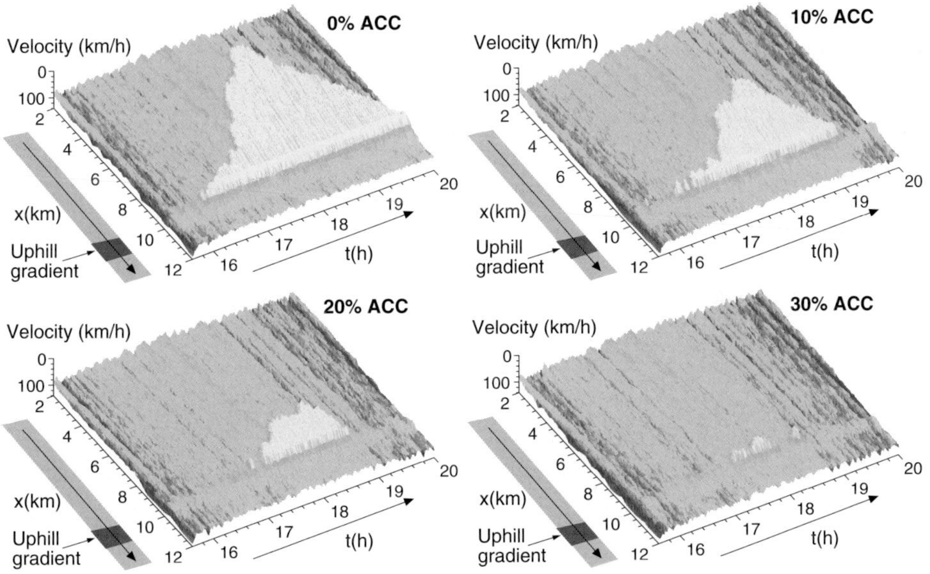

Fig. 3. Spatiotemporal dynamics of the simulation of a three-lane freeway with an uphill gradient at location $x = 10$ km, which produces a bottleneck effect. The diagrams show the lane-averaged (inversely displayed) velocity as a function of space and time for different proportions of ACC vehicles. The simulations illustrate the impact of the traffic-adaptive driving strategy for different proportions of ACC vehicles

demonstrating the great efficiency of the proposed ACC driving strategy. The improvement of the traffic efficiency scales non-linearly with the proportion of ACC vehicles. A gradual increase of ACC vehicles in the mixed traffic flows has a significant impact on the maximum free flow, and, thus, on the traffic efficiency. Note that the strong sensitivity of the scaling function at small equipment levels is crucial for a successful market introduction of the traffic assistance system.

3.2 Impact of ACC Systems on Travel Times and the Quality of Service

For matters of illustration, let us now consider the travel time as the most important variable for a user-oriented quality of service. While the travel time as a function of simulation time reflects mainly the perspective of the drivers, the cumulated travel time is a performance measure of the overall traffic system. The latter quantity can be associated with the economic costs of traffic jams. As indicated in Fig. 4, traffic breakdowns have a strong effect on the travel times. For example, the cumulated travel time without ACC vehicles amounts to about 3800 h, whereas the scenario with a fraction of 30% ACC vehicles results in approximately 2500 h. Therefore, the traffic breakdown leads to an increase of the overall travel time by 50%, compared to free flow conditions. In comparison, the travel time of individual drivers at the peak of congestion ($t \approx 18{:}45\,\mathrm{h}$) is even tripled compared to the uncongested situation. Increasing the proportion of ACC vehicles reduces the travel times significantly due to a delayed breakdown of traffic flow and, consequently, a reduction in the length of the traffic jam.

Fig. 4. Current and cumulated travel times for different ACC equipment levels. The diagrams show the strong impact of a traffic breakdown on the travel times, which is the most important measure of traffic quantity for the drivers. During the peak of traffic congestion, the travel time is approximately tripled compared to the travel time of approximately 8 min under free flow conditions. The cumulated travel time indicates the impact of congestion on the overall system. A proportion of 30% ACC vehicles prevents the traffic breakdown completely

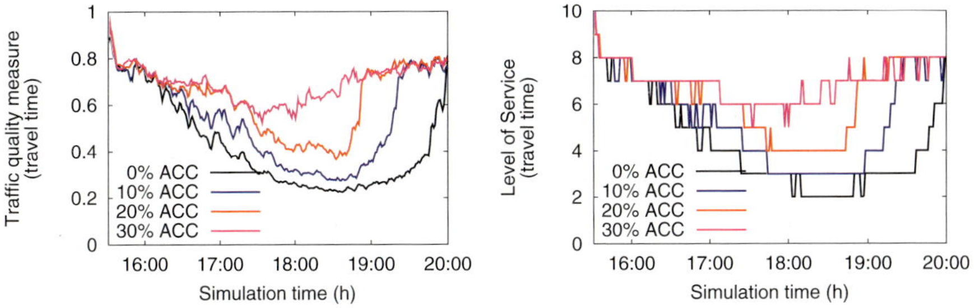

Fig. 5. Time series of the quality of service $y = \tau_0/\tau$ based on the travel time and resulting level of service index. The reduction of traffic congestion by vehicles equipped with our proposed 'traffic assistance system' leads to a considerable better traffic quality compared to the scenario without ACC-equipped vehicles

Finally, let us consider the travel time τ as the most important measure for the user-oriented quality of service. As the desired velocity $v_0 = 120\,\text{km/h}$ of car drivers (see Table 2) together with the length of the considered road section of 13 km determine a reference travel time $\tau_0 = 6.5\,\text{min}$, we define the quality measure by

$$y = \frac{\tau_0}{\tau}. \tag{9}$$

By rounding Ny with $N = 10$ to integer values, we define a discrete quality index in the range from 1 to $N = 10$, where 10 indicates the best quality value. Figure 5 shows the time-dependent quality of service $y(t)$ and the resulting discrete index time series for different proportions of vehicles equipped with the traffic-adaptive ACC system.

4 Inter-Vehicle Communication

The detection model presented in Sec. 2.2 is exclusively based on *local information*. For a more advanced vehicle-based traffic state estimation, non-local information must be additionally incorporated in order to improve the detection time and quality (cf. Fig. 2). For example, a short-range communication between vehicles (inter-vehicle communication, IVC) is a reasonable extension providing up-to-date information about *dynamic* up- and downstream fronts of congested traffic, which cannot be estimated without delay by local measurements only [6, 7, 8, 9, 19]. In contrast to conventional communication channels, which operate with a centralized broadcast concept via radio or mobile-phone services, IVC is designed as a local service. Vehicles, equipped with a short-range radio device, broadcast messages which are received by all other equipped cars within the limited broadcast range. The message transmission is not controlled by a central station, and, therefore, no further communication infrastructure is needed. Wireless local-area networks (WLAN) have already shown their suitability for IVC with typical broadcast ranges of 100–300 m. In addition, the short-range broadcast technology allows also for a roadside-to-vehicle communication, e.g., while passing a stationary sender.

In the context of freeway traffic, messages normally have to travel upstream in order to be valuable for their receivers. In general, there are two strategies, how a message can be transported upstream via IVC as displayed in Fig. 6: Either the message hops from an IVC car to a subsequent IVC car within the same driving direction ('longitudinal hopping'), or the message hops to an IVC-equipped vehicle of the other driving direction, which takes the message upstream and delivers it back to cars of the original driving direction ('transversal hopping'). A problem of the longitudinal hopping process is that it may not work properly for low equipment rates due to the short broadcasting range. However, the latter mechanism with vehicles of the opposite driving direction serving as relay stations is operational for a reliable and fast information propagation in case of low equipment levels of some percent of the vehicle fleet. For example, even for an equipment rate of 5% only, a traffic-information message will be passed 1 km upstream with a probability of 50% within 36 seconds [6].

4.1 Dynamic Congestion Front Detection

Since our focus is to use the traffic information as input for traffic-adaptive ACC systems, the crucial events to be detected and transmitted are the positions of jam fronts. In most cases, jam front positions can only be exactly detected by cars passing the location. Figure 7 illustrates 3 examples of different congestion fronts:

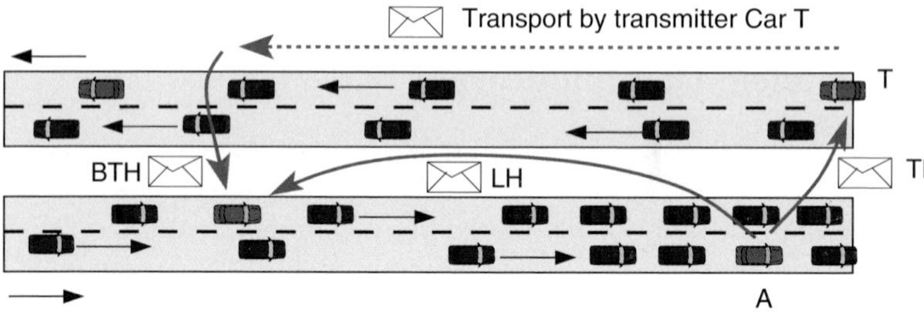

Fig. 6. Transport of a traffic-related information: The car A enters a traffic jam and broadcasts a corresponding message. The message is either received by a subsequent car via longitudinal hopping (LH) or by an equipped transmitter car T of the other driving direction via transversal hopping (TH). The message can travel with the transmitter vehicle further upstream until the message hops back to the original driving direction (BTH). The transversal hopping process is much more efficient for a fast and reliable message propagation in upstream direction

1. A downstream front of a traffic jam is pinned at some bottleneck, e.g., at the location of an incident referring to the straight line in Fig. 7.
2. An upstream jam front is moving with an propagation speed that depends on the upstream traffic flow.
3. A downstream dissolution front of congested traffic is propagating with a characteristic speed of about $-15\,\mathrm{km/h}$ [20, 21, 12].

In the following, we present a model for the detection of jam fronts based on floating-car data. In order to detect reliably these acceleration and deceleration processes, and to minimize the number of 'false alarms', each IVC vehicle smoothes the time series of its velocity, $v(t)$, using an exponential moving average (EMA), Eq. (2), with a relaxation time $\tau = 10\,\mathrm{s}$. An upstream jam front is determined via Eq. (6) with $\Delta v_{\mathrm{up}} = 15\,\mathrm{km/h}$, while a downstream front is identified by the condition (7) with $\Delta v_{\mathrm{down}} = 10\,\mathrm{km/h}$. When a congestion front is detected, a corresponding message containing position, time, and jam-front type is generated. This message is repeatedly broadcasted until it is discarded after 10 minutes. In order to reconstruct and predict the jam fronts locally by the received messages from other vehicles, each car sorts the messages according to the reported jam front type. Then, starting from the time, when the most recent generated message has been generated, a time interval of 120 s is considered. Only messages that are not older than 120 s compared to the most recent message are chosen for further evaluation. In case of two or more remaining messages, the best linear approximation for the jam front in the space-time plane is calculated by linear regression. This simple approximation approach is applicate with respect to the properties of jam fronts.

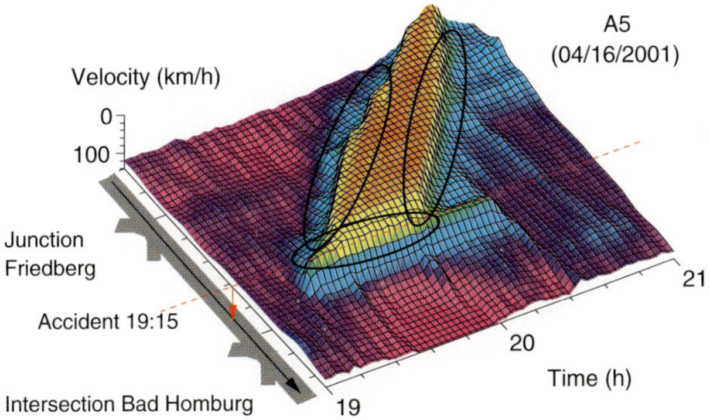

Fig. 7. Example for the spatiotemporal dynamics of congested traffic illustrating different types of jam fronts (cf. main text). This traffic jam on the German freeway A5 between Kassel and Frankfurt in direction South was caused by a blockage of the most-right of altogether three lanes after an accident occurred as noted in the sketch of the freeway

The following simulation demonstrates the local prediction algorithm. We assume a freeway with two driving directions and two lanes in each direction. In one driving direction, a 'moving localized cluster' [21, 12] is triggered while traffic is free in the other driving direction. We consider an IVC equipment rate of 3%. The resulting trajectories and the sending and receiving processes via transversal hopping are illustrated in Fig. 8. Note that the distance of the equipped vehicles exceeds the broadcast range of $R = 250\,\mathrm{m}$, even in the region of congested traffic. As a result from the simulation, the considered vehicle receives the first message about the upcoming traffic congestion already 2 km before encountering the traffic jam. Further received messages from other equipped vehicles are used to confirm and update the predicted downstream traffic situation. In order to measure the quality of the jam front prediction, we consider the difference e between the predicted and actual jam front. Figure 9 illustrates that the prediction error e decreases while the considered vehicle approaches the jam front, denoted by D. For details we refer to Ref. [22].

5 Adaptive Control of Traffic Signals

Decentralized approaches can also be applied to the adaptive control of traffic signals at intersections in urban road networks [23, 24]. When the arrival flows of vehicles are low, it is known that the first-in-first-out principle or the right-before-left principle without any further traffic regulation work well. For moderate flows, rotary traffic has shown to be efficient, reaching only small delays in travel times. However, for high traffic volumes, it is better to bundle cars with the same or compatible directions and to serve them group-wise rather than one by one. This implies an oscillatory mode of service, since it saves clearance times to serve many cars with the same direction.

A large amount of effort has been spent on optimizing traffic light control in the past. The classical approaches require vast amounts of data collection and processing as well as huge processing power. Centralized control concepts, therefore, imply a tendency of overwhelming the control center with information, which cannot be fully exploited online. Furthermore, today's control systems have difficulties responding to exceptional events, accidents, temporary building sites or other changes in the road network, failures of information channels, control procedures, or computing centers, natural or industrial disasters, catastrophes, or terrorist attacks. These weaknesses could be overcome by a decentralized , adaptive approach, which can utilize local information better. The independence from a central traffic control center promises a greater robustness with respect to localized perturbations or failures and a greater degree of flexibility with respect to the local situation and requirements. A decentralized control approach reduces the computational complexity and the sensitivity to far remote traffic a lot. As there is usually only one globally optimal solution, but a large number of nearly optimal solutions, the idea is to find the one which fits the local situation and demand best.

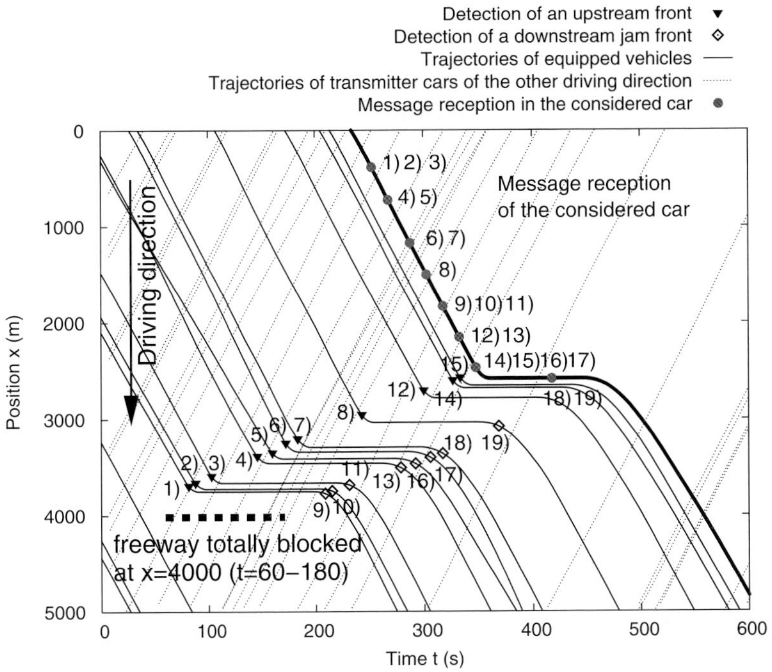

Fig. 8. Space-time diagram of our traffic simulation and the transmission of traffic-related messages by IVC-equipped vehicles moving in the opposite driving direction. Only the trajectories of IVC-equipped vehicles are shown by solid or dotted lines. In the first driving direction *(solid lines)* a temporary blockage of the road triggers a stop-and-go wave *(moving traffic jam)*. When cars encounter the propagating localized cluster, they broadcast messages about the position and time of the upstream jam front and the following downstream jam front. The subsequent messages are represented by numbers. The receipt of these messages by the considered vehicle *(thick solid line)* is indicated by the same numbers

In a pending patent [25], we have described an autonomous adaptive control based on a traffic-responsive self-organization of traffic lights, which leads to reasonable operations, including synchronization patterns such as green waves. In particular, our principle of self-control is suited for irregular (i.e. non-Manhattan type) road networks with counter flows, with main roads (arterials) and side roads, with varying inflows, and with changing turning or assignment fractions.[4] This distinguishes our approach from simplified scenarios investigated elsewhere [26, 27, 28].

Our approach to the problem was inspired by pedestrian flows at bottlenecks [29, 30, 23]: One can often observe oscillatory changes of the passing direction, as if the pedestrian flows were controlled by a traffic light (see

[4] See our website `http://www.trafficforum.org/trafficlights`.

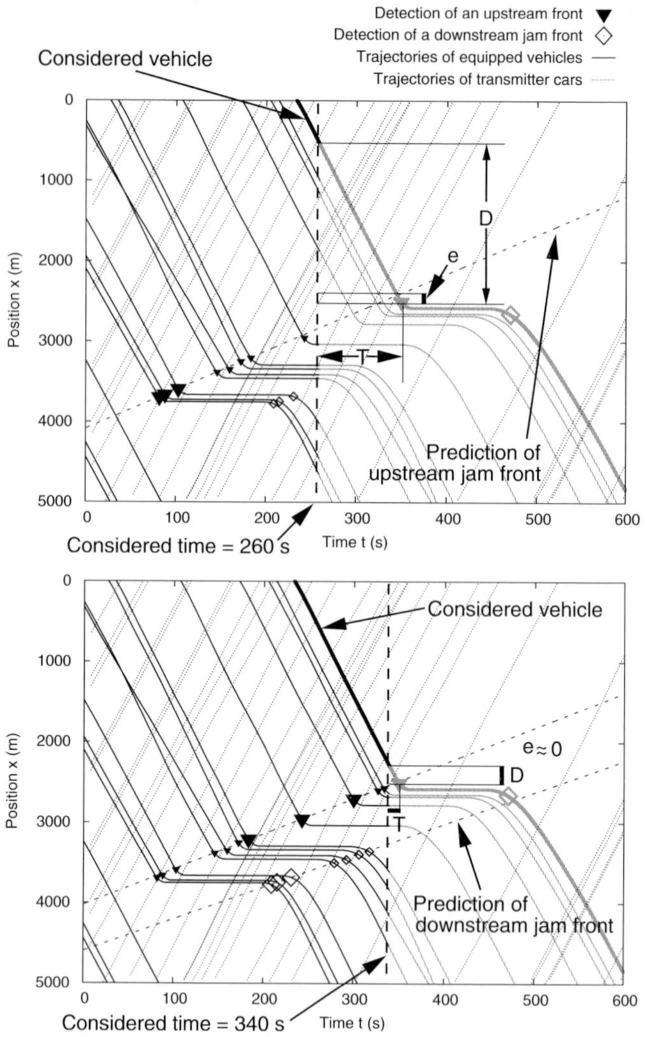

Fig. 9. Subsequent snapshots of our congestion front prediction, as the considered vehicle *(thick solid line)* approaches the jam front. Large symbols denote a jam front detection event that has been used for the jam front prediction algorithm, while small symbols correspond to outdated or future messages that have not been used. The prediction quality is measured by e, which is the distance between the actual and the predicted position of the jam front. The prediction error e decreases with the vehicle distance D from the actual jam position

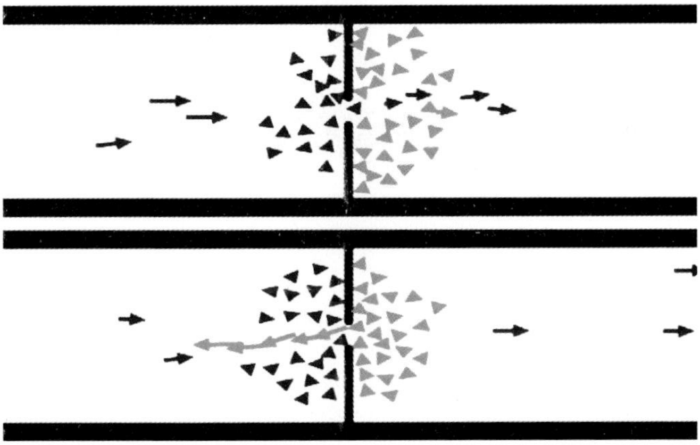

Fig. 10. Alternating pedestrian flows at a bottleneck. These oscillations are self organized and occur due to a pressure difference between the waiting crowd on one side and the crowd on the other side passing the bottleneck (after [29])

Fig. 10). Therefore, we extended this principle to the self-organized control of intersecting vehicle flows. Oscillations are an organization pattern of conflicting flows which allows to optimize the overall throughput under certain conditions [31]. In pedestrian flows, the mechanism behind the self-induced oscillations is as follows: Pressure builds up on that side of the bottleneck where more and more pedestrians have to wait, while it is reduced on the side where pedestrians can move ahead and pass the bottleneck. If the pressure on one side exceeds the pressure on the other side by a certain amount, the passing direction is changed. Transferring this self-organization principle to urban vehicle traffic, we define red and green phases in a way that considers 'pressures' on a traffic light by road sections waiting to be served and 'counter-pressures' by subsequent road sections, when these are full and green times cannot be effectively used. Generally speaking, these pressures depend on delay times, queue lengths, or potentially other quantities as well. The proposed control principle is self-organized, autonomous, and adaptive to the respective local traffic situation. It provides reasonable control results (see Fig. 11).

Our proposed autonomous, decentralized control strategy for traffic flows has certain interesting features: Single arriving vehicles always get a green light. When the intersection is busy, vehicles are clustered, resulting in an oscillatory and efficient service (even of intersecting main flows). If possible, vehicles are kept going in order to avoid capacity losses produced by stopped vehicles. This principle bundles flows, thereby generating main flows (arterials) and subordinate flows (side roads and residential areas). If a road section cannot be used due to a building site or an accident, traffic flexibly re-organizes itself. The same applies to different demand patterns in cases of mass events, evacuation scenarios, etc. Finally, a local dysfunction of sensors

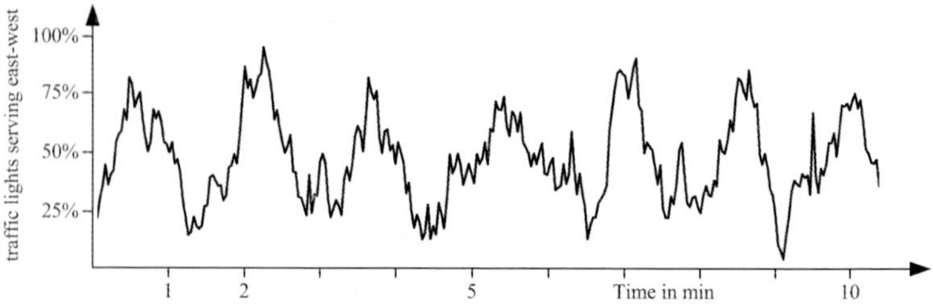

Fig. 11. The fraction of traffic lights in a regular road network serving a particular direction oscillates irregularly. The travel direction of green waves is permanently changing in order to adapt to the local traffic demands. A movie on our website `www.trafficforum.org/trafficlights` demonstrates this effect

or control elements can be handled and does not affect the overall system. A large-scale harmonization of traffic lights is reached by a feedback between neighboring traffic lights based on the vehicle flows themselves, which can synchronize traffic signals and organize green waves. In summary, the system is self-organized based on local information, local interactions, and local processing, i.e. decentralized control.

6 Summary and Outlook

We have presented two decentralized, vehicle-based approaches of traffic control: adaptive cruise control (ACC) and self-organized traffic light control. First, we have focused on a new ACC system that adapts its driving strategy to the respective traffic situation, which is autonomously detected. In order to enhance this system by non-local information, we have proposed a concept based on inter-vehicle communication (IVC), which is capable of the estimation and prediction of congestion fronts. Further studies of the impact of such decentralized traffic control strategies show that already a small percentage of vehicles equipped with such systems can have significant positive effects on the overall traffic situation in terms of increasing the stability and capacity of traffic flows. Our simulations indicate that an equipment rate of only 20% could get rid of most traffic jams that we fact today.

There are also manifold implications of decentralized approaches for future adaptive traffic light control. Our self-organizing signal control is adapting to local traffic demands instead of dominating it by pre-determined traffic light schedules. This is the key for a more effective usage of the network capacities and also implies reduced travel times for the individual drivers. Our traffic light operation distinguishes several regimes: (i) At low traffic demand, each vehicle gets a green light upon arrival. (ii) At higher demand, conflicts become more likely and delay times are unavoidable. These delays, however, impose a bundling effect resulting in a more efficient usage of the green times. Moreover, suitably delayed green phases give rise to the emergence of 'green waves'. (iii) If the demand exceeds capacity, some turning directions will be prohibited. This leads to an even more efficient service at the intersections, although some routes become a little longer. (iv) In extreme cases, where the demand is considerably above capacity and heavy congestion is unavoidable, our control generates 'green waves' for gaps. The goal of this operation scheme is to balance the load equally in the road network.

Acknowledgments

The authors would like to thank Hans-Jürgen Stauss for the excellent collaboration and the Volkswagen AG for partial financial support within the BMBF project INVENT. D.H. and S.L. kindly acknowledge partial financial support from the DFG project He 2789/5-1.

References

1. European Commission: Energy & Transport, White Paper European transport policy for 2010: time to decide. COM (2001) 370 final (2001)
2. Sussman, J.S.: Perspectives on Intelligent Transportation Systems (ITS). Springer (2005)
3. Chowdhury, M.A., Sadek, A.: Fundamentals of Intelligent Transportation Systems Planning. Artech House (2003)
4. Kesting, A., Treiber, M., Schönhof, M., Helbing, D.: Extending adaptive cruise control (ACC) towards adaptive driving strategies. Transportation Research Record: Journal of the Transportation Research Board (2007) *in print*
5. Kesting, A., Treiber, M., Schönhof, M., Kranke, F., Helbing, D.: Jam-avoiding adaptive cruise control (ACC) and its impact on traffic dynamics. In: Traffic and Granular Flow '05. Springer, Berlin (2007) 633–643
6. Schönhof, M., Kesting, A., Treiber, M., Helbing, D.: Coupled vehicle and information flows: message transport on a dynamic vehicle network. Physica A **363** (2006) 73–81
7. Yang, X., Recker, W.: Simulation studies of information propagation in a self-organized distributed traffic information system. Transportation Research Part C: Emerging Technologies **13** (2005) 370–390
8. Wischhof, L., Ebner, A., Rohling, H.: Information dissemination in self-organizing intervehicle networks. IEEE Transactions on intelligent transportation systems **6** (2005) 90–101
9. Wu, H., Fujimoto, R., Riley, G.: Analytical models for information propagation in vehicle-to-vehicle networks. In: Vehicular Technology Conference. Volume 6. (2004) 4548–4552
10. Green, M.: 'How long does it take to stop?' Methodological analysis of driver perception-brake times. Transportation Human Factors **2** (2000) 195–216
11. Kranke, F., Poppe, H., Treiber, M., Kesting, A.: Driver assistance systems for active congestion avoidance in road traffic (in German). In Lienkamp, M., ed.: VDI-Berichte zur 22. VDI/VW Gemeinschaftstagung: Integrierte Sicherheit und Fahrerassistenzsysteme. Volume 1960. Association of German Engineers (VDI), Wolfsburg (2006) 375

12. Schönhof, M., Helbing, D.: Empirical features of congested traffic states and their implications for traffic modeling. Transportation Science 41 (2007) 1–32
13. Bertini, R.L., Lindgren, R., Helbing, D., Schönhof, M.: Empirical analysis of flow features on a German autobahn. Transportation Research Board Annual Meeting, Washington, D.C. (2004)
14. Lindgren, R., Bertini, R., Helbing, D., Schönhof, M.: Toward Demonstrating the Predictability of Bottleneck Activation on German Autobahns. Transportation Research Record **1965** (2006) 12–22
15. Daganzo, C., Cassidy, M., Bertini, R.: Possible explanations of phase transitions in highway traffic. Transportation Research B **33** (1999) 365–379
16. Treiber, M., Hennecke, A., Helbing, D.: Congested traffic states in empirical observations and microscopic simulations. Phys. Rev. E **62** (2000) 1805–1824
17. Kesting, A., Treiber, M., Helbing, D.: MOBIL – A general lane-changing model for car-following models. Transportation Research Record: Journal of the Transportation Research Board (2007) *in print*
18. Kesting, A., Treiber, M., Lämmer, S., Schönhof, M., Helbing, D.: Decentralized approaches to adaptive traffic control and an extended level of service concept. In Gürlebeck, K., Könke, C., eds.: Proceedings of the 17th International Conference on the Applications of Computer Science and Mathematics in Architecture and Civil Engineering. Bauhaus University Weimar (2006)
19. Wu, H., Lee, J., Hunter, M., Fujimoto, R., Guensler, R.L., Ko, J.: Efficiency of simulated vehicle-to-vehicle message propagation in Atlanta, Georgia, I-75 Corridor. In: Transportation Research Record: Journal of the Transportation Research Board. Volume 1910., Transportation Research Board of the National Academies (2005) 82–89
20. Kerner, B., Rehborn, H.: Experimental features and characteristics of traffic jams. Phys. Rev. E **53** (1996) R1297–R1300
21. Helbing, D., Hennecke, A., Treiber, M.: Phase diagram of traffic states in the presence of inhomogeneities. Phys. Rev. Lett. **82** (1999) 4360–4363
22. Schönhof, M., Kesting, A., Treiber, M., Helbing, D.: Autonomous detection of jam-fronts and their anticipation from messages propagated by inter-vehicle communication. Transportation Research Record: Journal of the Transportation Research Board (2007) *in print* preprint (physics/0611261)
23. Helbing, D., Lämmer, S., Lebacque, J.P.: Self-organized control of irregular or perturbed network traffic. In: Deissenberg, C., Hartl, R.F., eds.: Optimal Control and Dynamic Games. Springer, Dortrecht (2005) 239–274
24. Lämmer, S., Kori, H., Peters, K., Helbing, D.: Decentralised control of material or traffic flows in networks using phase-synchronisation. Physica A **363** (2006) 39–47
25. Helbing, D., Lämmer, S.: Verfahren zur Koordination konkurrierender Prozesse oder zur Steuerung des Transports von mobilen Einheiten innerhalb eines Netzwerkes (pending patent DE 10 2005 023 742.8) (2005)
26. Brockfeld, E., Barlovic, R., Schadschneider, A., Schreckenberg, M.: Optimizing traffic lights in a cellular automaton model for city traffic. Physical Review E **64** (2001) 056132
27. Huang, D., Huang, W.: Traffic signal synchronization. Physical Review E **67** (2003) 056124
28. Fouladvand, M.E., Sadjadi, Z., Shaebani, M.R.: Optimized traffic flow at a single intersection: traffic responsive signalization. Journal of Physics A: Mathematical and General **37** (2004) 561–576

29. Helbing, D., Molnár, P.: Social force model for pedestrian dynamics. Physical Review E **51** (1995) 42824286
30. Helbing, D., Farkas, I., Vicsek, T.: Simulating dynamical features of escape panic. Nature **407** (2000) 487–490
31. Helbing, D., Schönhof, M., Stark, H.U., Hołyst, J.A.: How individuals learn to take turns: Emergence of alternating cooperation in a congestion game and the prisoner's dilemma. Advances in Complex Systems **8** (2005) 87–116

Critical Infrastructures Vulnerability: The Highway Networks

Limor Issacharoff[1], Stefan Lämmer[1], Vittorio Rosato[2,3], Dirk Helbing[4,5]

[1] Dresden University of Technology, A.-Schubert-Str. 23, 01062 Germany limor|laemmer@vwi.tu-dresden.de
[2] ENEA, Casaccia Research Centre, Via Anguillarese 301, 00123 S.Maria di Galeria, Roma, Italy rosato@casaccia.enea.it
[3] Ylichron S.r.l., Casaccia Research Centre, Via Anguillarese 301, 00123 S.Maria di Galeria, Roma, Italy
[4] Chair of Sociology, in particular of Modeling & Simulation, ETH Zurich, UNO D11, Universitätstrasse 41, 8092 Zurich, Switzerland dhelbing@ethz.ch
[5] Collegium Budapest – Institute for Advanced Study, Szentháromság utca 2, H-1014 Budapest, Hungary

1 Introduction

The analysis of the vulnerability of critical infrastructures has become a key issue in the last few years, particularly in view of the relevance of Critical Information Infrastructures Protection (CIIP) policies for economic development and security. CIIP studies have been originally focussed on Information Technology, Telecommunications and Energy supply systems but the importance of other sectors has been also recognized and the list of issues that such policies are expected to cover has grown over time. In particular, vehicular transportation systems and, specifically, road, rail, air and waterways transport networks, are now formally recognized as a crucial part of their critical sectors [1]. In this work we focus on the road transportation system and, in specific, on highway networks.

Highways should allow a fast and easy access to cities in normal and in emergency situations. They play, moreover, an important role for trading activities and for allowing a fast and reliable supply to industry, being intensively used for goods transportation. With their rapidly growing size and number of users, highways have naturally become complex both from the point of view of their topology and the traffic dynamics which takes place on them. A first perception of the growing complexity of this matter can be obtained by recalling that in Germany, France and Italy, the highway networks (in terms of total length) have grown of about a factor five in less than 50 years: in 1960 there were 2515 km of German highways, in 2004 there were 12174 km. The more complex the system, the more difficult is to predict its behavior and to ensure an adequate control. As a consequence of this rapid growth, highway critical infrastructures networks (HCIN) are more prone to failure; main causes of vulnerability should be comprised and managed in order to prevent (or mitigate) the effects which failures are likely to produce in other CIs to which they are functionally interconnected. Any network system, in fact, should be thought as being intrinsically vulnerable i.e. its functionality could be significantly reduced due to some failure produced by internal or external unexpected events. When dealing with systems at a national scale such as the road network, "...an unexpected reduction in functionality can have a serious impact on the health, safety, security or economics well-being of citizens or the effective functioning of governments ..." [1] Gaining a better understanding of what makes the network vulnerable and to what extent is thus a major interest of both governments and CI operators as a number of European Projects suggest [2]. In the present work, we employ methods and techniques developed in the context of Complexity and Transportation Science to address the problem of defining and measuring highways vulnerability.

This chapter is organized as follows. In Sec. 2 we give a short review of studies from the field of complexity science which have addressed the problem of unveiling network's properties from their topological structures. In Sec. 3 we introduce a methodology which, combining the results of topological analysis (complexity science) and those of dynamic flow simulations (transportation science), enables the evaluation of vulnerability properties of highway road networks. Using this methodology, we compare, in Sec. 4, data resulting from the analysis of three national highway networks, the Italian, the German and the French ones. Section 5 will be mainly devoted to draw some conclusions and to propose issues for future works.

2 Vulnerability Studies

We will first recall results of previous studies on road's networks vulnerability, from the Complexity and the Transportation Science perspectives.

Complexity Science started to address the issue of network's vulnerability in the late 90s [3]. Its general approach is to first "reducing" the components of a complex system into abstract mathematical structures such as the elements (nodes and arcs) of a graph. The reduction of the structure of a complex system into a graph allows the study of its topological properties which provides useful information on the "quality" of the network and keys for predicting its efficiency. It has been ascertained, in fact, that the topology of a network plays an important role in determining its function, tells us a lot about the causes and the mechanisms of its specific growth, and ultimately determines, to some extent, its vulnerability [4, 5].

---

[1] From the definition of CI given by the Commission of the European Communities in 2004 [1]

Different graph-theoretical measures (such as the clustering coefficient, the distribution of nodes degree, the network entropy etc.) have been applied to characterize the robustness of many networks [3, 22, 7, 8]. Several kind of networks have been analyzed under this perspective: among others, the world wide web [9], electricity grids [10, 11, 12], metabolic networks [13] and others (for a recent comprehensive review of this matter, see [7]).

Seminal papers have demonstrated that there are many unifying topological properties which associate networks of diverse origins. The most striking is the sharing of the functional form of the distributions of nodes degree (a "power law"), which allowed to call "scale-free" the topological class to which those networks have been attributed. It has been found that networks with power-law node's degree distribution are highly robust with respect to random node failures, but very sensitive to intentional attacks on the so-called "hubs" (high degree nodes [5, 14]).

Vehicular transportation networks have received less attention but recently there seems to be a growing interest also in this direction. Differently from most of the networks object of study of Complexity Science, road networks do not exhibit a peculiar topological structure. Due to "physical" constraints (roads junctions cannot have an arbitrarily large number of connections), they are expected to share more similarities with electrical grids (for instance an exponential-decaying distribution of nodes degree) than with the web or Internet network (which, in turn, present a "scale free" distribution). Power grids and road networks are also constrained by geography and high costs of physical equipment, that limits the possibilities of spreading. These limitations usually turn systems to show different distributions than power laws [15]. With the example of the web and the US highways, Gastner and Newman [16] show the topological differences between the structure of non-geographical and geographical networks and give an explanation for these differences in terms of costs and benefits of transportation. The Indian railway network, for example, obeys an exponential cumulative node degree distribution [17]. Also urban street networks have been shown to have unique topological properties, e.g. it was shown for German cities, that the relation between the size of the neighbourhoods and the traveling time scales with typical exponents [18].

Several approaches can be used to map a road network into a graph [19]. The first, a "direct" approach, the more intuitive one, represents roads as links and road intersections as nodes. In the second, a "dual" approach, in turn, roads are mapped into nodes and links represent road intersections. Both approaches have advantages and limits, as demonstrated in a recent study on the Polish transportation system [20] where both approaches have been applied and compared. In the mathematical representation form, links are differentiated by their relative weights. When links are chosen to represent roads, the weight of links can refer to their capacities, i.e. that is the maximum vehicle flow they can sustain per unit time. Recently, there has also been a growing interest in the study of weighted networks [21].

In a study of the topological characteristics of weighted graphs, for the case of the world-wide airport networks it has been shown that "the inclusion of weight and traffic provides evidence for the extreme vulnerability of complex networks to any targeted strategy and need to be considered as key features in the finding and development of defensive strategies" [22]. This implies that, aside to structural vulnerability, one must also consider the "dynamical" vulnerability of a network, which is the response of the network functioning to a given perturbation (i.e. the removal of some structural component such as a node or an arc). The combination of topological and functional analysis of a complex infrastructure will thus ensure to obtain a complementary set of data enabling to understand and predict the more vulnerable elements of a given network. Indeed, since the early nineties, traffic flow models have become a major research subject of physicists and extensive reviews are available in the literature [23, 24]. Most of these works are aimed at understanding some peculiarities of traffic flow, such as stop-and-go regimes and traffic jams; to date, however, we are not aware of tools which combine the two approaches, i.e. the topological and the dynamical analysis.

We turn now to vulnerability studies in Transportation Science. Though related subject such as reliability and risk analysis studies are big sub-fields in transportation science, vulnerability studies of road network was introduced in this field only in recent years. One of the first to approach this subject was Berdica who wrote a rather comprehensive "Introduction to road vulnerability: what has been done, is done and should be done" [25]. In the what "has been done" Berdica quoted that most studies are only "vulnerability-related" such as the case in many reliability studies. In engineering, reliability indicates the degree of stability of the quality of service which a system NORMALLY offers. Extreme cases and other cases which are "abnormal" such as the unforseen closure of road, are in general excluded, being events with low probability of occurrence. Hence, a new methodology for the assessing of roads' vulnerability still needs to be developed. Some attempts for developing such methods where done by Lleras-Echeverri et al. [26], D'este [27, 28] and others [29, 30] but still there seems to be no generally accepted method. Moreover, these studies seems to ignore the relevance, in this context, of a correct topological analysis of the network's graph.

In the following section we introduce a simple method in which both approaches, the topological-based and the flow-based analysis are combined. This method will then be used to make a qualitative vulnerability assessment of the highway networks of three large European countries.

3 The Proposed Approach

In this work, we attempt to combine Complexity Science and Transportation Science issues to perform a qualitative assessment of the vulnerability of three of the largest european highways systems.

Let us start with the description of the different actions which compose the approach used for the assessment.

The first action consists in reducing the highway networks into weighted graphs. We adopt the "direct" mapping where road segments are links and roads' junctions are nodes. We first analyze the graph's topology and determine networks elements which are more critical from the topological point of view. Then, we implement a "4-steps" approach to input traffic data and set up the simulation of traffic flow. This is determined by finding the resulting Wardrop equilibrium. After having determined the "physiological" traffic flow which establishes on the highway, in different traffic conditions, we attack the vulnerability study by selectively removing a number of links (i.e. closing a number of road segments) and evaluating the resulting values of purposely-defined estimators of highways Quality of Service. In the following sections we will detail all these actions.

3.1 General Definitions

A system is said to be vulnerable if its functioning can be significantly reduced by intentional or non intentional means. Thus, in order to assess a systems' vulnerability, we need to evaluate its functioning and to examine its response to failures where failure means the improper functioning or disfunction of one or more of the system's elements. This can be done by defining some measurable quantity f indicating the functionality of the system such as efficiency or quality of service and then measuring the relative change of these quantities in case of a failure (big changes will indicate high vulnerability). The level of vulnerability L of a system with functionality f could be indicated by

$$L = \frac{df}{du} \tag{1}$$

where u stands for the extent of the failure. Failures can be originated by internal events, such as tiredness of material or improper operation (human mistakes), or by external events such as natural disasters and intentional attacks. It is possible to classify failures according to some common characteristic of the individual elements which malfunction e.g. the type of elements or the failure magnitude (number of elements that failed) and the level of vulnerability can be measured for different classes of failure such as to indicate to which failure class it is more vulnerable. In the highway network, however, all these events will result in the closure (full or partial) of one or more of the highway road segments. Since we treat the highway as a network, closing a road corresponds to removing one or more links and partially closing the road corresponds to reducing the roads capacity. From the topology point of view, the amount of damage caused by the removal of one particular link depends on the position of that link in the network, on its degree, on its values of centrality properties etc. The functional damage caused to the roads network will be also related to these properties and to the level of traffic, its specific demand etc. In the following sections, we define, on one side, the topological quantities which could be usefully measured on the network's graphs and, on the other side, a dynamic model for the steady-state traffic flow simulation on the given network. Results of topology analysis and of the traffic flow simulation will then be used to extract information about the vulnerability of the network under study.

3.2 Topological Quantities

Let us assume the highway network as being represented by a "direct" approach: $G = G(N, E)$ with N nodes and E arcs. In the following, we will also assume that the graph is weighted and we will define the main topological properties accordingly. We distinguish between two different types of nodes: nodes where traffic can enter and exit the network and junction nodes. Junctions represent highway intersection (for example a biforcation) where no exit or entrance are present. Entrance and exit nodes are, in turn, the points where trips begin and end. Differently from other networks where nodes are relevant elements (i.e. in the Internet network where they represent, for instance, the position of active devices such as Autonomous Systems routers), for the highways graphs they assume only an ancillary role whereas links represent the basic feature of the graph. Therefore, rather than to node-related quantities, our analysis will focus on link-related quantities. We will focus on some of the so-called "centrality" measures and in particular on the quantities called "Betweenness Centrality" and "Information Centrality" (see for a current definition, [7], whose definition will be recalled in the following).

The Betweenness Centrality b_{ij} of the link connecting the nodes i and j (link ij) is expressed as:

$$b_{ij} = \frac{n_{ij}}{(N-1)(N-2)} \tag{2}$$

where n_{ij} is the number of shortest paths between each pair of node containing link ij, and N is the total number of nodes in the graph.

The topological *efficiency* of the network $E_T[G]$ provides an estimate of the average efficiency with which the network ensures that all nodes are reachable and can be defined as

$$E_T[G] = \frac{1}{N(N-1)} \sum_{i,j \in G} \frac{1}{d_{ij}} \tag{3}$$

where d_{ij} is the shortest path between nodes i and j. This definition is extremely general and holds also for non-connected networks (whereas for some i and j, $d_{ij} \rightarrow \infty$). Low values of $E[G]$ indicate that many couples i,j have a difficult connection (i.e. d_{ij} is large).

From the definition of the network efficiency $E[G]$ we can evaluate the *Information Centrality IC* of the generic link ij which indicates the change in the network's efficiency when that link is missing. The IC is expressed as:

$$IC_{ij} = \frac{\Delta E}{E} = \frac{E[G] - E[G']}{E[G]} \tag{4}$$

where $E[G]$ and $E[G']$ are the *efficiency* of the unperturbed network and that of the network after the removal of the link ij, respectively. The larger the IC_{ij}, the higher the importance of the ij link in the global efficiency of the network (i.e. that link allows to have, in general, reasonably small values of d_{kl}, for all the nodes k,l in the network).

It is evident that the evaluation of the IC of all nodes will give a first insight of the links which are mostly relevant for the graph's vulnerability.

Relating the *efficiency* to the functionality of the system f and the removal of the link to the failure u, eq.(4) provides the form of the level of vulnerability L (eq.(1)). We can thus define the topological vulnerability L_T as

$$L_T = \frac{\Delta E}{E} = \frac{E[G(0)] - E[G(u)]}{E[G(0)]} \tag{5}$$

where $G(u)$ indicates the graph formed after the failure u and $G(0)$ the graph with no failure. We should point out here that in eq.(2) the element n_{ij} depends on the shortest path between all possible pair of nodes; similarly the sum in eq.(3) is extended over all possible pair of nodes. This is because in other type of networks such as the internet or the humans social network, nodes usually belong to the same type (routers, humans etc.) and the connection between each pair of nodes has similar meaning. In a highway network, in turn, a trip can begin only in an entrance node and end up in an exit node. Trips between junctions are meaningless. We thus define a sub-set of node pairs containing the pairs of nodes which are either origin and/or destination, i.e. the set of all pairs which could act as entrance and exit points and we denote it as D. The Betweeness and the *efficiency* of a highway network should be then related to this set; when calculating the topological level of vulnerability, the efficiency is evaluated with the sum over the set D rather then the set of all nodes pairs.

3.3 Flow-dependent Quantities

The function of the highway critical infrastructure is to allow vehicles to move between different geographical locations in a reasonable time. The system's functioning thus depends, other than its topological structure, on how many vehicles use it (are present on the network) in a given time period (the flows), and the routes used by the vehicles when traveling from one point of the network to another. In order to describe the distribution of traffic flows in the highways and to estimate the associated traveling times we will use a static traffic model. We therefore consider the network to be in a stationary state. Although the traffic volume changes within one day, we assume it to stay constant during one particular hour of several successive days, e.g. the rush hour. The stationary state will be reached on a long time-scale after all drivers got informed about the closure of a road segments and had the chance to discover alternative routes. This state is referred to as the Wardrop-Nash equilibrium [31, 32, 33] and results from the assumption that each driver minimizes individually its traveling time.

An estimate of the traffic flow on each network at equilibrium, can be obtained with classical distribution models. We chose the 4-step-model [31] which will be described briefly. It consists in four different actions which allow to input the traffic demand and output the equilibrium traffic flow which originates from the input conditions and results from the application of the Wardrop-Nash equilibrium. The four actions are:

1. the trip generation which calculates the amount of traffic entering or leaving the network within one hour. A demand vector V_i represents then the outgoing flow from the source node (entrance node) i. We further assume that outgoing and incoming flows are equal and the total traffic demand is then given by $V = \sum_i V_i$.
2. the trip distribution which chooses the origin-destination relation for each pair of cities. To determine how the outgoing flow distributes among the different cities, that is the magnitude of each origin-destination pair, we assume that the probability of going from one city to another is inversely proportional to the time required to cover the travel. In more details, if P_{ij} is the probability of going from origin i to destination j, then $P_{ij} \propto e^{-\beta T_{ij}}$ where T_{ij} is the traveling time from i to j and β a suitable parameter. This distribution is obtained by solving a set of non-linear equations which are also referred to as the Logit model equations [31].
3. the modal split; this is required when considering different kinds of transports e.g. public transport, commercial etc. In our model we omit this step because we will only consider the case of private vehicles and omit to account for the use of public transportation.
4. the traffic assignment, where the traffic is distributed among the different roads of the network. This is the central step as it is used to determine the model output, i.e. the resulting flow distribution in the different road segments. To obtain the flows on each link for a given traffic demand V and a given flow distribution of origin-destination pairs, we have to evaluate the traffic flux by using an iterative procedure which allows the finding of the Wardrop equilibrium. The iteration runs as follows:
   a) find shortest path from origin to destination in terms of time
   b) assign flows to the shortest path
   c) recalculate the traveling time for the new distribution of flows
   d) if there is no change in time finish, else go to step 1.

   For static traffic distribution the time is usually calculated using the Capacity Constraint Function $T_l(q_l)$ of the specific road l depending on the traffic flux q_l

$$T_l(q_l) = T_l^0 \left[a(1 + (q_l/k_l)^b\right] \tag{6}$$

where T_l^0 is the "free" travel time on link (road segment) l, estimated as the ratio between the road's length and the road's maximum allowed velocity, k_l is the capacity of road l (in terms of vehicles per hour), a and b are network parameters. We also assume all roads to have the same capacities and same maximum velocities. The traveling time T_{ij} between the generic origin-destination pair ij will be thus a function of the traffic flux q_l on each road segment which composes the path

$$T_{ij}(q_l) = \sum_l T_l(q_l) \tag{7}$$

By applying the 4-step model to a given network, we can obtain the distribution of flows and traveling times on the links for various values of traffic demand. These values are used to evaluate the networks flow dependent efficiency eq.(10). To quantify the efficiency with which an highway is able to fulfill its function, for a given traffic demand, we must introduce a "Cost Function" C allowing to measure the performance of the highway with respect to some given reference. There are many possible choices to specify the cost function since the damage or the cost of a failure is not an absolute value. A one hours delay in arrival could be a nuisance but not highly significant event if looked from a private driver's point of view, while for a factory one hours late of all personal can have serious implication on the whole productivity of the factory. The interest of this study however is not to minimize costs for individual drivers or specific sectors but more to observe how the network reacts to perturbation for different traffic demands. It is thus useful to take a classical form of the cost function that depends only on the traffic demand and the resulting traveling time. Traffic demand is the number of trips per unit time between each pair of nodes defined on the D set. In a more formal way the cost funciton C for a given demand is:

$$C = \sum_{ij} C_{ij} = \sum_{ij} D_{ij} T_{ij} \tag{8}$$

C_{ij} is the cost of all trips between nodes i and j. D_{ij} is the flow (number of trips per unit time) between node i and node j and T is the flow-dependent travel time matrix with elements T_{ij} indicating the traveling time between origin i and destination j. The lowest possible cost value for a given demand, $C_{\min}$, is taken as a reference point for our analysis. Its value depends on the specific function used to calculate the traveling time T. Using this reference point, we can define for the system its *Quality of service* which will indicate the functionality of the system eq.(1). A commonly used quality of service function in this context is of the form

$$Q = \frac{\mathrm{C}_{\min}}{\mathrm{C}}, \tag{9}$$

where $0 \leq Q \leq 1$ and the value of one corresponds to the best possible functioning of the system. However, this definition is quite limiting since it is not well defined for the case in which the system is disconnected. Entellus and Madison [34] solved this problem by defining a separate *importance* function. Here, instead, we choose to solve this problem by using the topological efficiency definition eq.(3). We convert it into a flow-related form by replacing the value in the denominator d_{ij} (which can be thought as a topological cost function), with that of the Cost Function previously defined, for each origin-destination pair. Flow dependent efficiency E_F is then:

$$E_F = \frac{1}{N(N-1)} \sum_{ij \in D} \frac{1}{C_{ij}} \tag{10}$$

with

$$C_{ij} = q_{ij} T_{ij}. \tag{11}$$

Analogously, we derive the maximum efficiency from the cost function; the flow-dependent quality of service becomes

$$Q = \frac{E_F}{E_{F_{max}}} = \frac{\sum_{ij} \mathrm{C}_{ij}{}^{-1}}{\sum_{ij} (\mathrm{C}_{ij}^{\min})^{-1}} \tag{12}$$

where we introduce the maximum network efficiency $E_{F_{max}}$ as reference point for calculating Q. We can define the maximum flow dependent efficiency E^F_{max} by taking the limit of E^F for an infinite capacity of all roads

$$E_{F_{max}} = \frac{1}{N(N-1)} \sum_{ij \in D} \lim_{k \to \inf} \frac{1}{C_{ij}} = \frac{1}{\mathrm{D}_{ij} T^0_{ij}} \tag{13}$$

Finally we define the flow-dependent Level of vulnerability $L_F^{(u)}$ in accordance with eq.(1)

$$L_F^{(u)} = 1 - \left\langle \frac{Q(u)}{Q(0)} \right\rangle, \tag{14}$$

where $Q(0)$ and $Q(u)$ are the quality of service for the non-perturbed and the perturbed system, respectively. (L_F ranges between the values one and zero. One indicating the highest level of vulnerability).

The normalization of both Q and L_F allows the comparison between different networks, though having different sizes or traffic demand.

4 Analysis of some European highways

As test cases, we will apply the proposed methodology to compare three highway networks with different geographical features: the German, the Italian, and the French highway network, shown in Fig. 1. Each network is represented by a graph $G(N, E)$ with N nodes and E links. These networks are directed and weighted. Links are the highways road segments and their capacity is the

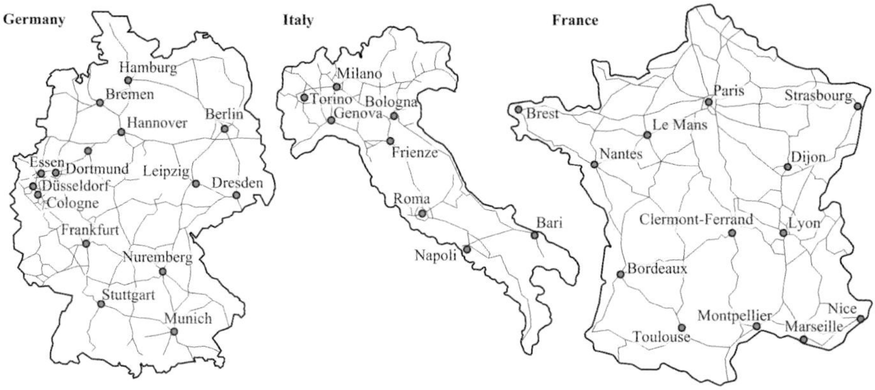

Fig. 1. The highway networks of Germany, Italy, and France. Some relevant nodes are highlighted

links weight. The nodes are the entrance and exit point from the highways as well as the intersection points of different roads. Each of the national highways contain hundreds of exit and entrance points; here we restricted the network to a small subset of nodes and links connecting the $N_{\text{city}} = 29$ largest cities of each country. All roads in this set are bidirectional i.e. they are composed of two segments (links) one for each direction; thus a road is represented by two links.

In Table 1 we report general properties and data of the chosen networks' subsets.

Further information on the network can be obtained by evaluating the Topological vulnerability Level L_T eq. (5) and the Betweenness centrality of links in the different networks. The Betweenness Centrality b_i of link i can be estimated by setting $N = N_{\text{city}}$ in eq.(2).

Table 1. General properties of the three networks used in the test. Population indicates the total population of the 29 largest cities considered. $< b_{ij} >$ and $<$ degree $>$ represent the average Betweenness Centrality eq.(2) and the average node's degree, respectively

	Italy	Germany	France
population	10701491	17366502	8381434
nodes	43	43	52
road's number	57	65	71
total road's length (km)	4946	5453	8194
average road's length (km)	86	83	115
$<$ degree $>$	2.6	3.0	2.7
$< b_{ij} >$	0.0451	0.0328	0.0332

The average value of b for all network's links is reported in table 1 and the normalized distributions reported in Fig. 2. These results show that German highway is characterized by relatively low values of b indicating a sharp distribution of link's centrality (whose maximum value is as low as $b = 0.1$ and the ability of network to make an "homogenous" use of all its links. The French and Italian network, in turn, have a quite broader distributions with maximum values which are double of that of the German one. This indicates that there are links which will be intensively stressed by traffic flow as they will contain minimum paths joining several origin-destination pairs.

Data are reported on Fig. 2. Also for this quantity, it is evident that the Italian and Frech networks suffer of a high average IC values. This means that the networks should be prone to be heavily damaged in the case of a specific link's removal. In sec. 3.2 we have defined the Topological vulnerability $L^T(u)$ eq.(14). We now present the result of this measure applied to the three networks under study. L was measured for five different magnitudes of failure. The failure magnitude λ is the size of the set u (i.e. the number of road segments simultaneously closed). For $\lambda = 1, 2$, L was measured by taking the average of all the possible choices of road segments removal. For $\lambda = 3, 4, 5$ L_T has been averaged over 10,000 different contributions of road segments simultaneously removed. Fig. 3 reports the average value of L for each set of measures.

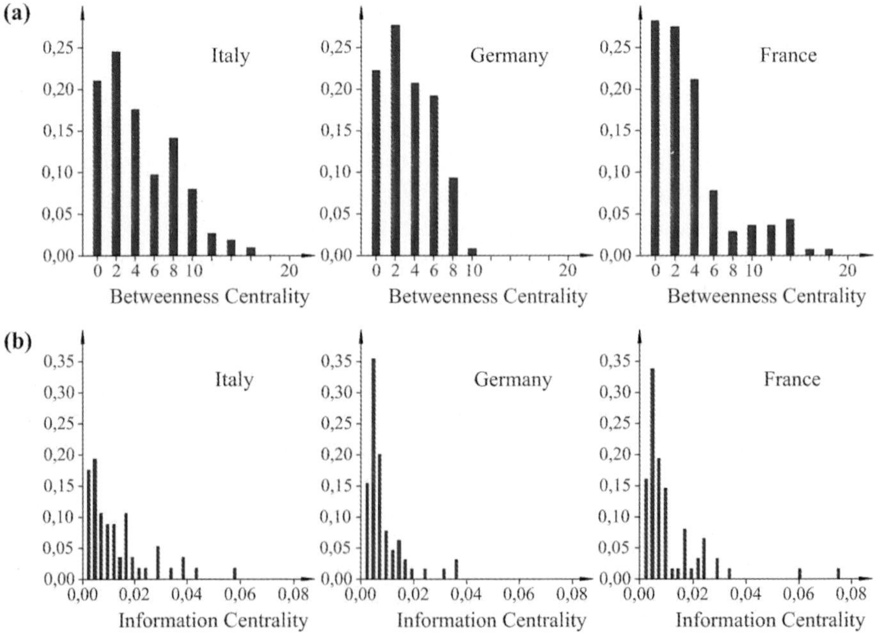

Fig. 2. The normalized distribution of **(a)** the betweenness centrality and **(b)** the information centrality of Italy, Germany, and France

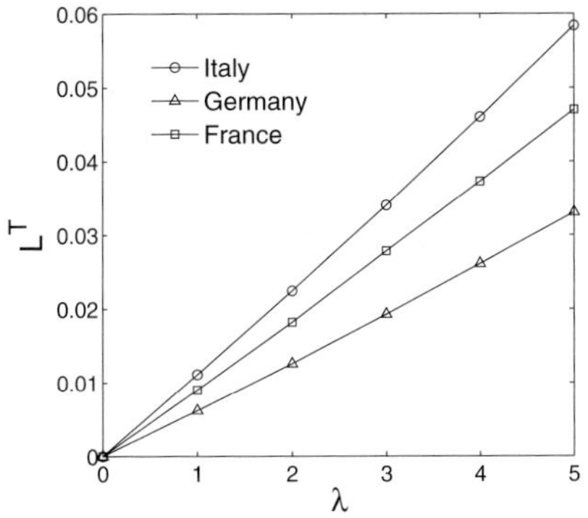

Fig. 3. The topological vulnerability level for Germany, Italy, and France for different magnitudes $\lambda = 1, 2, ..., 5$. For $\lambda = 1, 2$ the result is averaged over all possible cases. For $\lambda = 3, 4, 5$ results are averaged over 10,000 different scenarios

It is clear from Fig. 3 that there is a linear dependence between the topological vulnerability and the magnitude of perturbation. The larger the slope, the higher is the extent of topological damage that roads' closure is able to induce on the network. The average loss of efficiency can be as high as 6% in the case of the simultaneous closure of $\lambda = 5$ road segments in the Italian network.

4.1 Flow-related Vulnerability Analysis

The application of the "4-steps" model for the evaluation of the equilibrium traffic flow which establishes in the networks provides, as a function of the total traffic demand, a quality of service plot reported in Fig. 4 for the the three highway networks.

Results show that the German network can handle a larger traffic demand, better then both the Italian and France networks. There are several reasons which might help explaining this result. First of all, the large scatter of the cities population and the cumulated size of the three major italian cities (Rome, Milan, Naples) which aggregates a large traffic demand on the North-South direction, affecting a large fraction of the whole network. On the other side, the Betweenness Centrality of the links of the Italian network, as reported on Fig. 2 is higher, in average, then those of the other networks; this implies that there are links which will receive a large amount of traffic which will thus reach a congested situation prior then in other cases. However, whereas for Italy and France (which have a similar amount of population considered in this example, see Table 1) there should be a similar average traffic demand, Germany has a largest total population; this would imply that its average network's quality of service (Q) should be referred to a total traffic demand higher than that of the Italian and French cases.

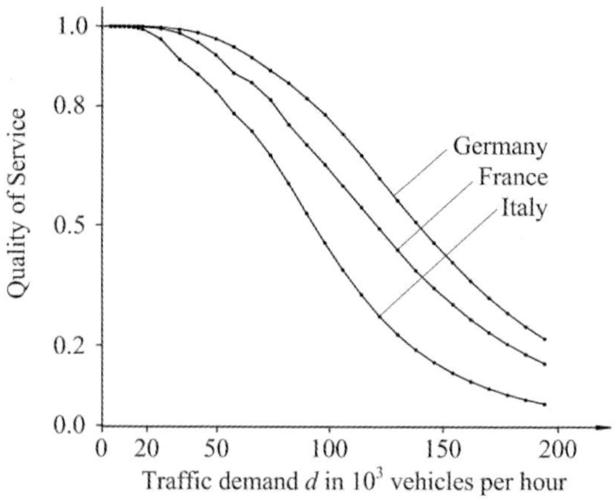

Fig. 4. The quality of service (Q) for Germany, Italy, and France for different total traffic demands d. Higher traffic demand leads to higher traveling times and to a resulting smaller Q value

To evaluate the flow-related vulnerability we chose all networks to be in the same initial functional state that is to have the input traffic demand such that $Q = 0.5$. Then we perturb the network by removing links from the graph (closing roads) and measure $Q(u)$. Recalling that λ represents the number of elements composing the set of removed links u, we report the results of the perturbation for $\lambda = 1, 2, 3$. For $\lambda = 1, 2$, level of vulnerability L_F was measured by taking the average of all the possible choices of road segments removal. For $\lambda = 3$, L_F has been averaged over 10,000 different contributions of road segments simultaneously removed. Results are reported in Fig. 5.

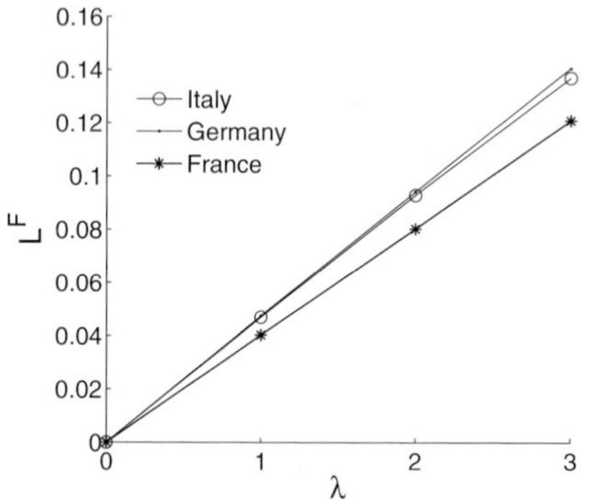

Fig. 5. The network level of vulnerability for Italy, Germany and France. Networks were perturbed from an initial state of $Q = 0.5$. Perturbations are of order $\lambda = 1, 2, 3$. For $\lambda = 1, 2$ the result is averaged over all possible roads removal. For $\lambda = 3$ results are averaged over 10,000 different randomly chosen sets of removed road segments

Similarly to what was observed for the topological vulnerability, there is a linear dependence between the magnitude of perturbation and the level of vulnerability. The obtained result presents a German network which, although being characterized by an higher topological quality of service compared to the French and Italian ones, here exhibits a large vulnerability similar to that of the Italian network. This effect is a consequence of the complex interplay existing between topology and flux distribution. It is a further evidence that flow analysis must be performed in order to validate the results based on a mere topological analysis of the network graph.

5 Summary and Outlook

We have attempted to obtain a qualitative vulnerability assessment of selected sub-networks of three out of the major European highway networks. To this purpose, we have proposed an approach which combines a topological analysis of the graphs representing the networks with the evaluation of the steady-state traffic flow on the network though a Wardrop-Nash equilibrium search. After having evaluated the traffic flow in normal conditions, we have perturbed the network by simultaneously removing an increasing number of the links and measured a suitable function which describes the associated loss in the Quality of Service. Results indicate that, whereas topological analysis of the networks provide first keys to evaluate the network's elements which could decrease the system's robustness, the evaluation of the flow-dependent quality of service issued upon network perturbation provides a comprehensive assessment of the effects.

The proposed method can be applied to networks of larger extent, representing both highways and urban roads. To follow this aim, we are implementing a computer program which will allow the treatment of very large and complex graphs with a number of nodes ranging over $N \in 10^4$.

Acknowledgement

This study has been partially supported by the **ESF Cost action P10 "Physics of Risk"** and by the "Cooperative Center for Communication Networks Data Analysis", a NAP project sponsored by the Hungarian National Office of Research and Technology under grant No. KCKHA005. The authors would also like to thank Martin Treiber for valuable discussions and suggestions.

References

1. Abele-Wigert, I. and Dunn, M.: International CIIP handbook 2006. ETH, Swiss Zurich (2006)
2. Integrated Risk Reduction of Information-based Infrastructure Systems (IRRIIS), http://www.irriis.org/
3. Albert, R. and Barabási, A.–L.: Rev. Mod. Phys. **74** (2002)
4. Callaway, D.S., Newman, M.E.J., Strogatz, S.H. et al: Phys. Rev. Lett. **85**, 25 (2000)
5. Albert, R., Jeong, H., Barabási, A.-L.: Nature (London) **406**, 387 (2000)
6. Newman, M.E.J.: SIAM Rev. **45** 2, (2003)
7. Boccaletti, S., Latora, V., Moreno, Y. et al: Phy. Rep. **424**, 4–5 (2006)
8. Demetrius, L. and Manke, T.: Phys. A **346** 3–4 (2005)
9. Pastor-Satorras, R. and Vespigniani, A.: Evolution and structure of the internet, Cambridge University Press 2004
10. Albert, R., Albert, I., Gary L. Nakarado: Phys. Rev. E **69** 025103(R) (2004)
11. Crucitti, P. *et al.*, Physica A **338** 92 (2004)
12. Rosato, V., Bologna, S., Tiriticco, F., Elect. Pow. Sys. Res. **77** (2007) 99–105.
13. Hughes, T.R., Marton, M.J., Jones, A.R. et al: Cell **102** 1 (2000)
14. Buhl, J. et al.: The European Physical Journal B **42**(1) 123–129 (2004)
15. Amaral, L.A.N., Scala, A., Barthelemy, M., Stanley, H.E.: Proc. Nat. Ac. Sci **97** 11149 (2000)
16. Gastner, M.T. and Newman, M.E.J., The European Physical Journal B **49**(2) 247–252 (2006)
17. Sen, P. *et al.*: Phys. Rev. E **67** (2003) 036106
18. Lämmer, S., Gehlsen, B. and Helbing, D.: Physica A **363**(1) (2006) 89–95
19. Porta, S., Crucitti, P. and Latora, V.: Physica A **369**(2) (2006) 853–866
20. Sienkiewicz, J. and Holyst, J.A.: Phy. Rev. E 72 (2005) 046127
21. Wu, Z., Braunstein, L.A., Havlin, S. et al: Phys. Rev. Lett. **96** 148702 (2006)
22. Dall'Asta, L., Barrat, A., Barthelemy, M. et al.: J. Stat. Mech. (2006) P04006
23. Helbing, D.: Rev. Mod. Phys. **7**3 (2001) 1067
24. Chowdhury, D. *e*t al.: Physics Reports **3**29 (2000) 199
25. Berdica, K.: Transport Policy **9** (2002) 117–127
26. Lleras-Echeverri, G. and Snchez-Silva M.: Proceedings of the Institution of Civil Engineers, Transport **147** (2001) 1–14
27. D'Este, G.M. and Taylor, M.A.P.: Journal of the Eastern Asia Society for transportation studies **4**(2001) 1–14
28. D'Este, G.M. and Taylor, M.A.P. In: M.G.H. Bell and Y.Iida (eds.) The network reliability of transport. Pergamong (2003)
29. Chen, A, Kongsomsaksakul, K. and Zhou, Z.: Annual Meeting of the TRB, Washington, DC (2007)
30. Matisziw, T.C., Murry, A.T. and Grubesic, T.H.: Annual Meeting of the TRB, Washington, DC (2007)
31. Hensher, D.A. and Button, K.J.: Handbook of Transport Modelling Pergamon (2002)
32. Maerivoet, S. and De Moor, B.: Technical report 05–155 Katholieke Universiteit Leuven (2005)
33. Van Woensel, T. and Vandaele, N.: Asia-Pacific Journal of Operational Research (2006) *in press*
34. Jenelius E. and Mattsson, L.–G.: Nectar Cluster 1 Seminar, Molde, Norway, May (2006)

Part IV

Critical Infrastructures and Systemic Risks

Trade Credit Networks and Systemic Risk

Stefano Battiston[1], Domenico Delli Gatti[2], Mauro Gallegati[3]

[1] ETH Zurich, Kreuzplatz 5, 8032 Zurich, Switzerland sbattiston@ethz.ch
[2] Catholic University of Milan, Milan, Italy domenico.delligatti@unicatt.it
[3] Universita' Politecnica delle Marche, Ancona, Italy gallegati@dea.unian.it

1 Introduction

Credit is extended by banks to firms (loans), by one bank to another (interbank credit) and by one firm to another (trade credit). As a result, there is a *network* of credit relationships among firms, among banks and between firms and the banking system.

Credit relations create value but also financial dependency. Therefore, for a node in the credit network having many links is a way to diversify risk but it is also the ground for the so called financial contagion [1].

In particular, some of these credit relationships are between firms or institutions in different countries and thus connect national credit networks in a world wide network. The possibility of a systemic crisis affecting the whole or a significant part of a credit network raises growing regulatory concern and it is the responsibility of policy makers to ensure that adequate fire walls are in place in order to prevent the spill over of crisis across institutions and firms [18].

An important and open debate, with major policy implications, concerns whether or when higher network density (in other words more links in the network) leads to lower or higher systemic risk (in the sense of probability of joint failures causally related).

The dominant neoclassic approach in economics typically assumes (1) equilibrium and/or (2) indirect interaction through price. The failure of coordination which is likely to arise in a decentralized market economy is simply assumed away. [6].

This approach to economic theory has been recently challenged by the approach based on heterogeneous interacting agents, which conceives the economy as a complex system. The starting point of this approach is that while prices surely play a fundamental role, the price mechanism can work well only if information is perfect and markets are complete. If this is not the case, i.e., if the future is uncertain, it is not possible to ignore direct interactions and co-ordination mechanisms that arise in spatio-temporal way – i.e. supply chains, communication, imitation, learning, trust and credit relationships. In this context, complex patterns of heterogeneous agents' interactions at the micro level lead to the emergence of statistical regularity at the macroeconomic level through a self-organised process. A pioneering work applying a complex systems approach to the macroeconomic impact of a production network dates back to the early 90' [2]. There, it was shown that local uncorrelated fluctuations can nevertheless generate, through interaction, large aggregate output fluctuations. Concerning credit networks, the view resulting from the dominant approach tends to see more dense networks as more stable. In this chapter, we show how the interacting agents approach results in a different view.

While banks-firms credit relationships have been extensively studied since long ago in the economic literature (for an overview, see [5]), a recent interesting line of research has analysed phenomena of financial contagion in interbank credit [1, 18]. Finally, trade credit , is less investigated but yet an important part of the network of credit relationships. It represented, for instance, one half of the short term liabilities of the corporate sector in 2004 in the U.S. [3]. Moreover, trade credit is largely used as collateral in bank borrowing, especially by small and medium sized firms. In the U.S., lines of credit secured by accounts receivables represented approximately one quarter of total bank loans in 1998 [8]. In Italy, loans secured by receivables were 22% of total loans and 54% of short term loans in 2002 [12]. In the theoretical literature, [7] emphasize the role of trade credit as a propagation mechanism (the so called *balance-sheet contagion*) , while the dynamics of credit chains has been investigated by [3].

From the point of view of complex systems, few important works have applied the concept of self-organized criticality (see also below) to the context of interbank markets [15, 19].

However, the issue of systemic risk in credit networks remains to some extent underresearched, both at the theoretical and empirical levels.

In this chapter, we present a model recently introduced in [14, 16] and we discuss the features of a networked economy in which N firms are organised in M production levels. Each firm at a certain level is supplied by a subset of firms in the upper level (suppliers) and supplies a subset of the firms in the lower level (customers). The bottom level consists of retailers, i.e., firms that sell in the consumer market. The top level consists of firms that provide primary goods to the other firms. Firms are connected by means of two mechanisms: (i) the output of supplier firms is an input for customer firms; (ii) supplier firms extend trade credit to customers (as it is typically the case in reality).

However, in the model, the trade credit contract is only implicitly sketched: we neither design the optimal trade credit scheme nor look for the optimal amount of trade credit a customer firm should require. Instead, we focus on the mechanisms of propagation of bankruptcy . When a firm is unable to reimburse debt, it goes bankrupt. This may happen as a result of one of (or any combination of) three mechanisms: (1) there is a production default in the firm (production is lost and so is profit, while the firm still has to pay for input and processing); (2) some customers are not able to pay; (3) some suppliers are not able to deliver the agreed input (in this case the firm does not bear the cost of input but still does bears some cost due the fact the resources were allocated in view of processing that input).

Thus, the failure to fulfill debt commitments by a customer may hamper the solvency of the supplier, who may become unable in turn to pay its own suppliers located in the upper level, which may lead to a chain of similar failures (domino effect) and in extreme cases result in bankruptcy avalanches. When a firm goes bankrupt, in fact, the probability of bankruptcy in connected firms increases, yielding clustered fluctuations in the number of failing firms. In other words, a single bankruptcy may have systemic repercussions through an avalanche of bankruptcies.

In this context, having many customers and many suppliers is a way for the firm to diversify the risk of defaulting payment or delivery. If the network is dense enough, the default of a firm in paying its debt doesn't cause any other default. For instance, if every firm has k customers (with similar volume of orders), the default of one customer in paying causes a unexpected relative decrease in profit of order $\frac{1}{k}$. The larger k the smaller the unexpected loss.

However, in presence of externalities , the loss caused by a defaulted payment or delivery may be amplified through the network. Some multi-agent models of financial fragility have been able to account for this effect. In [4] a single bankruptcy may have systemic repercussions: in fact, the banking system reacts to the bankruptcy by restraining the supply of credit and pushing up the interest rate to all firms. The increase in the interest rate may cause some other bankruptcies and thus trigger an avalanche of bankruptcies. Such models incorporate only the indirect interaction among firms that takes place through the endogenous determination of the interest rate on bank loans.

In the present model, instead, direct interaction among firms takes place through supply and extension of trade credit which is also subject to an interest rate. If the interest rate is dynamic and depends on the change of growth rate of the firm itself and its neighbours, then losses can be amplified through credit relations. Under such conditions, increasing the network density, while decreasing the shocks to individual firms, it may also increase the systemic risk, thus inducing a trade-off between individual risk diversification and global instability.

This result is consistent with a recent work on failure avalanches in complex networks [17] which has pointed out the role of the interplay of two opposing mechanisms: diffusion and contagion. On one side, when energy diffuses from a node to its k neighbours, the energy received by each neighbour is of order $\frac{1}{k}$ of the initial one. On the other side, in a contagion process, nodes have discrete states and with a certain probability switch from one to the other when a neighbour has changed state. Therefore, if both mechanisms are at work at the same time, then increasing the density of the network, the impact of a change of state of a given node on the neighbours first decreases due the diffusion and then increases due to the contagion. The work by [17] bridges two strains of work in the physics literature on complex systems: the models on cascading phenomena on one side and those on epidemic spreading on the other side.

Avalanches of events in networks have been studied extensively in the context of self-organised criticality (SOC) and in particular in models inspired to the sand pile model [9] and the fiber bundle model [13]. In all these models, an event on a node of the network (a "toppling") transfers energy to neighbouring nodes, possibly triggering their toppling. Each node is associated with one state variable, which depends on the toppling of the neighbours and causes the node to topple when it reaches a given threshold. In the SOC models there is a slow build-up mechanism (flow of sand, increase of the force on the bundle) acting everywhere in the system and decreasing over time the distance of the state variable of the nodes from the toppling threshold. Without this build-up mechanism the system would not become critical and the network density would increase the resilience of the system. On the other hand, the works on epidemic spreading have shown that if the network is dense or there are hubs the onset of the epidemic phase is facilitated [11].

Overall, the investigation on failure propagation in the context of credit networks of firms deserves more attention. Besides the work of [14, 16] presented here, [10] have recently studied a similar model where, however, there is no credit and cost is only proportional to delivered input, so that bankruptcy occurs only as result of production defaults. In fact, such model addresses a different issue related to the emergence of activity patterns in geographical economics.

The rest of the chapter is organised as follows. In section 2 we describe a modelling framework for networks of firms engaged in supplier-customer relations. We discuss the properties of a specific model in section 2.10, reporting some analytical results and some computer simulations. Some conclusions are drawn in section 3.

2 The Model

2.1 Economic Environment

The economy consists of N firms organised in M production levels. We will denote firms with indices $i, j, k, l, ...$ and levels with indices $J, K, L, ...$ We adopt the convention that production takes place along the vertical axis in downwards direction. The structure of the connections defines the production network as in the example shown in figure 1, in which arrows represent supply of goods (supply proceeds downwards, while money moves upwards).

In the example, each node has the same number of links k, but in general this could be from any distribution and, besides, the number of incoming links

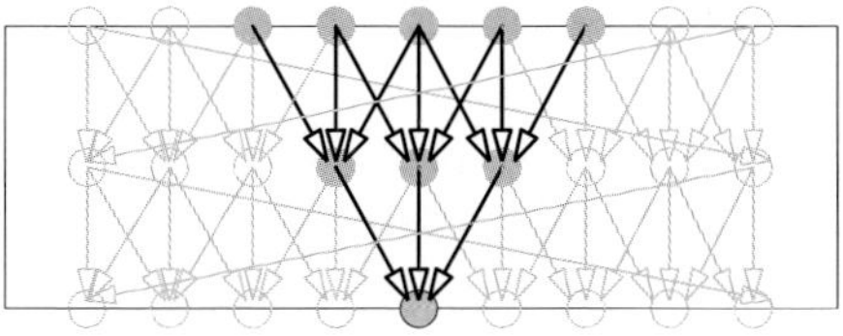

Fig. 1. Example of structure for the production network. The direction of production is from top to bottom. Each firm in a level receives goods from a subset (3 in this case) of firms from the upper level. The top level consists of primary producers. Layer 3 (from the top) consists of firms that sell in the consumer market (retailers). We have highlighted in dark gray the set of all suppliers upward from a given retailer (*in green*)

do not have to even outgoing links. Each firm in a level K is supplied by a subset of firms in the upper level $K-1$ and in turn supplies a subset of the firms in the lower level $K+1$. The bottom level $K = M$ represents firms that sell in the consumer market (retailers). The top level $K = 1$ represents primary producers. Firms are connected to each other through two mechanisms:

1. A firm asks for inputs from the suppliers in order to produce output.
2. A firm asks for payments from the customers in order to realize profit.

.

The output of each level K is produced by processing the input from the previous level $K-1$. Output is qualitatively different from input. For the sake of simplicity, we assume the following linear technology:

$$Y_i^{(K)} = \sum_{j \in V_i^S} Q_{ij}^{(K,K-1)} Y_j^{(K-1)} \quad (1)$$

where Y_i is the output of firm i, S_i is the set of suppliers of firm i, and Q_{ij} represents the fraction of the total output of firm j that firm i uses to produce its own output. In other words, Q is the input-output matrix and for any K, it follows that:

$$\sum_{i \in \text{ level } K} Q_{ij}^{(K,K-1)} = 1, \forall j \in \text{level } K-1 \quad (2)$$

2.2 Timing

We have to model the fact that over time, firms decide their desired amount of production, send orders, produce, deliver to customers and pay suppliers. We assume that time is discrete and divided into periods, each period including the following events for all firms: At the beginning of each period (or time step) t, orders flow upwards; then production and delivery flow downward. At the end of the period, money flows upward.

In greater detail, at the beginning, all firms in the bottom level M determine their *desired output*, based on the demand they face on the market and their *production capacity*, and then send orders to the upper level $M-1$. Afterwards, all firms in level $M-1$ determine their desired output based on the demand they face from their customer firms in level M. One after another, all levels do the same, up to level 1 (primary producers). Once the desired output is known, firms can compute their *expected output*, based on the expected output of the suppliers, which they communicate to the firms downward. This allows customer firms to allocate the necessary resources and premises to process the inputs they will receive.

At this point, production starts in level 1 and proceeds downward one level after the other, as each firm needs the input from its suppliers in order to produce. Output produced by a firm is delivered to customers on the basis of full trade credit; we rule out the possibility of inventory accumulation. When production reaches the bottom level, products are fully sold in the consumer market.

At the end of the period, a sequence of payments proceeds upwards from the retailers up to the primary producers. At each level, each firm pays its suppliers upstream only after having been paid by its customers. If costs exceeds revenues, the firm goes bankrupt and does not pay the suppliers in the current period. Moreover, the firm stops production for a number τ of periods in the future, after which it is replaced by a new firm endowed with an assigned initial value of production capacity. During those τ periods, the suppliers of that firm do not receive orders from it, nor do the customers receive production from it. Therefore, bankruptcy at the end of period t results not only in disruption of payments but also in a temporary local disruption in the production chain which is repaired in period $t+\tau+1$.

2.3 Remarks

The structure of the connections does not change during the process. This means that when a firm goes bankrupt, its customers do not create new links with other suppliers. This follows from the assumption of prohibitively high costs of establishing relations with new suppliers. So far, we have described a general framework, while the mechanisms involved can be specified in several ways (for example, we have to specify the dynamics of price, profit and net worth). However, some of the results presented in this paper do not depend on the specification of such mechanisms. Therefore, the present structure is a candidate for a class of models sharing similar behavior, in particular, concerning the conditions for the occurrence of avalanches of bankruptcies which are analysed in section 2.10. In the following, we provide a detailed description of a simple version of the model and a discussion of its limitations. In any period t each firm i is endowed with a level of real net worth $A_i(t)$, defined as the stock of the firm's assets in real terms, that has been financed only through net profits (we assume complete equity rationing).

2.4 Desired Output

Firm i at level K determines at time t its *desired output*, $Y_i^{(d,K)}$. This depends on the orders received from level $K+1$, with the constraint of its production capacity that we assume to be proportional to net worth $A_i^{(K)}$ by a constant $\theta > 0$ (as stated in eq. 3). Therefore, capacity is financially constrained as, for instance, in Greenwald and Stiglitz,(1993) and in related work by Delli Gatti et al., (2005). As in Greenwald and Stiglitz we can conceive of $\theta A_i^{(K)}(t)$ as the optimal (i.e., maximizing expected profit) output in the presence of bankruptcy costs.

Hence, desired output is defined as follows:

$$Y_i^{(d,K)}(t) = \min\{\theta A_i^{(K)}(t), \sum_{j \in V_i^C} O_{ij}^{(K,K+1)}(t) Y_j^{(d,K+1)}(t)\} \tag{3}$$

In the equation above, V_i^C is the set of customers of firm i, $O^{(K,K+1)}$ is the order matrix describing the orders from level $K+1$ to K, and in particular $O_{ij}^{(K,K+1)}$ is the fraction of the total supply needed by firm j, that firm j orders to firm i. In matrix notation we can write:

$$Y^{(d,K)}(t) = \min\{\theta A^{(K)}(t), O^{(K,K+1)} Y^{(d,K+1)}(t)\} \tag{4}$$

For level M, we assume that at each time step the consumer market absorbs the whole production and therefore:

$$Y^{(d,M)}(t) = \theta A^{(K)}(t) \tag{5}$$

2.5 Expected and Effective Output

Once the desired output is known at all levels, firms compute their *expected output*, based on the expected output of the suppliers. Here, "expected" has nothing to do with "expectation value" in statistical sense. A firm i may not be able to fulfill the orders of its customers, either because they exceed its production capacity or because the input from its suppliers is insufficient. As a result, supply can be smaller than the ordered quantity and therefore the *expected output* of firm i, $Y_i^{(e)}$, can be smaller than the desired one $Y_i^{(d)}$. In this version of the model, firms have a fixed set of suppliers (the network structure is static) and they cannot look for new suppliers. However, there is some freedom in the way firms decide to place orders to their suppliers, in other words, the way $O_{ij}^{(K,K+1)}$ are determined. This is discussed later on, and plays an important role. The production function of firms is assumed to be linear so that the output of a firm in level K is a linear combination of the input received from the suppliers in level $K-1$. This yields:

$$Y_i^{(e,1)}(t) = Y_i^{(d,1)}(t)$$
$$Y_i^{(e,K)}(t) = \sum_{j \in V_i^S} Q_{ij}^{(K,K-1)}(t) Y_j^{(e,K-1)}(t) \tag{6}$$

For firms at level 1, the expected output coincides with the desired one, as they do not have suppliers. V_i^S is the set of suppliers of firm i, $Q^{(K,K-1)}$ is the input-output matrix describing the transformation of input from level $K-1$ into the output of level K. Each entry $Q_{ij}^{(K,K-1)}$ represents the fraction of the total output of firm j that firm i uses to produce its own output. Firms in level 1 are primary producers and do not need any supply, therefore $Y_i^{(e,1)} = Y_i^{(d,1)}$. In matrix notation, the output of any level can be expressed as a function of the output of the first level as follows:

$$Y^{(e,K)}(t)=Q^{(K,K-1)}(t)Y^{(e,K-1)}(t)=Q^{(K,K-1)}(t)\cdot \ldots \cdot Q^{(2,1)}(t)Y^{(e,1)}(t) \quad (7)$$

The expected output is communicated downward to customers. Any two firms engaged in a supplier-customer relation agree on this amount to be delivered and paid at the end of the period. Customer firms allocate the necessary resources and premises to process the expected input they will receive from suppliers.

At this point, we include in the model some occasional production failures (due, for instance, to technical problems). At each period t, with probability q, the production of firm i is lost during the processing and no output is delivered to customers. This event occurs independently of the financial state of firms i and this failure lasts only one period. Therefore, we have to rewrite the *effective output* of i, $Y_i^{(K)}(t)$, as:

$$Y_i^{(K)}(t) = Y_i^{(e,K)}(t)S_i(t) \quad (8)$$

where $S_j(t) = 1$ with probability q and $S_j(t) = 0$ with probability $1-q$.

2.6 Production Costs

The output produced by firm i is sold to the customer at the price $P_i(t)$ (no inventory accumulation). We can think of the price of a firm's output in level K as $P_i(t) = P^{(K)}(t)u_i(t)$ where $P^{(K)}(t)$ is the general price at level K and $u_i(t)$ is the relative price for the output of the single firm. We assume that $u_i(t)$ is a random variable, uniformly distributed in $[1-\delta_P, 1+\delta_P]$ and independent of $P^{(K)}(t)$. Therefore, firm i incurs the following cost to get its supply of inputs from level $K-1$:

$$\tilde{C}_i^{(s,K)}(t) = \sum_{j \in V_i^S} Q_{ij}^{(K,K-1)} P^{(K-1)}(t)u_j(t)Y_j^{(K-1)} \quad (9)$$

The cost of inputs in real terms is obtained by dividing nominal costs by the level of prices in the level K:

$$\begin{aligned} C_i^{(s,K)}(t) &= \frac{P^{(K-1)}(t)}{P^{(K)}(t)} \sum_{j \in V_i^S} Q_{ij}^{(K,K-1)}(t)u_j(t)Y_j^{(K-1)}(t) \\ &= c_s \sum_{j \in V_i^S} Q_{ij}^{(K,K-1)}(t)u_j(t)Y_j^{(K-1)}(t) \end{aligned} \quad (10)$$

where c_s is defined as the ratio of the price levels at level $K-1$ and K and we assume it to be the same for all K.

Firm i also incurs a cost associated with the resources used in processing the input (labour and premises). As for the supply cost, this cost is assumed to be proportional to the expected output through a constant $c_r > 0$. We assume that the resources allocated by the customers of i to process its expected output cannot be dis-allocated within the current time period. Therefore, in case of a production failure of i, its customers run a cost proportional to the expected output and not to the effective output:

$$C_i^{(r,K)}(t) = c_r Y_i^{(e,K)} = c_r \sum_{j \in V_i^S} Q_{ij}^{(K,K-1)}(t) Y_j^{(e,K-1)}(t) \tag{11}$$

Of course, in the case of a production failure by i, the customers of i do not incur any supply cost. On the other hand, firm i not only does not receive any payment but has also to pay for the input from its suppliers. The production of firm i resumes at the next time step, if it has survived the shock. In conclusion, the production cost of firm i is the sum of the two terms defined above:

$$C_i^{(K)}(t) = C_i^{(s,K)}(t) + C_i^{(r,K)}(t) \tag{12}$$

2.7 Profit and Bankruptcy

In each period, when output is sold in the consumer market and payments start, some firms may realize sales revenue smaller than their supply costs. If this loss is high enough, firms go bankrupt and do not pay their suppliers . Therefore, we have to distinguish between the output delivered by firm i to its customers, $Y_i(t)$, and the output $Y_i^s(t)$ that is actually paid for ("s" for "sold"), at price $u_i(t)$, to firm i by its customers. Profit in real terms is equal to the difference between revenues and costs in real terms.

$$\pi_i^{(K)}(t) = u_i(t) Y_i^{(s,K)}(t) - C_i^{(K)}(t) \tag{13}$$

Profit, which can be negative or positive, incrementally changes the real net worth of the firm:

$$A_i^{(K)}(t+1) = \rho A_i^{(K)}(t) + \pi_i^{(K)}(t) \tag{14}$$

where $1-\rho$ measures a depreciation rate.

We assume that firms go bankrupt when the ratio of profit and net worth becomes smaller than a negative threshold value:

$$\pi_i^{(K)} < -\beta A_i^{(K)} \tag{15}$$

with $1 > \beta > 0$. If a firm goes bankrupt at time t, it stops supplying customers and paying suppliers for a number τ of time steps (referred to as "inactivity time" in the following). During these time steps, neighboring firms are not allowed to look for alternative customers or suppliers, as the network structure is static. Firms can however (at least in some of the scenarios considered in the following) adjust their orders as a function of the production capacity of the suppliers. As this is proportional to net worth, it means that customers order less and less when a supplier's net worth decreases. Once the inactivity time elapsed, the bankrupt firm is replaced by a new firm with the same links as its predecessor and its net worth is re-initialized: $A(t+\tau+1) = A_{entry}$.

2.8 Strategies for Placing Orders and Delivery

Although the network is static in this version of the model, and therefore the set of suppliers of a firm is fixed, still there are many possible ways to allocate orders to the suppliers. Consistently with our bounded rationality framework, we consider simple strategies for placing orders and one strategy for delivering. Firm i places orders evenly:

$$O_{ij}^{(K,K+1)}(t) = \frac{1}{|V_i^S|} \tag{16}$$

where $|V_i^S|$ is the cardinality of the set of suppliers j of firm i (notation is consistent with equation 3).

Firm i delivers to each customer j in proportion to its order:

$$Q_{ij}^{(K,K-1)}(t) = \frac{O_{ij}^{(K,K+1)}(t)Y_j^{(K-1)}(t)}{\sum_{l \in V_j^C} O_{lj}^{(K,K+1)}(t)Y_j^{(K-1)}(t)} \tag{17}$$

The equation above satisfies the condition of equation 2. For other possible strategies see ([14])

2.9 Generic Properties of the Model

A number of specific models can be investigated within the framework presented so far. In particular, in the presence of delayed payments (trade credit) and costs due to failures in supply (as assumed above), in this model it is possible to have avalanches of bankruptcies originating locally and spreading both upstream and downstream.

If bankruptcies can propagate simultaneously in both directions, then, and only then, are they "reflected" diagonally at each level and the result is a net horizontal propagation, that is perpendicular to the direction of production (figure 2, c-d). The horizontal propagation is important because it is a necessary condition for the spreading of an avalanche to a significant part of the network if the number of layers is much smaller than the number of site positions. This is typically the case in several sectors and was one of the weak points among the condition for the emergence of Self Organized Criticality

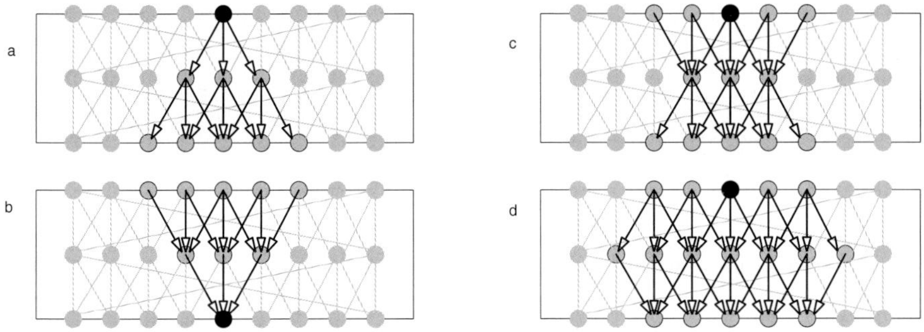

Fig. 2. Different modalities of failure propagation. Edges through which failure propagate are in darker gray. The firm triggering the avalanche is represented by the node in dark gray. **a.-b.** Downward and upward propagation of failures. **c.-d.** Horizontal propagation occurs when each level transmits downward but also reflects upwards. In panel c failures have propagated up to two degrees of separation from the initial firm; in panel d up to three degrees

in the work of [2]. In particular, the horizontal axis could also represents a geographical or technological space.

In the following we will speak of *horizontal bankruptcy propagation* to mean the situation in which bankruptcies can propagate potentially to the whole network and not only to the downward/upward cone of firms.

On the contrary, previous models ignore local interaction so that the propagation of bankruptcies is activated only by means of global coupling: the more firms fail, the higher the interest rate for all, hence the more they fail.

The model, as presented so far, reproduces qualitatively important properties of a production network:

1. Spatio-temporal correlation of output, growth and bankruptcies
2. Exponential growth
3. Oscillations of de-trended aggregate output
4. Heterogeneous firm size distribution
5. Exponential probability distribution of aggregate growth (right side)

For a detailed discussion of this properties, see [14]. Varying allocation strategies, dynamics on prices and other parameters one can investigate the role of the main factors involved in models of financial fragility and address the following issues:

1) The role of trade-credit relationships in the propagation of bankruptcies

2) The role of interest rate and policies to prevent the occurrence of large avalanches

3) The role of the structure of the network of interactions

4) Policies to make such structure more robust against large avalanches

In the rest of this chapter, we will focus on the role of the network density in combination with the dynamics of the interest rate, an issue discussed more in depth in [16].

2.10 Analysis of a Specific Setting

In order to isolate the impact of network density on the dynamics of avalanches, we will now consider a specific case of the model in which there is no accumulation of net worth (we set $\rho = 0$ in eq. 14). As a result, aggregate output is no longer growing exponentially and, moreover, firm size distribution does not evolve in time into a skewed distribution. Consistently, we also set $\beta = 0$ in eq. 15 : as firms do not accumulate net worth, they go bankrupt when profit is not positive. Therefore, in this setting, a production default implies also bankruptcy.

Concerning prices, we assume, as discussed in [14], that prices are stochastic and independent, distributed according to a uniform distribution in $[1 - \delta_P, 1 + \delta_P]$. Eq. 15 implies that a firm goes bankrupt if the price falls below a critical value, which can be approximated as:

$$u_i^*(t) = \frac{(c_r - \beta/\theta)Y_i^e(t) + c_s Y_i(t)}{Y_i^s(t)} \tag{18}$$

Because it is $Y_i^s \leq Y_i \leq Y_i^e$, the probability of bankruptcy increases the smaller are, with respect to the expected output Y_i^e, Y_i and Y_i^s (see section 2.7).

Density Decreases Systemic Risk

For the sake of simplicity, we assume that firms have the same number k of suppliers and customers, and we study the impact of different values of k. We consider M production layers, each including the same number n of nodes. If $k = 1$ the network is actually composed of isolated chains, while if $k = n$ all possible links are realized. Clearly, increasing k reduces the fluctuations of input from suppliers due to production default and therefore, the probability of bankruptcy. Consider for simplicity, a two-layer network; for $k = 1$, with probability q a customer is not delivered at all (because the supplier experiences a production default with probability q) and goes bankrupt. For large n, input delivered to each customer approaches the fraction $1 - q$ of the input requested, and thus, if $\frac{c_r}{1-q} + c_s < 1$, the probability of causing bankruptcy of customers to go bankrupt as a result of a production default among suppliers is zero. [4]

[4] Of course, one could assume that firms incorporate probability if defaults in their decision, by ordering $\frac{1}{1-q}$ so to be delivered the exact requested quantity. Instead, as before, for the purpose of this work we assume agents are boundedly rational and do not take into account production defaults in their decisions.

We can make precise predictions on the bankruptcies caused by a production default by computing the profit of the neighbouring firms. In a two layer network, after a production default in a supplier, the profit of each customer is:

$$\pi_j(t) = pY_i^s(t) - c_s Y_i(t) - c_r Y_i^e(t) = (p - c_s - \frac{k}{k-1} c_r) Y_i(t) \tag{19}$$

Similarly, after a production default in one customer, the profit of each supplier is:

$$\pi_j(t) = pY_i^s(t) - c_s Y_i(t) - c_r Y_i^e(t) = (\frac{k-1}{k} p - c_s - c_r) Y_i(t) \tag{20}$$

In both cases, the increase of k increases the profit and makes a bankruptcy less probable. We can also estimate the profit of firms in case of multiple defaults in the neighbourhood, and thus compute the expected profit in the general case [16]. Overall increasing network density reduces the probability of bankruptcies of individual firms, as well as the probability of joint bankruptcies, in other words, it reduces the systemic risk.

Systemic Risk in Presence of Positive Feedback

However, if there is a positive feedback of the probability of bankruptcy of a firm i on the cost i faces (namely, that the more a firm is likely to fail the higher the cost it faces), then the effect of network density can be to increase the instability of the system. In fact, in a very dense network, an increase in the average probability of failure would increase the cost of all firms, thus increasing in turn their probability of failure. In a sparse network, the coupling is only local so that the probability of failure may increase somewhere while decreasing somewhere else. In order to investigate quantitatively this issue, we now, consider the cost of the firm to be dependent on the financial state of the firm. The rationale for this is that firms pay an interest rate on the supply they receive (trade credit) and/or on the funds used to pay wages or processing (loan). The interest rate a firm is charged by other firms or by the bank increases (at least within a range) with the financial fragility of the firm itself as its partners need to compensate their risk in extending credit to it (however, when the interest rate is very high, creditor usually don't have an incentive to increase it further, see [5] (chapter 5). In order to capture this effect, we assume that an increase in bankruptcy risk (a decrease in profit) leads to an increase in interest rate and therefore in production costs. As a consequence, the production cost for firm i is multiplied by a factor η evolving in time as follows:

$$\eta_i(t+1) = \eta_i(t) + \alpha \cdot \mathrm{sign}(\sum_{j \in V_i} P_j(t) - P_j(t-1))$$
$$\eta_i \in [\eta^{\min}, \eta^{\max}] \tag{21}$$

where V_i is the neighbourhood of i including i itself, while α is a parameter. The range of variation of η_i is bound, corresponding to a minimum and maximum interest rate. The equation above implies that whenever profit decreases/increases among neighbours, cost increases/decreases by a fixed quantity α. Other functional dependencies are possible and reasonable, but the important feature here is that the net average change of profit in the neighbourhood causes a discrete change in the cost.

Understanding the Dynamics in a Simplified Model

A similar dynamics has been recently introduced in [17] in the context of cascades in complex networks . There, agents are associated with a state variable, representing their fragility, that evolves as a function of the neighbours. At each time step, the fragility of each agent receive an i.i.d. shock (through a normalized stochastic variable $\xi(t)$, with standard deviation equal to 1), which is shared with the neighbours. If, at time t the fragility of agent i exceeds a given threshold θ, the agent fails and the quantity a, representing the damage associated with its failure, is distributed to the neighbours (by incrementing their fragility), which may in turn fail. All the toppling events following the initial failure occur at time scale faster than the one for fragility, in other words they all occur before the next time step $t+1$. In formulas the dynamics reads:

$$\phi_i(t+1) = \sum_{j \in V_i} W_{ij}(\phi_j(t) + \sigma\xi_j(t)) + \alpha \cdot \mathrm{sign}(\sum_{j \in V_i} W_{ij}(\phi_j(t) - \phi_j(t-1))) \tag{22}$$

where W is a matrix representing interaction among agents, with $\sum_j W_{ij} = 1 \forall i$, and σ is a parameter. Additionally, if $\phi_r(t+1) \geq \theta \; \exists r$, then:

1. For all neighbours of each node r:
   $\phi_s \to \phi_s + aW_{sr}$
2. For all such r, $\phi_r \to 0$

3. Repeat until $\phi_i < \theta$ for all i in the system.

In order to understand the onset of instability we analyse the dynamics above in a mean field approximation, in the case $W_{ij} = \frac{1}{k}$. We consider the average fragility in case of large k:

$$\Phi(t+1) = \frac{1}{N}\sum_i \phi_i(t+1) \simeq \Phi(t) + \frac{\sigma\xi(t)}{\sqrt{k}} + \alpha \cdot \mathrm{sign}(\Phi(t) - \Phi(t-1)) \tag{23}$$

In the regime where $\frac{\sigma\xi(t)}{\sqrt{k}} \gg \alpha$, the process is dominated by the first term of eq. 24 and it is approximated by a random walk with step of amplitude

$\frac{\sigma}{\sqrt{k}}$, thus decreasing with k. In particular, if the first term dominates, the difference $\Phi(t) - \Phi(t-1)$ is positive or negative with equal probability and thus the second term does not contribute any systematic drift. In this regime, increasing k makes the step of the random walk smaller and thus decrease the probability to hit the threshold.

If $\frac{\sigma}{\sqrt{k}} \ll \alpha$, then expressing $\Phi(t)$ in terms of $\Phi(t-1)$, we have:

$$\text{sign}(\Phi(t) - \Phi(t-1)) = \text{sign}(\frac{\sigma\xi(t)}{\sqrt{k}} + \alpha(\text{sign}(\Phi(t-1) - \Phi(t-2)))) = \\ +\alpha(\text{sign}(\Phi(t-1) - \Phi(t-2)))) \quad (24)$$

The expression above is always true if the distribution of $\xi(t)$ has a limited support and α is larger than the right limit of such support, otherwise it is true with a certain probability that can be computed. Therefore, in this regime the average fragility tends to keep moving in the direction it is already moving. Because Φ is repelled from 0 by construction, in the limit of large k, it moves upwards with constant slope and periodically it hits the threshold and is then reset to 0. When the average fragility hits the threshold and the individual fragility trajectories are sufficiently close, then one or few failures cause an avalanche involving the whole system.

The argument suggests therefore that, increasing the density of network in the system described by eq. 22, the probability of failures first decreases and then increases. In other words there is a trade-off between diversifying the risk by sharing the shocks with many other agents and the systemic risk resulting from the synchronization of the fragility trajectories. For a more formal analysis see [17].

The argument above suggests that a similar result should also hold for the economic model presented in this chapter. The reader may notice that we have inferred an important property of a fairly complicated economic model from a basic argument, based on a mean field approximation of a simplified model that captures some essential dynamical features of the original model. In the next section we will examine the results of computer simulations of the original model.

2.11 Results of Computer Simulations

In this section, we compare the time evolution of some quantities measured on a network of 2 production layers, with three different values of connectivity degree, $k = 1, 5, 20$. Unless specified otherwise, results reported in this work are obtained with constant price (interval width $\delta_P = 0$), production default probability $q = 0.03$, $c_s = c_r = 0.3$, inactivity time $\tau = 1, 2, 3$ with equal probability. Firms are endowed with constant value of net worth $A_i(t) = A_{init} = 1 \, \forall i$. As explained above the bankruptcy threshold is set as $\beta = 0$ and the depreciation factor is set as $\rho = 0$ (which yields a depreciation rate $1 - \rho = 1\%$). Finally, the value of α in the dynamics of the cost factor $\eta_i(t)$ is set to 0.05.

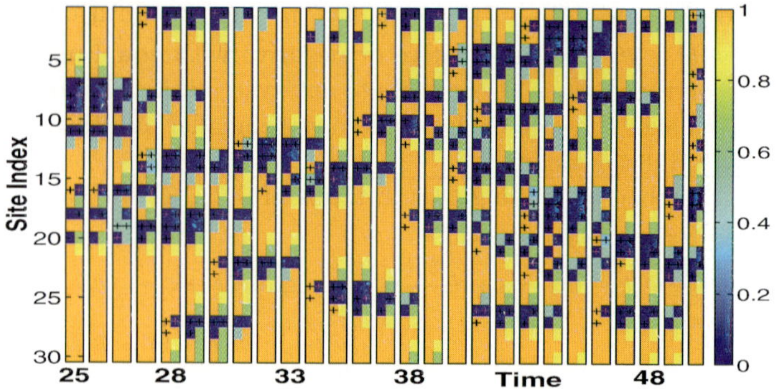

Fig. 3. Time evolution of output of the production network with degree $k = 2$. Zoom on a time interval. Each frame represents the production network at a given time step with index of the node on the y axis and production layers on the x axis. The layer of primary producers is now the left column of each frame, while the layer of retailers in the consumer market is the right column of each frame. Output is normalized in each frame by its maximum value in order to emphasize the relative spatial distribution and is represented by a color scale as specified by the color bar. Magenta crosses indicate production defaults occurring stochastically with probability q, while black crosses indicate bankruptcy (see text for more details)

In figures 3, 4, 5, the evolution of output over the production network is shown in an interval of 25 time steps. In order to follow the propagation of bankruptcies, we choose a represent different from the one used in fig. 1. Each frame represents the production network at a given time step with index of the node on the y axis and production layers on the x axis. The layer of primary producers is now the left column of each frame, while the layer of retailers in the consumer market is the right column of each frame. Output is normalized in each frame by its maximum value in order to emphasize the

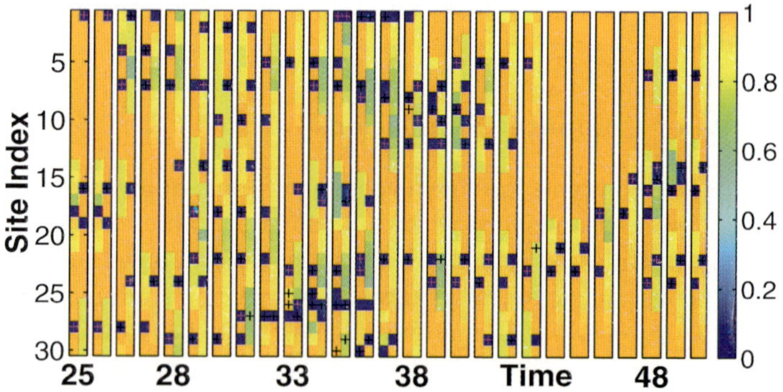

Fig. 4. Time evolution of output of the production network with degree $k = 5$. The figure is constructed in the same way as figure 3

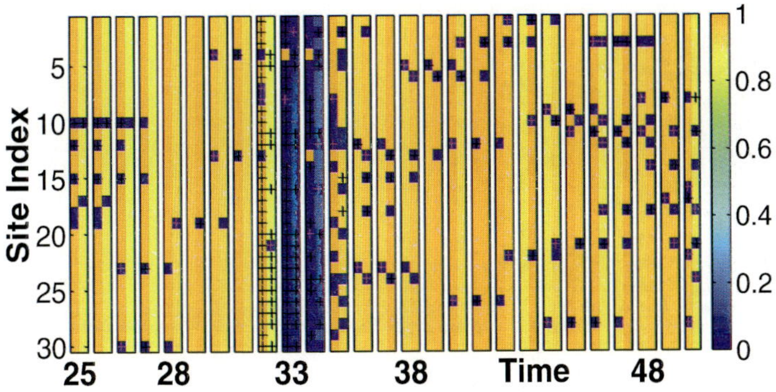

Fig. 5. Time evolution of output of the production network with degree $k = 5$. The figure is constructed in the same way as figure 3

relative spatial distribution and is represented by a color scale as specified by the color bar. Magenta crosses indicate production defaults occurring stochastically with probability q, while black crosses indicate bankruptcy. Following the position of the crosses from one frame to the next, it is possible to observe the propagation of bankruptcies over time. With the parameters chosen in this specific setting, production default in a firm also implies its bankruptcy, although it is not true in general.

With degree $k = 1$ (not shown), production is organized in chains, which are obviously very fragile to shocks, as the production default of a supplier/customer implies also the bankruptcy of its only customer/supplier. With the chosen values for cost, $c_s = 0.3$ and $c_r = 0.3$, and with degree $k = 2$, the default of a supplier is very likely to cause, the bankruptcy of its two customers, which in turn do not pay their suppliers, causing two additional bankruptcies. Overall, five firms go bankrupt in such an event, while with degree $k >= 3$ instead, the default of a supplier is very likely not to cause any bankruptcy. Simulations shown in figures 3, 4 confirm this estimate, although some deviations are possible, due the internal dynamics of the cost factor. With high degree ($k = 30$, 5) the feedback mechanism prevails on the risk diversification and larger avalanches occur (at time 32 in the figure).

The effect can be seen also in figures 6, 7, 8, where the evolution of output $Y_i(t)$ and cost factor $\eta_i(t)$ is shown over 100 time step. In order to emphasize the spatio-temporal patterns we use now another representation with respect to figures 3, 4, 5. The x axis is time, while the y axis represents the index of the nodes from 1 to N. In other words, positions from 1 to n on y (from to the top) represent the nodes of the first layer ($n = 30$ in this example), while positions from n to $2n$ (from to the top) represent the nodes of the second layer. We chose n and time interval relatively small to make the patterns visible. Output and cost factor are represented by a color scale as specified by the color bar. Magenta crosses indicate production defaults (occurring

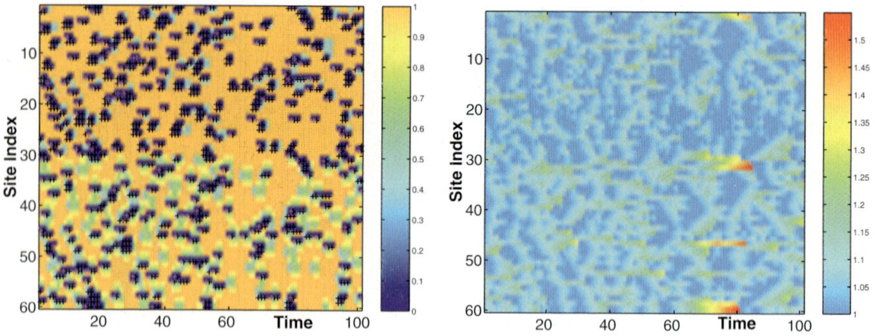

Fig. 6. Time evolution of output $Y_i(t)$ (**left**) and cost factor $\eta_i(t)$ **right** with degree $k = 2$. The x axis is time, while the y axis represents the index of the nodes from 1 to N. Output and cost factor are represented using a color scale specified by the color bar. Magenta crosses indicate production defaults, while black crosses indicate bankruptcy

stochastically with probability q), while black crosses indicate bankruptcy. Following the position of the crosses from one frame to the next, it is possible to observe the propagation of bankruptcies over time.

An important aspect of the phenomenon investigated here is that the instability induced at high k is also visible at the aggregate level. In figure 9 the aggregate output of the network is shown for 200 time steps. Going from degree $k = 2$ (blue curve) to $k = 5$ (green curve) and to $k = 30$ (magenta curve), fluctuations first decrease and then increase.

A more detailed investigation of the trade off between individual risk diversification and systemic risk is out of the scope of this work and can be found in [16].

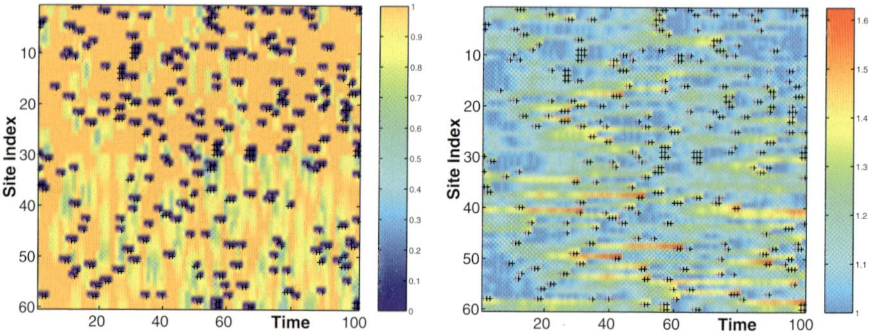

Fig. 7. Time evolution of output $Y_i(t)$ (**left**) and cost factor $\eta_i(t)$ **right** with degree $k = 5$. The figure is constructed in the same way as figure 6

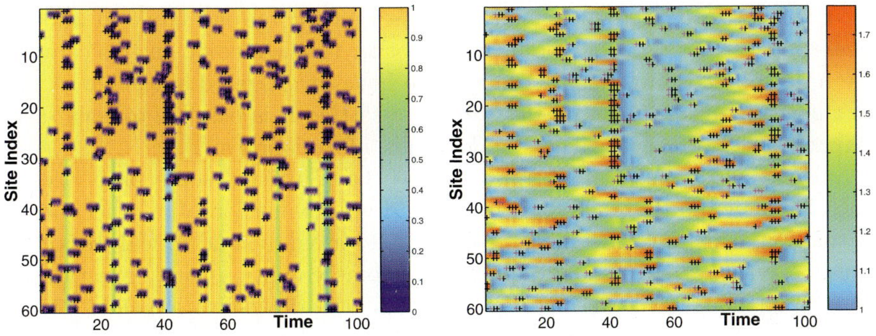

Fig. 8. Time evolution of output $Y_i(t)$ (**left**) and cost factor $\eta_i(t)$ **right** with degree $k = 30$. The figure is constructed in the same way as figure 6

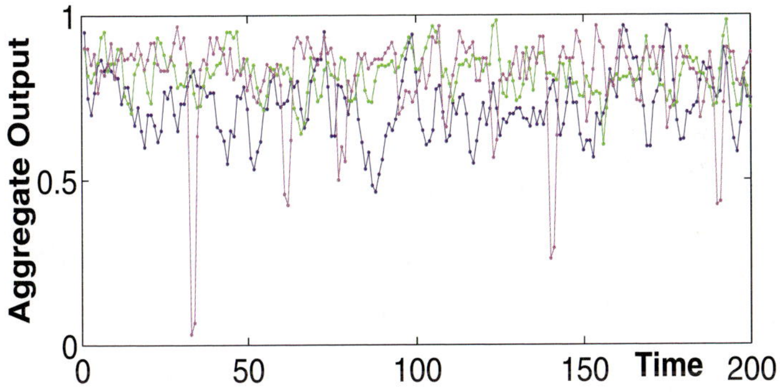

Fig. 9. Time evolution of aggregate output of the production network with different value of degree k: $k = 2$ (blue), $k = 5$ (green) and $k = 30$ (magenta)

3 Conclusion

In this chapter we have examined the impact of network density on the systemic risk in networks of credit relations. A node in a credit network may form several links in order to diversify its risk, but this may also induce a form of financial contagion.

Systemic risk raises today growing regulatory concern and policy makers would like to know how to ensure adequate fire walls in order to prevent the spill over of a crisis across institutions and firms. Avalanches of failures in networks have been studied extensively in the Complex System literature in the context of SOC and epidemic spreading, but, outside such two contexts, the investigation deserve more attention. In the economic literature, there is a growing body of work on systemic risk in credit networks, although there seems to be a dominant view on the positive role of the network density on the systemic risk.

Credit is extended by banks to firms (loans), by one bank to another (interbank credit) and by one firm to another (trade credit). In this chapter, we have focused on the last case and we have investigated a specific setting of the model introduced by [16]. In such a setting it is possible to isolate the impact of network density on the dynamics of avalanches.

Using a complex system approach, we have inferred some properties of the economic model based on a mean field approximation of another model, actually much simpler, that captures some essential dynamical features of the original one. We have shown how, under some conditions on the parameter chosen, a trade off emerges between individual risk diversification and systemic risk. This result is in line with recent finding in the economic literature, but it is in contrast with the dominant view. This works contributes to the debate on what are the appropriate regulations to ensure the robustness of credit networks.

References

1. Allen, F. and Gale, D.: Fincancial contagion. Journal of Political Economy **108** (2001) 1–33
2. Bak, P., Chen, P., Scheinkman, J. and Woodford, M.: Aggregate fluctuations from independent sectoral shocks: Self-organized criticality in a model of production and inventory dynamics. Ricerche Economiche **47** (1993) 3–30
3. Boissay, F.: Credit chains and the proagation of financial distress. Working paper, European Central Bank **573** (2006)
4. Delli Gatti, D., Di Giulmi, C., Gaffeo, E., Giulioni, G., Gellati, M. and Palestrini, A.: A new approach to business fluctuations: heterogeneus interacting agents, scaling laws and financial fragility. Journal of Economic Behavior and Organization **56** (2005) 489–512
5. Stiglitz, J.E. and Greenwald, B.C.: Towards a New Paradigm in Monetary Economics. Cambrigde University Press, Cambridge (2003)
6. Hahn, F.H. and Solow, R.: A Critical Essay on Modern Macroeconomic Theory. MIT Press, Cambridge (1995)
7. Kiyotaki, N. and Moore, J.: Credit Chains. ESE Discussion Papers **118**, Edinburgh School of Economics, University of Edinburgh (1997) http://ideas.repec.org/p/edn/esedps/118.html
8. Klapper, L.: The uniqueness of short-term collateralization. Policy Research Working Paper, World Bank **2544** (2001)
9. Lee, D.–S., Goh, K.–I., Kahng, B. and Kim, D.: Sanpile avalanche dynamics on scale-free networks. arXiv:cond-mat/0401531 (2004)
10. Weisbuch, G. and Battiston, S.: From production networks to geographical economics. Journal of Economic Behavior and Organization *in press* (2007)
11. Vespignani, A. and Pastor-Santorras, R.: Epidemic spreading on scale-free networks. Physical Review E **65** (2002) 035108
12. Omiccioli, M.: Trade credit as collateral. Temi di discussione della Banca d'Italia **553** (2005)
13. Crucitti, P., Latora, V. and Marchiori, M.: Model for cascading failures in complex networks. Physcial Review E **69** (2004) 045104

14. Battiston, S., Delli Gatti, D., Gallegati, M., Greenwald, B.C.N. and Stiglitz, J.E.: Credit chains and bankruptcies avalanches in supply networks. Journal of Economic Dynamics and Control *in press* (2007)
15. Aleksiejuk, A., Holyst, J.A. and Kossinets, G.: Self-organized criticaly in a model of collective bank bankruptcies. Int. Mod. Phys. C. **13** (2001) 333
16. Battiston, S., Delli Gatti, D., Gallegati, M., Greenwald, B.C.N. and Stiglitz, J.E.: Trade credit network and systemic risk. *submitted*
17. Battiston, S., Peters, K., Helbing, D. and Schweitzer, F.: Cascades on networks. *submitted*
18. Freixas, X., Parigi, B.M. and Rochet, J.–C.: Systemic risk, interbank relations, and liquidity provision by the central bank Journal of Money, Credit and Banking **32** (2000) 611–638
19. Iori, S. and Jafarey, F.P.: Systemic risk on the interbank market. Journal of Economic Behavior and Organization **61** (2006) 525–542

A Complex System's View of Critical Infrastructures

Vittorio Rosato[1], Ingve Simonsen[3,4], Sandro Meloni[1], Limor Issacharoff[1,3], Karsten Peters[2], Nils von Festenberg[3], Dirk Helbing[5]

[1] ENEA, Casaccia Research Center, Computing and Modelling Unit, Roma, Italy rosato@casaccia.enea.it
[2] Institute of Transport and Economics, Dresden University of Technology, D-01086 Dresden, Germany simonsen|limor|festenberg@vwi.tu-dresden.de
[3] Department of Physics, The Norwegian University of Science and Technology, N-7491 Trondheim, Norway
[4] Institute for Logistics and Aviation, TU Dresden, 01062 Dresden karsten.peters@tu-dresden.de
[5] Chair of Sociology, in particular of Modeling & Simulation, ETH Zurich UNO D11, Universitätstrasse 41, 8092 Zurich, Switzerland dhelbing@ethz.ch

1 Introduction and Motivation

Our contemporary societies are examples of highly complex systems with many interacting constituents that are organized in ways that often are hard to grasp. Their organizational systems and infrastructures are time-dependent and highly interconnected. Thus, what may appear as different parts of our societies, do indeed depend on and influence each other.

Large Complex Critical Infrastructures (LCCIs) are national — or international — technological systems whose correct functioning has a high social impact. A current definition of a "Critical Infrastructure" is a large scale infrastructure which if degraded, disrupted or destroyed, would have a serious impact on health, safety, security or well-beings of citizens or the effective functioning of governments and/or economy [4]. This definition therefore allows to label many infrastructures that we are well-familiar with from our daily lives, as being "critical". Among them are, for instance, the networks for the transmission and the distribution of electrical power, those allowing communication to occur (in all its forms, from telephones to the Internet), transportation networks like roads, railways and sky-routes up to pipelines for drinking water, gas and oil, *etc.* LCCIs are thus strategic (in the wider sense of the term); as such, an enormous care should be taken to keep them operational and efficient, preventing their failure due to accidents or intentional attacks.

A further major issue comes from the high level of *interdependency*, *i.e.* the fact that each LCCI interacts (in a more or less explicit way) with one another. This may have the implication that a disturbance in one of them might affect the functionality of others. This renders the task of preventing failures and, in general, the same operational control, an extremely complex task. It is indeed desirable to have the best possible control on *single* infrastructures in order to prevent faults. However, optimizing and securing individual infrastructures independent of the presence of others, is often *not* sufficient to securing such interconnected system.

LCCIs are also intriguing technological objects. They are "complex", according to the current definition of complexity, as their behavior cannot be simply predicted on the basis of the behavior of their single components. Complexity triggers the *emergence* of new phenomena which cannot be predicted by usual means but only through a complete description of all its components altogether. Emergence of new phenomena occurs, *a fortiori*, when many LCCIs are functionally coupled together: also in the case of a weak connection, there is the seed for the emergence of further unpredictable behavior.

All this conceptual entanglement has attracted the interest of the Complexity Science (CS) community. This work intends to introduce some basic statements, show the CS methods and tools and some recent results of their application in the field of LCCIs. In this chapter, we intend to make a first recognition of some basic problems which can be tackled by making use of mathematical models and numerical methods, with the aim of producing results useful for the understanding of some fundamental questions related to their structure.

2 Why LCCIs Become/Behave More and More Complex

Historically, in Europe (at least), the LCCIs were often national monopolies typically owned and/or controlled by the national governments. Over the last decades, this situation has changed to a large extent; many LCCI sectors have been *deregulated* and thus the monopolistic state removed. This opened up for new market players that together with the former monopolists (of a given region) could compete. Notice that this situation was not (in principle) restricted to a geographical (national) region, but also international competition was encouraged by the market liberalization. For instance, one prominent example of the latter is the European power market. The formal basis of the deregulation of the European electricity market was laid out in the 1996 EU Directive 96/92. However, about three years later, on 19 February 1999, the electricity market in thirteen countries in the European Union (EU) and the European Economic Area (EEA) began to open up on an international basis. A competitive European Power market was born!

With the deregulation of the European LCCI sector, new challenges were created. Now (big) consumers could, say, buy their electrical power from any market participant. This implied that the backbone of the European transmission grid had to be fully interconnected, and that it should be able to handle rising loads. However, the European transmission grids were not designed for this purpose (and volume) in mind. Connections to neighboring states were typically built up for backup reasons, and to handling short term import-export scenarios. Hence, the new business model that was put in place (due to the liberalization) prompted some technically minded people to question the robustness of the ever more complex power transmission grid. This concern became strengthened by the increased terrorism threat as well as the recent large-scale Italian September 2003 blackout, and the similar previous cases from London, North America, Sweden and Denmark. For instance, the cause of the London blackout was traced back to a badly-installed fuse at a power station; indeed all the others happened for similar reasons. Furthermore, it was realized, by a careful analysis of the cause of events, that problems typically start at one place and propagate over large geographical distances, like a domino effect. For instance, the great 2003-blackout in New York initially was triggered by an event in the mid-west (Ohio) [1, 27].

Analogous problems must be faced in telecommunication (TLC) systems, where a large number of stakeholders crowd common infrastructures and compete for bandwidth and customers. TLC routes are constantly stressed by a constantly increasing traffic level.

Most LCCIs have grown in an unsupervised regime (there is not a general controller of worldwide Internet network) and needs to face a dramatic increase of their usage by adopting an "intrinsic" ability to adapt themselves according to changing external conditions. This seems to be a key point in this matter: Are technological networks able to autonomously react to external input in a way to adapt their functioning to constantly guarantee a reasonable efficiency? If so, which are the agencies that allow adaptive behavior to occur? What can LCCI managers reasonably do to let adaptation mechanisms run faster and more efficiently, and to better respond to mutated external conditions?

Complexity Science tries to answer these questions also by identifying common scenarios which subtend rather "universal" behaviors which take place in complex systems. This approach has allowed a flow of data and methods from diverse scientific fields and triggered the customization of ideas and methods, typical in one domain, to other domains. Living objects, for instance, as bacterial colonies, swarms and bird flocks do display a number of intriguing control strategies which, if properly understood, could be mutuated and used to analyse and control technological systems (*bio-mimetic* strategies).

This scenario has prompted calls for improved coordination between basic and applied research on the evaluation and the design of new tools for the analysis and the control of LCCIs at a multi-national level. For instance, many EU funded projects have been launched within this domain in the sixth Framework Program (FP6), and similar projects have received public funding in the US. Dedicated programs within this area are forecasted also for FP7.

In the remaining part of this chapter, we will introduce and discuss LCCIs from a complex network perspective. In particular, the *graphs*, which are the network's basic model and will be the object of the present study, will be introduced. They are mathematical objects onto which CS can deploy its methods whose results can provide measures of their ability of providing useful information on the networks. Various analyzing methods suitable for LCCIs will be proposed and discussed. Examples of results that can be obtained from such analysis will be given for some example LCCIs.

3 What is A Network?

The term network is used in every-day life, so most of us have an impression of what is meant by it. As we will attempt to reply to basic questions on these networks, also the metaphors which will be used to describe networks will be at a high level of abstraction. However, here we will put a specific meaning to that word. By a *network* we will mean a set of N objects, referred to as *nodes* or *vertices*, that are connected through what is typically known as *links*, *arcs* or *edges* (L). A network G will thus be indicated as a collection of objects $G = (N, L)$ (in these terms, the network can be also represented as a mathematical *Graph* which is indicated by explicitating the same entities). Some simple networks are sketched in Fig. 1. In this figure, the nodes are indicated by red (or grey) filled circles, while the links are black lines between the nodes. For instance, for an electrical power network, nodes correspond to power generators or distribution (or transmission) stations, while links represent the power transmission lines connecting the nodes.

It should be noticed that links do not have to be *physical* connections; they might also represent *logical* connections between nodes, such as in the case of a so-called social network. Here nodes are persons, and a link exists between two persons if they are considered (in some way) to be friends.

Networks represent the natural starting point for modelling infrastructures. As mentioned above, the investigation of networks has received an increasing attention in the last decade in the CS community. Genuinely, the field was embodied in mathematics as *graph theory*, which, due to progress in computer power and the growing consciousness of the relevance of network's structure in several fields, has prompted this topic to a wider scientific audience. The field experienced a strong push forward when the famous "small-world" paper by Watts and Strogatz was published in 1998 [2] by delivering examples of networks where seemingly distant nodes actually are surprisingly close to each other due to the peculiar network structure; this property has been called "small-world" as it perfectly reproduces the situation of a globalized world where local events can have a long-range impact . This property has also been fixed in the common language by the phrase "six degrees of

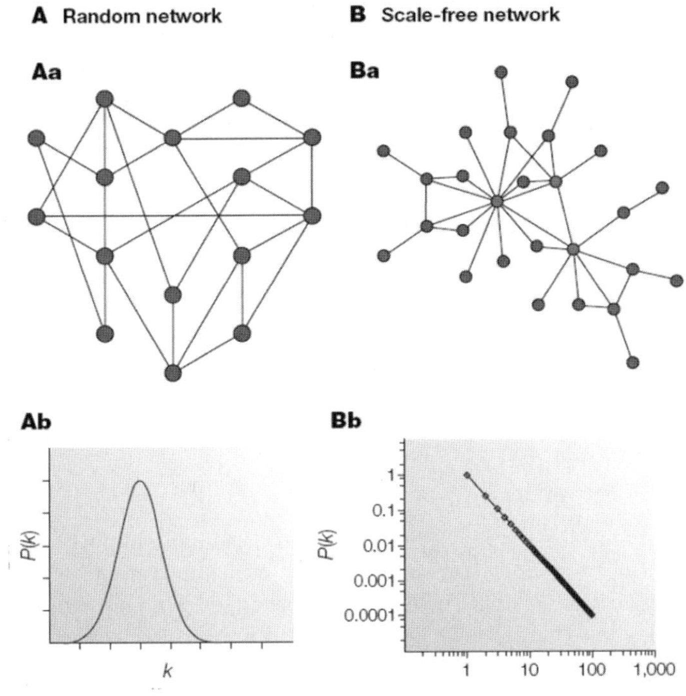

Fig. 1. Examples of complex networks. Fig. 1(**A**) depicts a *random network*, while a scale-free network is shown in Fig. 1(**B**). The typical degree distribution, $P(k)$ of each class of network is shown in the lower part of the figure, *i.e.* the distribution of the number of links, k that is associated with each node. Notice in particular the marked difference in topology that results from the change in the degree distribution. (After Ref. [26])

separation"[6]. It expresses that in a small-wold class network, two arbitrary chosen nodes can be connected in, on average, only six steps. For social networks, this effects (as well as the number of six) had already been known empirically since the 1960's due to some cleverly designed experiments conducted by the social psychologist Stanley Milgram [5]; with a small chain of friends and friends-of-a-friend, each of us can reach whatsoever other person in the world in (on average) six steps.

A major outcome of this new branch of theoretical disciplines is the recognition that diverse networks (from sociology, technology and biology) display a peculiar structure with clear small-world characteristics. This seems to be a property *emerging* from complexity and, as such, probably brings

[6] The phrase was made well-known outside scientific circles by John Guare's popular theater play of the same name (and later movie).

some added-value to the network's property. Much work has already been performed in order to show which are exactly the benefits engendered by such a topological structure.

Recently, several comprehensive reviews on network research (graph theory) have appeared in the literature [3, 6, 7, 22] displaying the current state of its application to real world networks. Most of the work, so far, has focused on static properties and behaviour of networks, *e.g.* the question of network robustness [8].

The main property of a network stems from its classification as belonging to a specific topological class. These are related to the specific form displayed by the distribution of the node's *degree*, k, of the network, $P(k)$ (degree distribution). The degree, k, of a node is defined as the number of nodes to which it is *directly* (physically or logically) connected. The most relevant topological classes are:

- Random networks
- Scale-Free networks

In the first case (see Fig. 1A), $P(k)$ has a Poissonian shape; the network is thus composed of *almost* equivalent nodes, with an average degree $\langle k \rangle$ and a given standard deviation. In the second case (see Fig. 1B), the situation is more complex, as $P(k)$ follows a power-law, *i.e.*

$$P(k) \sim k^{-\gamma}, \tag{1}$$

where γ is a real positive constant which has been found to take values typically in the range $2 < \gamma < 3$ [6]. This situation occurs when nodes are highly non-equivalent. Such networks have been named Scale-Free (SF hereafter) because a power-law has the property of having the same functional form at all scales. In fact, power-laws are the only functional forms $f(x)$ that remain unchanged, apart from multiplicative factors, under a rescaling of the independent variable x. They are the only solutions to the equation $f(\alpha x) = \beta f(x)$. SF-networks, having a highly inhomogeneous degree distribution, result in the simultaneous presence of a few nodes (the *hubs*) linked to many other nodes, and a large number of poorly connected elements (the *leaves*). Each of these network-classes occurs in specific cases; there are, however, other topological classes which will be referred to, in the following, when they will be eventually mentioned. Up to the eighties, the current opinion was that practically all networks representing real world structures (from social to technological networks) could be ascribed to the class of *Random* networks. After all, they were thought of as resulting from unsupervised growth processes and, as such, believed to be produced by a growth mechanism where new nodes stuck randomly to existing nodes (random-growth mechanisms). Relevant studies, at the end of the last decade, have shown the inadequacy of this scenario to represent the topological features of real networks: they have demonstrated that, although resulting from unsupervised growth processes, a large number of networks grow under the action of some *effective selective pressure* whose resulting effect is the realization of a structure more appropriately ascribed to the SF class [3].

From the knowledge of the network's graph, many different topological properties can be deduced which further specify the network's properties and characteristics. These data allow to design specific *growth mechanisms* able to design networks with desired topological properties. For comprehensive reviews on the proposed growth mechanisms to reproduce networks with different topological structures, the reader is referred to Refs. [3, 6].

4 Critical Infrastructures as Networks

In this work, we will attempt to analyze available data of several CIs by using the methods and the ideas of topology analysis. According to the definition of CIs given previously, the following technological infrastructures may be certainly ascribed to the CI set:

- Public power supply networks
- Telecommunication networks
- The Internet

In the following sections, we will apply the methods of graph analysis to the graphs resulting from the available data of the technological networks of the above mentioned CIs.

4.1 Public Power Supply Networks

The Power Grid

The public power supply network transmits power from generation to loads thereby providing the link between producers and consumers. The network connects large numbers of generators and loads together thus *(i)* improving the reliability of the power supply, *(ii)* reducing needs for reserve, peak, control and storage capacity, *(iii)* enabling more efficient and economic power production, and *(iv)* providing a necessary platform for the electricity market. The strengthening of the cross border transmission capacity has made the public power supply network increasingly international and spatially very extended. The power supply network is an essential, but often very international part of the national critical infrastructure.

The power supply network is hierarchically organized to transmission and distribution networks. Transmission networks cover very wide geographical areas, and have typically very high voltage levels and large power flows. Distribution networks, on the other hand, connect the loads and distributed generation with the transmission network. The distances are traditionally shorter and the voltage level lower than in the transmission network. Distribution networks are normally organized in a radial way, although redundancy is provided by a meshed network topology. Low voltage customers are connected to the distribution network via low voltage distribution networks.

About one third of the cost of power supply comes from the distribution of power. The power distribution network has also a much higher impact on the reliability and power quality than the power transmission network. Failures on the transmission network are relatively rare, but their impact spreads over much wider areas than those occurring on distribution networks.

A power network is characterized by the fact that it has very little buffering storage capacity and the physical balance of supply and demand must be maintained, otherwise the power transmission system will collapse. The deregulated electricity market is an important tool for finding a cost efficient initial solution for this balancing problem. Operation of the power network is highly and increasingly dependent on protection, automation, information and communication systems.

Distributed generation, intermittent generation from renewable energy sources (*e.g.* wind energy), pressures to cost reduction and power quality improvements, ever bigger generation and transmission units are expected challenges to the power network. This calls for significant changes in the power distribution systems and their automation and operation, but the power distribution systems have much inertia. The required lifetime of the power network related investments is very long. Thus rapid fundamental changes are seldomly possible.

Topological Analysis

From the topological point of view, the graphs representing electrical transmission networks cannot be properly ascribed to *random* nor to SF networks. In fact, as it happens for networks whose structure is constrained (*i.e.* by geographical reasons) or that cannot present an arbitrarily large node's degree (as for roads, for instance, where there is a very low maximum degree), electrical networks have a Gaussian shape , with a heavy exponential tail that drops the values of the highest degrees to smaller numbers (for electrical transmission lines the maximum degree of a node is usually of the order of 10) [10, 11, 12].

The electrical network which has been widely studied in recent years, and which will also be the object of the present analysis, is the Italian high-voltage (380 kV) electrical transmission network (HVIET hereafter). A graph of HVIET, as deduced from publicly available data, is depicted in Fig. 2, and it consists of $N = 310$ nodes and $L = 361$ links (transmission lines). In fact there are different node types; generators (117), loads (139), and junctions (54), but a distinction between them will not be made in our analysis. Moreover, 14 (of the total 361) links are *double* (transmission) lines.

Several topological analysis have been performed on the HVIET network. One of the relevant properties of the network, allowing to classify the topology of the network, is constituted by the distribution of the node's *degree* k (the degree is the number of links connecting each node to its nearest neighbors).

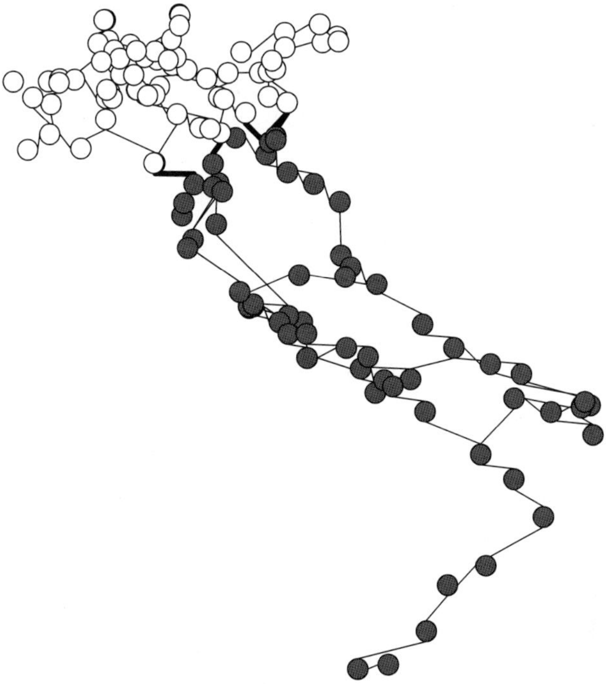

Fig. 2. Graph of the Italian high-voltage (380 kV) transmission network, where nodes are located at approximately correct geographical location. The 6 links located in the central region of Italy and represented by thick solid links correspond to the first critical section (min-cut selection) that divides the network into two almost equal-in-size parts (as indicted by the dark and white node symbols). Different node types (generators, loads and junctions) are not distinguished

The distribution of node's *degree* of HVIET is reported in Fig. 3 which confirms that HVIET does not show neither a clearcut SF nor a random character. The network has a limited number of hubs, whose maximum degree is $k_{max} = 11$. Another property which has been measured on the HVIET network is the average *clustering* coefficient[7] C, which measures the propensity of nodes to form small-scale communities (*c.f.* Refs. [3, 22] for additional details and formal definition). The clustering coefficient C is large when nodes, neighbors of a common node, are also neighbors of each other, *i.e.*, if node 1 is connected to node 2, and 2 to 3, then, if C is large, there is a relatively high probability that node 1 is also connected to node 3. Hence, we see that C measures in some sense the (relative) number of *connected* triangles in the network. In the HVIET network, the tendency to form connected triangles is rather small, and the clustering coefficient is as small as $C = 2.06 \times 10^{-2}$ (we will later see

[7] Notice that some authors refer to this same effect as network transitivity [22].

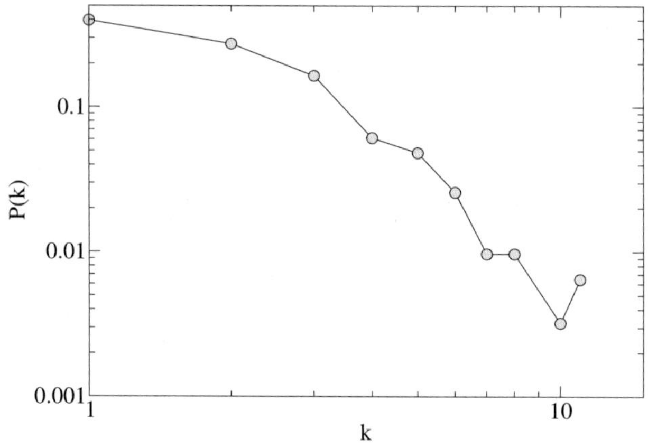

Fig. 3. The degree distribution, $P(k)$ *vs.* node degree, k, (in log-log scale) for the HVIET network depicted in Fig. 2

in Sec. 4.2 that, for instance, the clustering in the backbone of the Internet can be orders of magnitude higher).

An interesting result on HVIET has been evaluated by using the min-cut theorem associated with the spectral analysis of the so-called *Laplacian* $\mathcal{L}$. In order to define this matrix, let us start by introducing the adjacency matrix, $\mathbf{A}$, who's matrix elements, A_{ij}, take the value 1 if node i and j are connected, and 0 otherwise [22]. Then, in terms of $\mathbf{A}$ the (symmetric) Laplacian matrix is defined according to

$$\mathcal{L}_{ij} = \begin{cases} \sum_{k=1}^{N} A_{ik}, & \text{if } i = j \\ -A_{ij}, & \text{if } i \neq j \end{cases} \tag{2}$$

An interesting result that can be obtained from the spectral analysis of $\mathcal{L}$ can be stated as follows: The signs of the components of the eigenvector associated with the first non-vanishing eigenvalue of the Laplacian allow to optimally bisecate the network. As $\mathcal{L}$ is symmetric, the first eigenvalue is always vanishing. The n components of the eigenvector $\mathbf{v}_2^{\mathcal{L}} = (v_1, v_2, ..., v_n)$ associated with the second eigenvalue, solve the one-dimensional *quadratic placement* problem of minimizing the function

$$z = \frac{1}{2} \sum_{i=1}^{n} \sum_{j=1}^{n} (v_i - v_j)^2 A_{ij}. \tag{3}$$

The vector is subjected to the constraint $|\mathbf{v}| = (\mathbf{v}^T \mathbf{v})^{\frac{1}{2}} = 1$ [17]. The process allows to partition the graph $G = (N, L)$ into disjoint subsets G_1

and G_2 such that $L_{12}/(N_1 \cdot N_2)$ is minimized, where L_{12} is the number of links to be removed and N_1 and N_2 are the number of nodes in the two resulting subnetworks. It comes clear that this procedure allows for the "optimal" bisectioning of the graph, *i.e.* it forms the closest possible subnetworks G_1 and G_2 with the minimum amount of broken links L_{12}. If one applies the min-cut procedure, one gets the following bisection: the HVIET network is divided into two, connected, parts HVIET_1 and HVIET_2. The first is formed by $N(\text{HVIET}_1) = 195$, and the second by $N(\text{HVIET}_2) = 115$ nodes. The two parts are separated by *only* six links; The removal of these links allows to totally bisecate the network, which would separate it into two, non-communicating, parts (see Fig. 2). The ideal line, joining the location of the removed links, has been called *first critical section.* Indeed, there are several other lines of cut of the network, grouping a set of links whose removal produces a bisection of the network. These sets, although being composed (in some cases) by a lower number of links, do not minimize the function of Eq. (3) and, ultimately, are less efficient in separating the graph into two (almost) equal parts. This is a major outcome of the spectral analysis; this provides a way to locate the critical vulnerability lines of the network. The procedure can be iterated on the different components of the graphs, by creating *critical sections* of higher orders. If on a simple graph, the min-cut procedure can be almost done by visual inspection, for larger graphs it cannot be simply performed by any other mean.

Relevant information on the *robustness* of a network can be gained by simulations. Starting from the graph structure, for instance, one can evaluate what is the probability of *physically* disconnecting one (or more) nodes by disconnecting one (or more) lines. This will produce a qualitative evaluation of the *structural* robustness of the network. The knowledge of further technical details on the network (*i.e.* the electrical characteristics of lines) allows the formulation of a *dynamical* model for the power transport on the network. A recent work [13] has attempted to reproduce the flow on the HVIET produced by a given gauge of injected/extracted electrical power by/from the different nodes and by the real electrical admittance of the different electrical lines. The availability of such information opens the way to evaluate the so-called *flow* vulnerability. If one eliminates a given number of lines, the dynamical model allows to evaluate the new flux distribution; in case of overtaking given threshold of maximum flux on the lines, the flow equations are re-evaluated by starting from a different gauge of injected/extracted electrical power. When relevant lines are missed, the network must undergo a severe reduction of the injected power in order to be able to correctly sustain the power flux. If one associates the amount of power reduction to re-establish the flux to the specific removed line, one can classify the different lines as a function of the damage that their absence produce to the whole network. If applied to HVIET, this procedure allows to obtain a classification of the lines as a function of the damage which their absence is able to produce (which can be as large as 1.5GW, see Fig. 4).

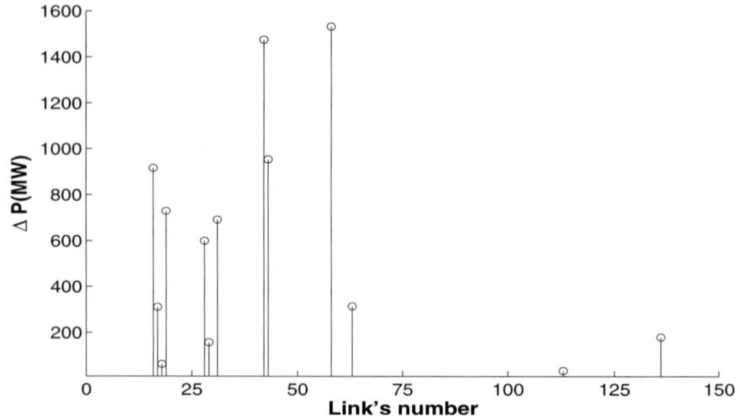

Fig. 4. Lines of HVIET whose removal is associated to the largest injected power reduction: The illustration shows on the abscissa the number of the power line, on the ordinate the amount of injected power (in MW) to be reduced to re-establish a correct power flux in the network

4.2 The Internet

Organizational Issues

The Internet should be known (and appreciated) by all of us, and therefore probably does not need any further introduction. In the following, by the term "the Internet" we will be referring to the network formed by the so-called *Autonomous System* (AS) router level [14]. An AS is a collection of IP networks and routers under the control of one entity (or sometimes more) that presents a common routing policy to the Internet. Therefore any sub-network appears as an AS, and the important difference between *Intra-AS* routing and *Inter-AS* routing must be introduced. The entity that controls an AS can choose the routing protocol to be used inside it, so in general AS can use different routing protocols. But in order to make interconnectivity between AS possible, each AS must employ one or more routers to interface with the "outer world", in order to informing it of the AS presence and topology. Usually there are specifically designated routers dedicated to accomplish this task — the so called *Border Routers.* Clearly these routers must adhere to the Internet rules and protocol set (explained further on). Thus, the AS-level routers form the backbone of the Internet which speaks the same language (*i.e.* adhere to the same protocol).

Since the first Internet connection was made on June 6, 1969, its size and complexity has grown dramatically. A recent paper examined the growth rate for nodes (g_n) and links (g_l) of such a network during the end of last decade [10], and it was found that $g_n \sim 140$ nodes/month and $g_l \sim 300$

links/month. It is worth recalling that a new AS-level router corresponds to the introduction of a new subnetwork which can also contain thousands of (internal) nodes.

Topological Data

To get *accurate* data on the topology of the Internet is difficult. In fact, the Internet should be measured "from its inside", since no one has the complete, up-to-date map of it. This need has prompted a number of large-scale projects aimed at "mapping" the Internet in the most accurate way. Examples of such projects are the DIMES [15] and the RouteViews Projects [16].

Data which will be referred to in the present work have been collected from the DIMES project funded by the EU. They refer to a snapshot of the map taken at a given date (July 2005). A repository of several snapshots, collected at different times, is also contained in the projects web site. These are useful in order to monitor the growth of the network (or, at least, its time variation); These data could be used to infer growth mechanisms underlying the time variation of the network's properties (size, degree, clustering *etc.*) [10].

Other less accurate data sets of the Internet, but covering a larger geographical region, can be found in *e.g.* Ref. [28].

Topological Analysis, and Network Growth Models

Topological analysis to evaluate the major topological properties have been performed on the DIMES network and, for comparison, on similar data taken from the repository of a US-funded project (RouteViews). Data collected on the two sets of data (DIMES and RouteViews) are reported in Table 1.

Several points of Table 1 need to be highlighted. First of all the small characteristic path length $\langle d_{ij} \rangle$ (which should be compared to its predicted value for a *random* network of a similar dimension $\langle d_{ij}^{rand} \rangle$ ($\langle d_{ij}^{rand} \rangle = \log N / \log \langle k \rangle$). This is a quite controversial point in the literature. Standing on their analysis, some authors have claimed an Internet characteristic

Table 1. Relevant topological properties of the DIMES and the RouteViews Internet network data. N denotes the number of nodes of the network, L the number of links, γ the exponent of the degree distribution (see Eq. (1)), C the clustering coefficient, k_{max} the maximum degree of a node (the largest hub of the network), $\langle k \rangle$ the average degree, d the diameter of the network (the largest inter-node distance) and $\langle d \rangle$ the average node's separation, $\langle d_{ij}^{rand} \rangle$ the average node's separation of a *random* network of equal N and $\langle k \rangle$

Data set	N	L	γ	C	k_{max}	d	$\langle d_{ij} \rangle$	$\langle d_{ij}^{rand} \rangle$
DIMES	14154	38928	2.41	0.41	1932	9	3.343	5.606
ROUTEVIEWS	11461	32730	2.35	0.35	2432	9	3.565	5.712

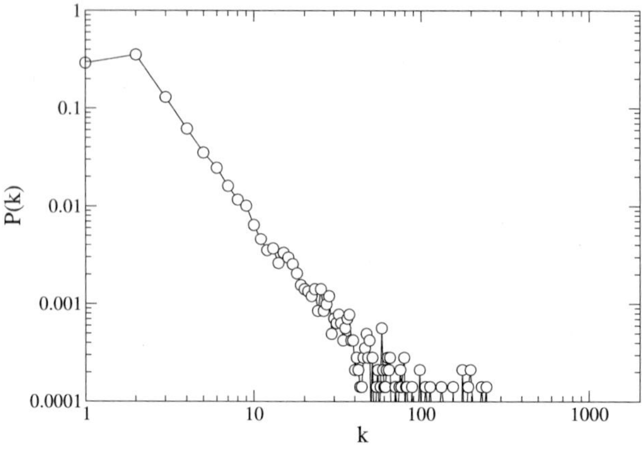

Fig. 5. The degree distribution, $P(k)$ *vs.* node degree, k, (in log-log scale) for the DIMES data

path length higher than that predicted for a random networks [18], others have measured a slightly lower diameter [3]. Then the large *clustering coefficient*, C, of the network which measures the propensity of nodes to form small-scale communities (Refs. [3, 22]). SF networks *do* not necessarily have large C values. Thus is a peculiar feature of the Internet network and of many social networks [3]; other Critical Infrastructures (such as Power grids) do not share this feature.

DIMES shows, as expected, a distribution of node's degree which properly fits a power-law with exponent $\gamma = 2.41$ (Fig. 5).

In order to summarize the observations made concerning the node's degree distribution and of the large *clustering* coefficient, we have attempted to define an "empirical" growth mechanisms allowing to reproduce the Internet topology. We succeded in this task by using a suitable combination of the Preferential Attachment (PA) [3] and the Triad Formation (TF) [9] mechanisms, the first allowing a SF network to be produced, the second able to account for an arbitrarily high value of the *clustering* coefficient. If, in a growth mechanisms where, starting from an initial set of n nodes, we wish to add a new node, we define that $P_{(n+1)\to j}$ is the probability that the new node $(n+1)$ sticks on the node j belonging to the network, it will be as follows:

$$P_{(n+1)\to j} = (1-\beta)PA^{\alpha} + \beta TF. \tag{4}$$

It means that the new node will stick with a probability $(1-\beta)$ with a modified PA algorithm (indicated as PA$^{\alpha}$ or with probability β with a TF mechanism).

The value of the parameters providing the best agreement with the DIMES data set are: $\alpha = 1.44$ and $\beta = 0.93$.

Our speculations follow a previous attempt made on the issue of modeling the Internet's large-scale topology [19]. The authors pointed on a modification of the PA mechanisms by introducing a further dependence on the *distance* among nodes: highly connected nodes are favoured if geographically close. With this assumption, links to far away nodes are discouraged while clustering is favored because node's proximity tends to enhance the establishment of links particularly among neighboring nodes.

The Random Walk Approach

In the previous sub-subsection, we saw that one could characterize the "clusteredness" of a network by, *e.g.*, the clustering coefficient C. However, given a network topology, how can one identify the nodes belonging to the same cluster? For large networks, like the Internet, this is a highly non-trivial (and often computationally daunting) task. Recently, several dedicated numerical algorithms have been proposed with this purpose in mind [3, 6, 20, 21, 22, 23, 24, 25, 29]. Here we do not intent to present a full overview of such *clustering-algorithms*, but instead outline a particular approach based on diffusion or random walkers.

To motivate this algorithm in simple terms, let us consider the following mental image; Assume the (very hypothetical) scenario that a car driver is located randomly somewhere in North-America, without the ability to gain information about direction from traffic signs, maps *etc.* Whenever he approaches a cross road, he randomly picks (with equal probability) one of the possible connecting roads. In this way, the driver randomly moves around on the road network without being assisted by any directional information that we all are so used to benefiting from. If the main aim of our "random driver" is to reach a given destination in, say, South America, you can probably easily guess, that the drivers strategy is far from being optimal. The driver will most probably find himself driving around in North America for a very long time, simply because there are relatively few roads "connecting" North and South America. In other words, the random driver will spend most of his time in the "northern" cluster where he started off. There is only a small probability that he will find his way through the bottleneck, here represented by Central America.

If there is not only one (random) driver, but instead a large number of them, one may ask for the relative fraction of drivers being at a particular node i at time t. This fraction, or density, is simply $\rho_i(t) = N_i(t)/N$ where $N_i(t)$ is the number of drivers at node i at time t, and N is the total number of drivers. If the system is evolving according to the random dynamics outlined above, one may suspect that the walker density in highly connected regions of the network, *i.e.* within a cluster (if any), will reach an almost constant value much faster than in not so highly connected regions of the network. It was this suspicious that, in the first place, lead us to consider it as a candidate for a clustering detection algorithm.

Given the underlying network topology, the process of the random drivers (or walkers) can easily be formulated mathematically, and the suggestions made above can be confirmed within a solid framework. The process is mathematically described by the "diffusion-like" equation [23, 24, 25]:

$$\partial_t \boldsymbol{\rho}(t+1) = \mathbf{D}\boldsymbol{\rho}(t) \tag{5}$$

where $\boldsymbol{\rho}(t)$ is the density vector of walkers at time t, and $\mathbf{D}$ a matrix that can be called the *diffusion matrix* (operator) for the system. This matrix is related to the adjacency matrix A_{ij} in the following way $D_{ij} = A_{ij}/k_j - \delta_{ij}$, where k_j refers to the degree of node j, and δ_{ij} is the Kronecker delta function. Notice that $\mathbf{D}$ is non-symmetric, unlike the adjacency and Laplacian matrices (2). The solution to Eq. (5) should be readily obtained as the linear combination of $\mathbf{v}^{(\alpha)} exp(-\lambda^{(\alpha)} t)$ where $\mathbf{v}^{(\alpha)}$ and $\lambda^{(\alpha)}$ are corresponding pairs of eigenvectors and eigenvalues, respectively, of the diffusion matrix. The index α is used to label the ordered sequence of eigenvalues so that $\alpha = 1$ corresponds to the largest one (that can be shown to be exactly one), $\alpha = 2$ to the next-to-largest one, and so on. Hence one realizes that the terms corresponding to increasing α's (where $\lambda^{(\alpha)} > 0$) correspond to faster-and-faster decaying modes of the system. The interpretation of this observation is that the largest α's different from one (the slowly decaying modes), can be related to the large scale topological features of the network. This has been demonstrated in recent publications [23, 24, 25] by plotting *e.g.* the *current of walkers*, $c_i^{(\alpha)} = \rho_i^{(\alpha)}/k_i$, leaving node i for an increasing number of modes α. For a given (small) mode $\alpha \neq 1$, the signs of the corresponding currents, $c_i^{(\alpha)}$, indicate a partitioning (into two parts) that may, or may not, correspond to a well-defined module or cluster for the network. To determine if a given partitioning can be characterized as a module we have used the so-called *modularity* measure. It is defined, given a (predefined) partition, as essentially the total number of links falling within modules minus the expected number of links for an equivalent network where links are placed at random [31, 32, 33, 30].

If the modularity for a given partitioning is large, one says that the partitioning represents a "good" modular structure, otherwise not. By repeating this process for higher and higher (diffusive) modes, α, a rather rich community structure can be identified (*cf.* Ref. [25] for additional details).

We will now analyze the topology of the Internet by this random walk current mapping technique. In the following we will consider an AS-data set obtained from Ref. [28]. It consists of about 6 500 nodes from various parts of the world. Fig. 6 shows the 2-dimensional current mapping of the network using the two slowest decaying diffusive modes, *i.e.* $\alpha = 2, 3$. All AS-systems have been labelled with black dots. Later all nodes from some selected nations have in addition been labeled differently for convenience. The star-like

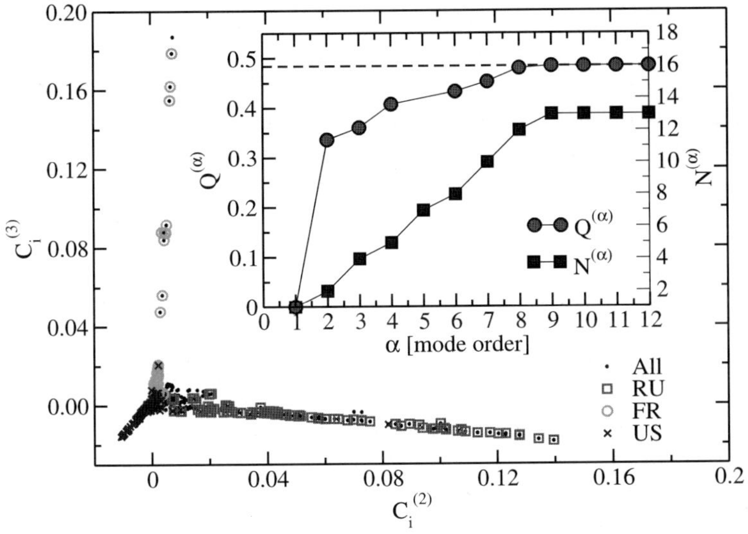

Fig. 6. The two-dimensional current mapping of an Autonomous System (AS) network [23, 28]. The symbols refer to the geographical location of the AS: Russia (□), France (○), USA (×). The inset shows the modularity, $Q^{(\alpha)}$, and number of detected communities, $N^{(\alpha)}$, for the optimal partitioning of the network at a given diffusive mode. (After Ref. [25])

structure indicates that there is a hierarchy of vertices where those located the furthest away from the origin of the current plot are the most peripheral vertices of the network. Furthermore, each hierarchy corresponds roughly to the national division of the AS network. Fig. 6 shows that the three legs of the star-structure correspond to Russia, the US and France. For the AS-network we identified 13 communities resulting in a modularity of about one-half (inset to Fig. 6).

This analysis indicates that the extreme edges of the Internet corresponds to US (blue crosses in Fig. 6) and Russian (red squares) AS. A closer investigation into the location and purpose of some of there "extreme systems" revealed that the "extreme" US sites corresponds to US military South Pacific (*i.e.* Hawaii). These systems, among all the American ones, are the least connected to the other systems of our analyzed data set.

Another peculiar observation to be made from Fig. 6 is the single American node located in the "Russian sector". It turned out that this node belonged to the Russian branch of the (American) car-maker Ford. This alone would probably not make it "Russian" since one still would expect that much of the routed traffic would be to other branches of Ford located outside Russia. According to the contact person for this AS, this system had, during the time period covered by the data set, been taken over by hackers. Our analysis therefore suggests that substantial parts of the traffic for this system therefore went over other Russian AS. This explains (partly) why this system, from the routing information alone, was classified alongside other Russian systems.

By applying the mapping technique in higher and higher dimensions, a more and more detailed clustering structure could be found. This is illustrated by the inset to Fig. 6 from which one sees that a total of 13 communities, mainly corresponding to a national location of the nodes, could be identified from the analyzed data set.

5 Conclusions and Outlook

Large Complex Critical Infrastructures (LCCIs) are national — or international — technological systems whose correct functioning is highly impacting our contemporary societies and the well-beings of their citizens. It is therefore essential that such systems have the highest level of protection and security. In order to make sure that this is the case, one is required to fully understand, and therefore analyze, their behaviors.

The science of Complex Systems can help in this regard. It is today widely recognized that most systems that grow in a not centrally optimized manner, including most LCCIs, do display common features. This might pave the way for a number of approaches aiming to migrate analysis and control systems tools from one system to another. Topological analysis of the graphs described by the (physical and logical) complexion of the infrastructural network might help in designing new networks and to increase their actual efficiency Functional models describing basic features of LCCIs can also disclose a number of complex phenomena taking place under appropriate conditions.

A major problem encountered in critical infrastructures is their interdependency. This defines the degree of correlation existing between two (or more) systems, allowing a perturbation acting on one of them to produce sizeable effects in others. Interdependency is the factor which produces large cascade effects when, for instance, electrical blackouts take place. Complexity Science can help in analyzing and understand these phenomena and in describing network interdependency.

Definitely intriguing is the possibility opened by the adaptation of biological control (and optimization) strategies to technological environments (*bio-mimetic* strategies); This and other recent fields are expected to foster a number of significant advancements in the field of infrastructures management.

Acknowledgments

The authors acknowledge all the partners of the EU-funded project IRRIIS (Integrated Risk Reduction of Information-based Infrastructure Systems) which have provided the starting point and the principal receptor of this analysis. In particular, the authors are indebted to Sandro Bologna, Ester Ciancamerla, Pekka Koponen, Gwendal Le Grand, Michele Minichino, Peter Popov, and Kizito O. Salato.

References

1. US-Canada Power System Outage Task Force, "Final Report of the August 14th Blackout in US and Canada", United States Department of Energy and National Resources Canada, April 2004.
2. Watts, D.J. and Strogatz, S.H. Nature **393** (1998) 440
3. Albert, R. and Barabasi, A.–L.: Statistical mechanics of complex networks. Rev. Mod. Phys. **74** (2002) 45
4. Schmitz, W.: Private Communication. (2006)
5. Milgram, S.: The small world problem. Psychology Today (1957) 60
6. Boccaletti S. et al.: Complex networks: Structure and dynamics. Phys. Rep. **424** (2006) 175
7. Dorogovtsev, S.N. and Mendes, J.F.F.: Evolution of networks. Adv. Phys. **51** (2002) 1079
8. Albert, R., Jeong, H. and Barabasi, A.–L.: Error and attack tolerance of complex networks. Nature **406** (2000) 378
9. Holme, P. and Kim, B.J. Grwoing scale-free networks with tunable clustering. Phys. Rev. E **65** (2002) 026107
10. Rosato, V., and Tiriticco, F. Growth mechanism of the as-level internet network. Europhys. Lett. **66** (2004) 471
11. Amaral, L.A.N., Scale, A., Barthelemy, M. and Stanley, H.E.: Classes of small-world networks. Proc. Natl. Acad. Sci. **97** (2000) 11149
12. Bakke, J.O.H., Hansen, A. and Kertész, J.: Failures and avalanches in complex networks. Europhys. Lett. **66** (2004) 471
13. Rosato, V., Issacharoff, L., Bologna, S.: Influence of the topology on the power flux of the italian high-voltage electrical network. Submitted to ENEA Technical Report CAMO/2007/1(editorial) (2006)
14. Pastor-Satorras, R. and Vespignani, A. Evolution and structure of the Internet: A stasticial physics approach. Cambridge University Press, Cambridge (2004)
15. DIMES is a EU–funded project for the worldwide mapping of the Internet (see *www.netdimes.org*).
16. RouteViews is the US–funded project for the worldwide mapping of the Internet (see *www.routeviews.org*).
17. Kahng, A.B. and Hagen L.: New spectral methods for ratio cut partioning and clustering. IEEE T. Compt. Aid. D. **11** (1992) 1074
18. Faloutsos, M. et al.: Proc. ACM SIGCOMM. Comput. Commun. Rev. **29** (1999) 251
19. Barabasi, A.–L., Yook, S.–H. and Jeong, H.: Modeling the internet's large-scale topology. Proc. Natl. Acad. Sci. USA **99** (2002) 13382
20. Kleinberg, J.: Authorative sources in a hyperlinked environment. J. ACM. **45** (1999) 604
21. Gibson, D., Kleinberg, J. and Raghavan, P.: Inferring web communities from link topology. In: Proc. 9th ACM Conference on Hypertext and Hypermedia (1998)

22. Newman, M.E.J.: The structure and function of complex networks. SIAM Rev. **45** (2003) 167
23. Eriksen, K.A., Simonsen, I., Maslov, S. and Sneppen, K.: Modularity and extreme egdes of the internet. Phys. Rev. Lett. **90** (2003) 147801
24. Simonsen, I., Eriksen, K.A., Maslov, S. and Sneppen, K.: Diffusion on complex networks: A way to probe their large scale topological structures. Physica A **336** (2003) 167
25. Simonsen, I.: Diffusion and networks: A powerful combination. Physica A **357** (2005) 317
26. Source: http://mit.edu/vdb/www/6.977/l-blondel.pdf
27. R. Weron and I. Simonsen, *Blackouts, risk, and fat-tailed distributions.* In H. Takayasu, editor, Practical Fruits of Econophysics. Springer Verlag, 215–219, 2006. (arXiv:physics/0510077)
28. The data set can be obtained from http://moat.nlanr.net/AS/.
29. Girvan, M. and Newman, M.E.J.: Community structure in social and biological networks. Proc. Natl. Acad. Sci. USA **99** (2002) 7821
30. Danon, L., Diasz-Guilera, A. and Arenas, A.: Effect of size heterogeneity on community identification in complex networks. J. Stat. Mech. (2006) 11010
31. Newman, M.E.J. and Girvan, M.: Finding and evaluating community structure in networks. Phys. Rev. E **69** (2004) 026113
32. Newman, M.E.J.: Analysis of weighted networks. Phys. Rev. E **70** (2004) 056131
33. Newman, M.E.J.: Modularity and community structure in networks. Proc. Natl. Acad. Sci. **103** (2006) 8577

Part V

Information Systems

Bootstrapping the Long Tail in Peer to Peer Systems

Bernardo A. Huberman[1], Fang Wu[2]

[1] HP Labs, Palo Alto, CA 94304 bernardo.huberman@hp.com
[2] HP Labs, Palo Alto, CA 94304 fang.wu@hp.com

1 Introduction

The provision of digitized content on-demand to millions of users presents a formidable challenge. With an ever increasing number of fixed and mobile devices with video capabilities, and a growing consumer base with different preferences, there is a need for a scalable and adaptive way of delivering a diverse set of files in real time to a worldwide consumer base.

Providing such varied content presents two problems. First, files should be accessible in such a way that the constraints posed by bandwidth and the diversity of demand is met without having to resort to client server architectures and specialized network protocols. Second, as new content is created, the system ought to be able to swiftly respond to new demand on specific content, regardless of its popularity. This is a hard constraint on any distributed system, since providers with a finite amount of memory and bandwidth will tend to offer the most popular content, as is the case today with many peer-to-peer systems.

The first problem is naturally solved by peer to peer networks, where each peer can be both a consumer and provider of the service. Peer to peer networks, unlike client server architectures, automatically scale in size as demand fluctuates, as well as being able to adapt to system failures. Examples of such systems are Bittorrent [4] and Kazaa, who account for a sizable percentage of all the use of the Internet. Furthermore, new services like the BBC IMP, (`http://www.bbc.co.uk/imp/`) show that it is possible to make media content available through a peer-to-peer system while respecting digital rights.

It is the second problem, that of an adaptable and efficient system capable of delivering any file, regardless of its popularity, that we now solve. We do so by creating an implementable incentive mechanism that ensures the existence of a diverse set of offerings which is in equilibrium with the available supply and demand, regardless of content and size. Moreover, the mechanism is such that it automatically generates the long tail of offerings which has been shown to be responsible for the success of a number of online businesses such as Amazon or eBay [2]. In other words, while the system delivers favorite mainstream content, it can also provide files that constitute small niche markets which only in the aggregate can generate large revenues.

In what follows we describe an efficient incentive mechanism for P2P systems that generates a wide diversity of content offerings while responding adaptively to customer demand. Files are served and paid for through a parimutuel market similar to that commonly used for betting in horse races. An analysis of the performance of such a system shows that there exists an equilibrium with a long tail in the distribution of content offerings, which guarantees the real time provision of any content regardless of its popularity. In our case, the bandwidth fraction of a given file offered by a server plays the role of the odds, the bandwidth consumed corresponds to bettors, the files to horses, and the requests are analogous to races.

An interesting consequence of this mechanism is that it solves in complete fashion the free riding problem that originally plagued P2P systems like Gnutella [1] and that in milder forms still appears in other such systems. The reason being that it transforms the provision of content from a public good into a private one.

We then analyze the performance of such a system by making a set of assumptions that are first restrictive and are then relaxed so as to make them correspond to a realistic crowd of users. We show that in all these cases there exists an equilibrium in which the demand for any file can be fulfilled by the system. Moreover this equilibrium exhibits a robust empirical anomaly which is responsible for generating a very long tail in the distribution of content offerings. We finally discuss the scenario where most of the servers are bounded rational and show that it is still possible to achieve an optimum equilibrium. We conclude by summarizing our results and discussing the feasibility of its implementation.

2 The System and its Incentive Mechanism

Consider a network-based file exchange system consisting of three types of traders: content provider, server, and downloader or user. A content provider supplies—at a fixed price per file—a repertoire of files to a number of people acting as peers or servers. Servers then selectively serve a subset of those files to downloaders for a given price. In a peer-to-peer system a downloader can also, and often does, act as a server.

If the files are typically large in size, a server can only afford to store and serve a relatively small subset of files. It then faces the natural problem of choosing an optimal (from the point of view of maximizing his utility) subset of files to store so as to sell them to downloaders.

Suppose that the system charges each downloader a flat fee for downloading any one file (as in Apple's iTunes music store), which we normalize to one. Since many servers can help distribute a single file, this unit of income has to be allocated to the servers in ways that will incentivize them to always respond to a changing demand.

In order to do so, consider the case where there are m servers and n files. Let b_{ij} be the effective bandwidth of server i serving file j, normalized to $\sum_{i,j} b_{ij} = 1$. Also, denote the bandwidth fraction of file j by $\pi_j = \sum_k b_{kj}$.

Suppose that when a downloader connects to the system, it starts downloading different parts of the file simultaneously from all available servers that have it. When it finishes downloading, it will have received a fraction of the file j

$$q_{ij} = \frac{b_{ij}}{\sum_k b_{kj}} = \frac{b_{ij}}{\pi_j} \qquad (1)$$

from server i. Our mechanism prescribes that *the system should pay an amount q_{ij} to server i as its reward for serving file j.*

Now consider the case when server i's reserves an amount of bandwidth b_{ij} as his "bid" on file j. Because we have normalized the total bandwidth and the total reward for serving one request both to one, the proportional share allocation scheme described by Eq. (1) can be interpreted as redistributing the total bid to the "winners", in proportion to their bids. Thus our payoff structure is similar to that of a pari-mutuel horse race betting market, where the π_j can be regarded as the odds, the bandwidth corresponds to bettors, the files to horses, and the requests are analogous to races (Fig. 1). It is worth pointing out however, that in a real horse race all players who have placed

Fig. 1. A parimutuel horse betting board

a bet on the winning horse receive a share of the total prize, whereas in our system only those players that kept the "winning" file and also had a chance to serve it get paid. In spite of this difference it is easy to show that when rewritten in terms of expected payoffs, the two mechanisms behave in similar fashion.

3 The solution

3.1 Rational Servers with Static Strategies and Known Download Rates

In this section we make three simplifying assumptions. While not realistic they serve to set the framework that we will utilize later on to deal with more realistic scenarios. First, every server is rational in the sense that he chooses the optimal bandwidth allocation that maximizes his utility, whose explicit form will be given below. Second, every server's allocation strategy is static, i.e. the b_{ij}'s are independent of time. Third, we assume that each file j is requested randomly at a rate $\lambda_j > 0$ that does not change with time, and these rates are known to every server.

Consider a server i with the following standard additive form of utility:

$$U = \mathbb{E}\left[\int_0^\infty e^{-\delta t} u(t) dt\right], \tag{2}$$

where $u(t)$ is his income density at time t, and $\delta > 0$ is his future discount factor. Let X_{j1} be the (random) time that file j is requested for the first time, let X_{j2} be the time elapsed between the first request and the second request, and so on. According to our parimutuel reward scheme, server i receives a lump-sum reward b_{ij}/π_j from every such request, at times $X_{j1}, X_{j1} + X_{j2}$, etc. Thus, the server i's total utility is given by

$$U = \sum_j \frac{b_{ij}}{\pi_j} \sum_{l=1}^{\infty} \mathbb{E}[e^{-\delta \sum_{k=1}^{l} X_{jk}}] \equiv \sum_j \frac{b_{ij}}{\pi_j} u_j. \tag{3}$$

which amounts to each server receiving a utility proportional to the fraction of the file that he serves. Notice that the sum of expectations in Eq. (3) (denoted by u_j) can be calculated explicitly. Because the X_{jk}'s are i.i.d. random variables with density $\lambda_j^{-1} \exp(\lambda_j x)$, it can be calculated that $u_j = \lambda_j/\delta$. If we let $\lambda = \sum_j \lambda_j$ be the total request rate and $p_j = \lambda_j/\lambda$ be the probability that the next request asks for file j, then we can also write $u_j = \lambda p_j/\delta$. Plugging this back into Eq. (3), we obtain

$$U = \frac{\lambda}{\delta} \sum_j \frac{p_j b_{ij}}{\pi_j}. \tag{4}$$

Since we assume that server i is rational, he will allocate b_{ij} in a way that it solves the following optimization problem:

$$\max_{(b_{ij})_{j=1}^{n}\in\mathbb{R}_{+}^{n}} \sum_j \frac{p_j b_{ij}}{\sum_k b_{kj}} \quad \text{subject to} \sum_j b_{ij} \leq b_i. \tag{5}$$

where b_i is the total upload bandwidth of user i. Thus we see that the servers are playing a *finite budget resource allocation game.* This type of game has been studied intensively, and a Nash equilibrium has been shown to exist under mild assumptions [8, 11]. In such an equilibrium, the players' utility functions are strongly competitive and in spite of a possibly large utility gap, the players behave in almost envy-free fashion, i.e. each player believes that no other player has received more than they have.

3.2 Rational Servers with Static Strategies and Unknown Request Rates

We now relax some of the assumptions made above so as to deal with a more realistic case.

It is usually hard to find out the accurate request rate for a given file, especially at the early stages when there is no historical data available. Thus it makes more sense to assume that every server i holds a *subjective belief* about those request rates. Let p_{ij} be server i's subjective probability that the next request is for file j. Then server i believes that file j will be requested at a rate $\lambda_{ij} = \lambda p_{ij}$. Eq. (5) then becomes

$$\max_{(b_{ij})_{j=1}^{n}\in\mathbb{R}_{+}^{n}} \sum_j \frac{p_{ij} b_{ij}}{\sum_k b_{kj}} \quad \text{subject to} \sum_j b_{ij} \leq b_i. \tag{6}$$

which is still a finite budget resource allocation game as considered in the previous section.

It is interesting to note that when m is large, b_{ij} is small compared to $\pi_j = \sum_k b_{kj}$, so that π_j can be treated as a constant. In this case, the optimization problem can be well approximated by

$$\max_{(b_{ij})_{j=1}^{n}\in\mathbb{R}_{+}^{n}} \sum_j \frac{p_{ij} b_{ij}}{\pi_j} \quad \text{subject to} \sum_j b_{ij} \leq b_i. \tag{7}$$

Thus, user i should use all his bandwidth to serve those files j with the largest ratio p_{ij}/π_j.

This scenario (7) corresponds to the so-called *parimutuel consensus* problem, which has been studied in detail. In this problem a certain probability space is observed by a number of individuals, each of which endows it with their own subjective probability distributions. The issue then is how to aggregate those subjective probabilities in such a way that they represent a good consensus of the individual ones. The parimutuel consensus scheme is similar to that of betting on horses at a race, the final odds on a given horse being proportional to the amount bet on the horse. As shown by Eisenberg and Gale [6], an equilibrium then exists such that the bettors as a group maximize the weighted sum of logarithms of subjective expectations, with the weights being the total bet on each horse.

Moreover a number of empirical studies of parimutuel markets [5, 7, 9] have shown that they do indeed exhibit a high correlation between the subjective probabilities of the bettors and the objective probabilities generated by the racetracks. Equally interesting for our purposes is the existence of a robust empirical anomaly called *the favorite-longshot bias* [5, 7, 9]. The anomaly shows that favorites win more frequently than the subjective probabilities imply, and longshots less often. This anomaly enhances the long tail, which is populated by those files which while not singly popular, in aggregate are responsible for a large amount of the traffic in the system.

3.3 Rational Servers with a Dynamic Strategy

We now consider the case where the rate at which files are requested can change with time. Because of this, each server has to actively adjust its bandwidth allocation to adapt to such changes. As we have seen in the last section, user i has an incentive to serve those files with large values of p_{ij}/π_j. Recall that $\pi_j(t)$ is just the fraction of total bandwidth spent to serve file j at time t, which in principle can be estimated from the system's statistics. Thus it would be useful to have the system frequently broadcast the real-time π_j to all servers so as to help them decide on how to adjust their own allocations of bandwidth.

From Eq. (1) we see that, by serving file j, user i's expected per bandwidth earning from the next request is $p_j q_{ij}/b_{ij} = p_j/\pi_j$. Hence a user will benefit most by serving those files with the largest "p/π ratio". However, as soon as a given user starts serving file j, the corresponding p/π ratio decreases. As a consequence, the system self-adapts to the limit of uniform p/π ratios. If the system is perfectly efficient, we would expect that $p_j/\pi_j =$ constant. Because p_j and π_j both sum up to one, this implies that $\pi_j = p_j$, or $\sum_k b_{kj} = \lambda_j/\lambda \propto \lambda_j$. In other words, the total bandwidth used to serve a file is proportional to the file's request rate.

This result has interesting implications when considering the social utility of the downloaders. Recently, Tewari and Kleinrock [10] have shown that in a homogeneous network the average download time is minimized when $\sum_k b_{kj} \propto \lambda_j$. This implies that in the perfectly efficient limit, our mechanism maximizes the downloaders' social utility, which is measured by their average download times.

Since in reality a market is never perfectly efficient, the above analysis only makes sense if the characteristic time it takes for the system to relax back to uniformity from any disturbance is short. As a concrete example, consider a new file j released at time 0, being shared by only one server. Suppose that every downloader starts sharing her piece of the file immediately after downloading it. Because there are few servers serving the file but many downloaders requesting the file, for very short times afterwards the upload bandwidth will be fully utilized. That is, during time dt, an amount $\pi_j(t)dt$ of data is downloaded and added to the total upload bandwidth immediately. Hence we have

$$d\pi_j(t) = \pi_j(t)dt. \tag{8}$$

which implies that $\pi_j(t)$ grows exponentially until $\pi_j(T) \sim p_j$. Solving for T, we find

$$T \sim \log\left(\frac{p_j}{\pi_j(0)}\right). \tag{9}$$

Thus the system reaches uniformity in logarithmic time, a signature of its high efficiency.

3.4 Servers with Bounded Rationality

So far we have assumed that all servers are rational, so that they will actively seek those files that are most under-supplied so as to serve them to downloaders. In reality however, while some servers do behave rationally, a lot of others do not. This is because even a perfectly rational server sometimes can make wrong decisions as to which files to store because his subjective probability estimate of what is in demand can be inaccurate. Also, such a bounded-rational server can at times be too lazy to adjust his bandwidth allocation, so that he will keep serving whatever he has, and at other times he might simply imitate other servers' behavior by choosing to serve the popular files. In all these cases we need to consider whether or not the lack of full rationality will lead to equilibrium on the part of the system.

As a simple example, assume there are only two files, A and B. Let $p = \lambda_A/\lambda$ be file A's real request probability, and let $1-p$ be file B's real request probability. Suppose the servers are divided into two classes, with α fraction rational and $1-\alpha$ fraction irrational, arriving one by one in a random order. Each rational server's subjective probability in general can be described by an identically distributed random variable $P_t \in [0,1]$ with mean p. Then with probability $\mathbb{P}[P_t > \pi(t)]$ he will serve file A, and with probability $\mathbb{P}[P_t < \pi(t)]$ he will serve file B. In order to carry out some explicit calculation below, we consider the simplest choice of P_t, namely a Bernoulli variable

$$\mathbb{P}[P_t = 1] = p, \quad \mathbb{P}[P_t = 0] = 1-p. \tag{10}$$

(Clearly $\mathbb{E}[P_t] = p$, so the subjective probabilities are accurate on average.) It is easy to check that under this choice a rational server chooses A with probability p and B with probability $1-p$.

On the other hand, consider the situation where an irrational server chooses an existing server at random and copies that server's bandwidth allocation. That is, with probability $\pi(t)$ an irrational server will choose file A.[3]

From these two assumptions we see that

$$\mathbb{P}[\text{server } t \text{ serves A}] = \alpha p + (1-\alpha)\pi(t), \tag{11}$$

and

$$\mathbb{P}[\text{server } t \text{ serves B}] = \alpha(1-p) + (1-\alpha)(1-\pi(t)). \tag{12}$$

The stochastic process described by the above two equations has been recently studied in the context of choices among technologies for which evidence of their value is equivocal, inconclusive, or even nonexistent [3]. As was shown there, the dynamics generated by such equations leads to outcomes that appear to be deterministic in spite of being governed by a stochastic process. In the context of our problem this means that when the objective evidence for the choice of a particular file is very weak, any sample path of this process quickly settles down to a fraction of files downloaded that is not predetermined by the initial conditions: ex ante, every outcome is just as (un)likely as every other. Thus one cannot ensure an equilibrium that is both optimum and repeatable.

In the opposite case, when the objective evidence is strong, the process settles down to a value that is determined by the quality of the evidence. In both cases the proportion of files downloaded never settles into either zero or one.

In the general case that we have been considering, there are always a number of servers that will behave in bounded rational fashion and a few that are perfectly rational. Specifically, when $\alpha > 0$, which corresponds to the case where a small number of servers are rational, the $\pi(t)$ will converge to p in the long time limit. That is, a small fraction of rational servers is enough for the system to reach an optimum equilibrium. However, it is worth pointing out that since the characteristic convergence time diverges exponentially in $1/\alpha$, the smaller the value of alpha α, the longer it will take for the system to reach such an optimum state.

4 Conclusion

In this paper we described a peer-to-peer system with an incentive mechanism that generates diversity of offerings, efficiency and adaptability to customer

[3] This assumption can also be interpreted as follows. Suppose a downloader starts serving his files immediately after downloading, but never initiates to serve a file. (This is the way a non-seed peer behaves within Bittorrent.) Then the probability that he will serve file j is exactly the probability that he just downloaded file j, which is $\pi_j(t)$.

demand. This was accomplished by having a pricing structure for serving files that has the structure of a parimutuel market, similar to those commonly used in horse races, where the the bandwidth fraction of a given file offered by a server plays the role of the odds, the bandwidth corresponds to bettors, the files to horses, and the requests are analogous to races. Notice that this mechanism completely solves the free riding problem that originally plagued P2P systems like Gnutella and that in milder forms still appears in other such systems.

We then analyzed the performance of such a system by making a set of assumptions that are first restrictive but are then relaxed so as to make the system respond to a realistic crowd. We showed that in all these cases there exists an equilibrium in which the demand for any file can be fulfilled by the system. Moreover this equilibrium is known to exhibit a robust empirical anomaly, that of the *favorite-longshot bias*, which in our case generates a very long tail in the distribution of offerings. We finally discussed the scenario where most of the servers are bounded rational and showed that it is still possible to achieve an optimum equilibrium if a few servers can act rationally.

The implementation of mechanism is feasible with present technologies. The implementation of a prototype will also help study the behavior of both providers and users within the context of this parimutuel market. Given its feasibility, and with the addition of DRM and a payment system, it offers an interesting opportunity for the provision of legal content with a simple pricing structure that ensures that unusual content will always be available along with the more traditional fare.

Acknowledgements

We benefited from discussions with Eytan Adar, Tad Hogg and Li Zhang.

References

1. Eytan Adar and Bernardo A. Huberman. Free Riding on Gnutella. First Monday October (2000).
2. Chris Anderson. The long tail. `http://longtail.typepad.com/the_long_tail/` (2005).
3. Jonathon Bendor, Bernardo A. Huberman and Fang Wu. Management fads, pedagogies and soft technologies. `http://www.hpl.hp.com/research/idl/papers/fads/fads.pdf` (2005).
4. Bram Cohen. Incentives build robustness in Bittorrent. Working Paper, Workshop on the Economics of P2P Systems (2003).
5. L. Coleman. New light on the longshot bias. Applied Economics, 36(4), 315–326 (2004).
6. Edmund Eisenberg and David Gale. Consensus of subjective probabilities: The parimutuel method. Annals of Mathematics Statistics, 30(1), pp. 165–168 (1958).

7. M. Gramm and D. H. Owens. Efficiency in Pari-Mutuel betting markets across wagering pools in the simulcast era. Southern Economic Journal, 72(4), 926–937 (2006).
8. L. Shapley and M. Shubik. Trade using one commodity as a means of payment. Journal of Political Economy, Vol. 85:5, 937–968 (1977).
9. Richard H. Thaler and William T. Ziemba, Parimutuel betting markets: racetracks and lotteries, Journal of Economic Perspectives, Vol. 2, No. 2, pp. 161–174 (1988).
10. Saurabh Tewari and Leonard Kleinrock. On fairness, optimal download performance and proportional replication in peer-to-peer networks. Proceedings of IFIP Networking 2005, Waterloo, Canada (2005).
11. Li Zhang. The efficiency and fairness of a fixed budget resource allocation game. ICALP (2005).

Coping with Information Overload through Trust-Based Networks

Frank E. Walter, Stefano Battiston, Frank Schweitzer

Chair of Systems Design, ETH Zurich Kreuzplatz 5, 8032 Zurich, Switzerland
fewalter|sbattiston|fschweitzer@ethz.ch

1 Introduction

1.1 Motivation

Over the recent decade, the Internet has conquered people's homes and life: they pursue an increasing amount of activities on the World Wide Web and this has fundamentally impacted the lifestyle of society. For example, people use their computers for communication with others, to buy and sell products on-line, to search for information, and to carry out many more tasks. Along this development, so far unknown ways of marketing, trading and information sharing are booming. This situation is made possible by a set of related emerging technologies centred around the Internet – just to mention a few: collaborative work and information sharing environments, peer-to-peer networks , and rating, recommendation, and reputation systems. At the economic level, the impact of these technologies is already very high and it is expected to grow even more in the future. The Internet has become a social network, "linking people, organisations, and knowledge" [2] and it has taken the role of a platform on which people pursue an increasing amount of tasks that they have usually only done in the real-world. An approach looking at these emerging technologies and their effects from a complex systems perspective can, as we will show in this chapter, be very useful.

1.2 Emerging Technologies

In the following, we will look at the particular technologies already mentioned – collaborative work and information sharing environments, peer-to-peer

* The model discussed in this chapter is based on our paper of a trust-based recommendation system on a social network, see [1]. For more formal and detailed descriptions of the model, the analysis, and the simulations, please also refer to this paper. For further materials on our research in this area, please see our website www.sg.ethz.ch/research.

networks, and rating, recommendation, and reputation systems – in more detail. We will demonstrate that collaborative work and information sharing environments are tools to create vast amounts of globally available information; in addition, peer-to-peer networks help to quickly spread this information over large distances. This leads to the situation that people are confronted with an *information overload*; one possible solution to this problem lies, as we will demonstrate, in rating, recommendation, and reputation systems.

Collaborative Work and Information Sharing Environments

Collaborative work and information sharing environments have created platforms where people are able to share knowledge, tastes, bookmarks etc. An example of such a system would be `wikipedia.org`, a free on-line encyclopedia which can be accessed and edited by anyone on the web. Over the recent years, Wikipedia has grown manifold and now is considered a real challenge for the established encyclopedias available both on-line or as books. Wikipedia is an example of a whole range of websites that act as the platform for people making information available to others – there are many more, for example `delicious.com`, a repository for the bookmarks of people, `citeulike.org`, where people can make their bibliographies and literature lists available, `ohmynews.com`, which is an online newspaper with the motto "every citizen a reporter" and where anyone can contribute articles, and many, many more.

Peer-to-Peer Networks

At the same time, peer-to-peer networks have become very popular because they enable users to share information, typically digital content. Peer-to-peer networks are inherently distributed in the sense that they do not require a central server which coordinates clients but rather that nodes self-organise and adapt to change. This makes it very difficult to attack peer-to-peer networks (i.e., this includes attempts to take them off the network). Furthermore, they reflect the structure of social networks in the real life. The simplicity to duplicate and share digital content combined with the ineffective implementations of digital rights management platforms has caused some to suggest the revision of the notion of intellectual property. Several approaches can be thought of in the framework of these emerging technologies: for instance, a rating system of the digital content would allow to compensate authors based on the aggregate rating of the items that they offer. Nonetheless, the core feature of peer-to-peer networks is that they provide a medium to spread information without boundaries in space and time.

Information Overload

Now, the technologies mentioned so far – collaborative work and information sharing environments as well as peer-to-peer networks – confront people with an *information overload*: they are facing too much data to be able to effectively filter out the pieces of information that are most appropriate for them. The exponential growth of the Internet [3] implies that the amount of information accessible to people grows at a tremendous rate. Historically, people have – in various situations – already had to cope with information overload and they have intuitively applied a number of *social mechanisms* that help them deal with such situations. However, many of these, including the notion of trust, do not yet have an appropriate *digital mapping* [4]. Finding suitable representations for such concepts is a topic of on-going research [6, 5, 7, 8, 9] across disciplines.

Rating, Recommendation, and Reputation Systems

The problem of information overload has been in the focus of recent research in computer science and a number of solutions have been suggested. The use of search engines [10] is one approach, but so far, they lack personalisation and usually return the same result for everyone, even though any two people may have vastly different preferences and thus be interested in different aspects of the search results. A different proposed approach are rating, recommendation, and reputation systems [13, 11, 12]:

- *Rating systems* allow users to post their rating on items, which are then ranked according to the aggregate rating in the system. An example would be `ciao.com`, a website which allows to do product and price comparisons. The obvious drawback of such systems in which the aggregate rating is made the benchmark is that users with preferences deviating from the average will find the rating unsatisfactory for them.
- *Recommendation systems* based on collaborative filtering suggest users items based on the similarity of their preferences to other users. For example, on `amazon.com` users are often presented the message that "people who bought [a particular] book also bought these other books" followed by a list of related books. This kind of recommendation system works quite well for low-involvement items such as books, movies or alike. Many scientific teams are working on the data mining aspect, but the few works based on complex systems theory seem particularly promising [26], [25]. Furthermore, the combination of collaborative filtering with trust is one the hot topics in computer science in the near future [28], and again a complex systems approach is proving to be quite successful [32], [1]. In such recommendation systems, the fact that information is processed in a centralised way raises scalability issues. However, more importantly, if ratings concerned high-involvement services, such as health care, insurance, or financial services, centralisation also raises confidentiality issues. As we will see in the following, these limitations can be overcome with trust-based networks.

- *Reputation systems* are used more and more in trading. Possibly the most prominent example is `ebay.com`, the Internet auction platform where both buyers and sellers have an associated reputation value which reflects their reliability, quality-of-service, and trustworthiness. Such notions of reputation are gaining visibility – even to the point that people post their `ebay.com` reputation value on their curriculum vitae when looking for a job. However, there are several unsolved game theoretic drawbacks to such systems, for instance the incentive to give good ratings in order to avoid retaliation.

Figure 1 illustrates the use cases of such recommendation systems along the example of `amazon.com`. In the example, a user is searching for a travel guide to Switzerland. The recommendation system is used to establish a ranking of potential books to be bought and to facilitate the decision making of the user.

1.3 Applications to Business and Society

As we have seen from the examples, these concepts have formed the basis for recently founded businesses all over the world. This demonstrates their high impact at the global economic level. Moreover, the current trend is that the sector is continuously expanding with the foundation of new start-ups. However, the impact is not limited to the business world, but also affects society. For the first time in history, a large-scale real-time self-organisation of citizens in previously unknown forms is possible. For instance, now it is more straightforward for consumers to reach and share ratings of products independently of the producers. It is also feasible, for example, that groups of consumers form buying groups that negotiate with firms the delivery of products or services with specific features. Market diversity , which, today, is a producer-driven process, could become a consumer-driven process, a major change of perspective. In particular, the market share for sustainable products and services could increase significantly. In particular, the application of recommendation systems and akin is not limited to targeted marketing. On the contrary, there is an unprecedented potential for empowering citizens to make more informed choices in their daily life in a vast range of domains, from grocery purchases to political support.

1.4 Role of Complex Systems Theory

The important aspect from the perspective of complex systems theory is that these developments give rise to large-scale collective dynamics. While computer science research in this field mainly focusses on aspects such as protocols, algorithms, security, and infrastructure, the theoretical understanding of the large-scale emerging properties is poor. Research from a complex systems perspective can and should give important contributions to better understand these developments with respect to collective dynamics.

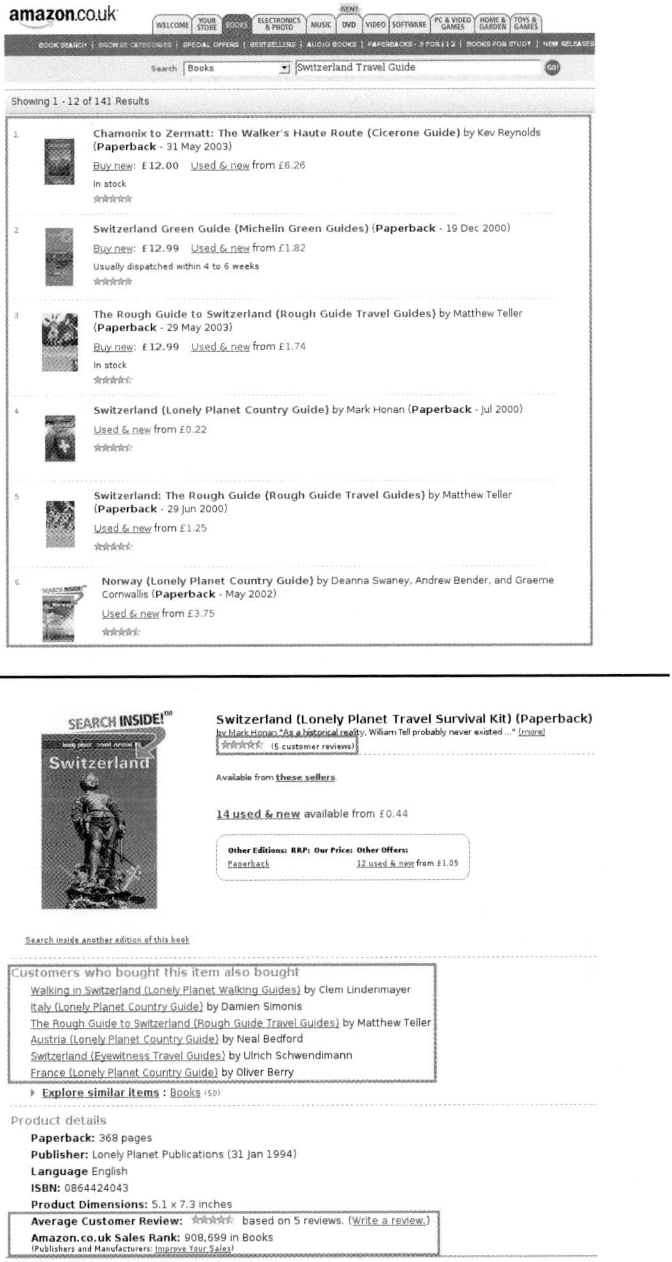

Fig. 1. Amazon, an example of a recommendation system. In the example on top, recommendations are used to *rank particular items in a category*, e.g. books that claim to be travel guides to Switzerland, and in the example at bottom, recommendations are used to *make choices based on ratings* that they provide, e.g. whether to buy/not to buy a particular book. Note the erroneous result on Norway in the list

1.5 Trust-based Networks

The complex systems approach offers a promising way to cope with all these mentioned challenges: *trust-based networks*. A trust-based network can be defined as an information processing system in which interconnected agents (citizens, firms, organisations) share knowledge in their domains of interest. Each agent has a set of neighbours – e.g., friends, partners, and collaborators – with which it decides to share lists of products, services, people, experts etc. together with ratings on these. Trust between neighbours is built up dynamically, based on the satisfaction experienced from the recommendations received by these neighbours.

Soon, paths of trust build up in the network, and each agent is able to reach and rely on – filtered – information, even if coming from another agent far away in the network. This emerging property has some reminiscence to the building of optimal paths in ant colonies [27, 29]. Some recent works have proven the overwhelming superiority of such trust-based recommendation systems over those based on the frequency the recommendations [32]. From the point of view of scalability, trust-based networks are inherently distributed in their nature and do not require centralised information.

A trust-based network can be regarded as an IT support tool for decision making shaped around the natural behaviour of individuals in society. Today's search engines allow the user to find a range of information/products/services from a centralised source, corresponding to a set of keywords. Through a trust-based network, an agent can, instead, search relevant items from specialised, distributed sources and evaluate the trustworthiness of the items with respect to its own preferences.

Subsequently, we present an example of a trust-based network by illustrating a model of a trust-based recommendation system. This system, in an automated and distributed fashion, filters information for agents based on the agents' social network and trust relationships.

The model that we are going to present enables a quantitative study of the problem and also provides a sketch for a solution in terms of a real Internet application/web service. The idea at the core of the model is that agents

- leverage their social network to *reach information*; and
- make use of trust relationships to *filter information*.

We describe the model and the results obtained through multi-agent simulations. To some extent, it is also possible to make analytical predictions of the performance of the system as a function of the preferences of the agents and the structure of the social network. In the following, we will refrain to go into all the details of the model and stay at the level of an informal treatise; for more formal and detailed descriptions of the model, the analysis, and the simulations, please also refer to [1].

The remainder of the chapter is organised as follows: in the following section, we put our work into the context of the related work. Subsequently, we describe an illustrative example of a situation in which a user could benefit from the use of a trust-based network. Then, we present our model of a trust-based recommendation system on a social network. This is followed by a summary of the results from computer simulations and analytical approximations as well as their interpretations. Subsequently, we outline an application of the model. Finally, we illustrate a number of extensions.

2 Related Work

Recent research in computer science has dealt with recommendation systems [11]. Such systems mostly fall into two classes: content-based methods suggest items by matching agent profiles with characteristics of products and services, while collaborative filtering methods measure the similarity of preferences between agents and recommend what similar agents have already chosen [38]. Interestingly, some of the achievements in this field come from the community of complex systems research [26, 25]. Often, recommendation systems are centralised and, moreover, they are offered by entities which are not independent of the products or services that they provide recommendations on, which may constitute a bias or conflict-of-interest.

Additionally, the diffusion of information technologies in business and social activities results in intricate networks of electronic relationships. In particular, economic activities via electronic transactions require the presence of a system of trust and distrust in order to ensure the fulfilment of contracts [4, 9]. However, trust plays a crucial role not only by supporting the security of contracts between agents, but also because agents rely on the expertise of other trusted agents in their decision-making.

Along these lines, some recent works have suggested to combine distributed recommendation systems with trust and reputation mechanisms [13, 12, 31, 39, 28]. Because of the fact that both building expertise and testing items available on the market are costly activities, individuals in the real world attempt to reduce such costs through the use of their social/professional networks.

Such complex networks, in particular their structure and function, are the subject of an extensive and growing body of research across disciplines [17]. In particular, it has been shown that the structure of social networks plays an important role in decision making processes [22, 23, 24].

In this chapter, we combine these three approaches – recommendation systems, trust, and social networks – along the lines of [1].

3 Illustrative Example

The situation we want to model could be illustrated by the following scenario: a person needs to buy a bottle of Swiss wine to accompany an evening with cheese fondue and, just having moved to the country, does not know which one to choose. Therefore, the person contacts its friends and asks them for advice. The friends either have a piece of advice or they pass the question on to their own friends. Let us assume that there are several brands of wine to choose from: $\{a, b, c, ...\}$. After some time, the person receives a number of recommendations, say 6 in number, for specific brands to choose from. For instance, there could be

- 3 recommendations that suggest brand a,
- another 2 that suggest brand b, and
- 1 that suggests brand c.

How is it possible to make the best use of the recommendations? One might choose brand a because it is the most frequently recommended, but it may also be that brand c has been recommended by a friend of a friend who is known to be an expert in wines. Now, there is a trade-off – should one rely on the opinion of the majority or on the opinion of an expert? For an person with average preferences, the opinion of the majority, i.e. the "average opinion" might do well. However, if the preferences of the person deviate from the average in its community, following the advice of an expert may be much more useful.

Let us assume, for the moment, that the person decides for brand a because it is the most frequently recommended choice. However, upon consumption, it discovers that this brand does not match its taste at all. Now, it may make sense that, at the next time when the person goes shopping for wine, it gives less importance to the recommendations of those agents that recommended brand a and that it may even try brand b or c. By following such a strategy, the person would, over time, learn which other people give reliable advice with respect to a particular context and which do not.

Note that in order for this system to work, all people concerned need to have identical definitions of the concept "wine": whenever they exchange recommendations on wine, they know that they all are talking of the same concept.

However, consider that the person now also requires a recommendation on which brand of cheese to buy for the fondue. By the same procedure as for the wine, it obtains a number of recommendations, some from the same people that also made recommendations for the wine. Should the experiences made with the former recommendations on wine influence the decision of which recommendation on cheese to follow? Certainly this must not necessarily be the case: for example, the expert on wines may give good recommendations on wine, but since he is not at all experienced in cheese, his recommendations on cheese may be completely useless. In other words, there may be some contexts in which people may follow recommendations by certain friends and other contexts in which they may not follow the recommendations by the same friends.

What people intuitively do in real life is to keep a mental mapping of the level of trust that they have towards the advice of friends in a particular context. However, this is a difficult task when the market offers thousands of product and service categories as well as dozens of brands in each category. Certainly, the recent developments in the field of information technology make it both desirable and possible to automate this process by means of a computer-assisted recommendation system.

4 Model Description

In the following, we describe an example of a trust-based network by illustrating a model of a trust-based recommendation system. The model deals with agents which have to decide for a particular item that they do not yet know based on recommendations of other agents. When facing to purchase an item, agents query their neighbourhood for recommendations on the item to purchase. Neighbours in turn pass on a query to their neighbours in case that they cannot provide a reply themselves. In this way, the network replies to a query of an individual by offering a set of recommendations. One way to deal with these recommendations would be to choose the most frequently recommended item. However, because of the heterogeneity of preferences of agents, this may not be the most efficient strategy in terms of utility. Thus, we explore means to incorporate knowledge of trustworthiness of recommendations into the system. In the following, we investigate under which conditions and to what extent the presence of a trust system enhances the performance of a recommendation system on a social network.

4.1 Agents, Objects, and Profiles

We consider a set S_A of N_A agents $a_1, a_2, a_3, ..., a_{N_A}$. The agents are connected in a *social network* such as, for example, a social network of people and their friends [15, 16, 17] that are recommending books to each other. Hence, each agent has a set of links to a number of other agents (which we call its neighbours). In reality, social networks between agents to evolve over time; in other words, relationships form, sustain, and also break up. In this chapter, we mainly focus on a static network while dynamic networks will be investigated more thoroughly in further work. At this stage, we assume the network to be described by a random graph [34].

Furthermore, there exists a set S_O of N_O objects, denoted $o_1, o_2, o_3, ..., o_{N_O}$. These objects represent items, agents, products, buyers, sellers, etc. – anything that may be subject to the recommendations, for example, books. We further assume that objects are put into one or more of N_C categories from S_C, denoted $c_1, c_2, ..., c_{N_C}$, where these categories are defined by the system and cannot be modified (i.e. added, removed, or redefined) by the agents. In a scenario where the recommendation system is on books, categories could be

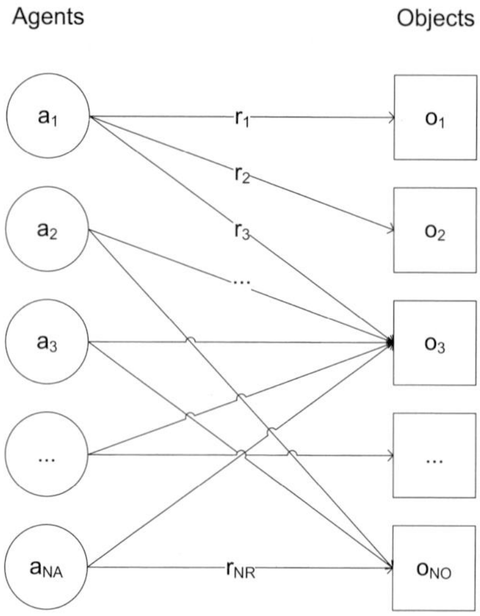

Fig. 2. Agents rating Objects: this is a bipartite graph with the agents on the left hand side and the objects on the right hand side, the ratings being the connections. The set of all possible ratings of an agent constitutes its respective profile [1]

books on 'epicurean philosophy', 'Swiss folklore', or 'medieval archery'. We denote the fact that an object o_i is in category c_j by stating $o_i \in c_j$.

Each agent a_i is associated to one certain preference profile which is one of N_P preference profiles in the system, where $S_P = \{p_1, p_2, p_3, ..., p_{N_P}\}$. In the following, we will use the terms 'preference profile', 'profile', and 'preferences' interchangeably. Such a profile p_i is a mapping which associates to each object $o_j \in S_O$ a particular corresponding rating $r_j \in [-1, 1]$, $p_i : S_O \rightarrow [-1, 1]$. This is illustrated in Figure 2. In the current version of the model, we only consider discrete ratings where -1 signifies an agents' dislike of an object, 1 signifies an agents' favour towards an object. In a future version of the model, this assumption can be relaxed; we chose to initially focus on a discrete rating scheme because most of the ones found on the Internet are of such type. We assume that agents only have knowledge in selected categories and, in particular, they do only know their own ratings on objects of other categories subsequent to having used these objects. Thus, each agent is and remains an expert only on a set of initially assigned selected categories.

4.2 Trust Relationships

In this model, we also consider trust relationships between agents: each agent a_i keeps track of a trust value $T_{a_i,a_j} \in [0, 1]$ to each of its neighbour agents a_j. These values are initialised to $T_{a_i,a_j} = 0.5$. It is important to stress that trust relationships only exist between neighbours in the social network; if two agents are not directly connected, they also cannot possibly have a trust relationship with each other. However, two such agents may indirectly be connected to each other through a path in the network. For example, agent a_i could be connected to agent a_j through agents a_k and a_l, should a_k and a_l, a_i and a_k, as well as a_l and a_j be neighbours. We can then compute a trust value along the path path(a_i, a_j) from a_i to a_j – in the example, path$(a_i, a_j) = \{(a_i, a_k), (a_k, a_l), (a_l, a_j)\}$ – as follows:

$$T_{a_i,\ldots,a_j} = \prod_{(a_k,a_l)\in \text{path}(a_i,a_j)} T_{a_k,a_l} \tag{1}$$

i.e. the trust value along a path is the product of the trust values of the links on that path. Figure 3 illustrates a part of such a social network of agents and a chain of trust relationships between two agents.

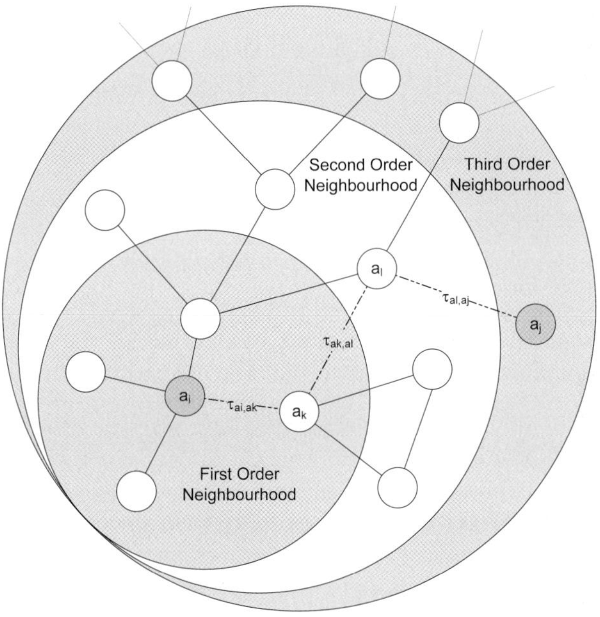

Fig. 3. Social Network of Agents and their Trust Relationships: a section of the social network around agent a_i, indicating a chain of trust relationships to agent a_j and ordering the neighbours according to their distance in hops ('orders of neighbourhood') [1]

4.3 Temporal Structure, Search for Recommendations

The model assumes a *discrete linear bounded model of time*. In essence, there are two possible types of search for a recommendation:

1. *Ranking within a category (RWC)*: agents query for a particular category and search recommendations for several objects in this category in order to decide for one of the recommended objects in the response from the network – typically the best one.
2. *Specific rating for an object (SRO)*: agents query for a particular object and search recommendations on this very object in order to decide for or against using it, based on the response from the network.

Of these, the RWC is a superset of the SRO; a system which can provide a RWC can trivially be extended to provide SRO, too. Hence, in the following, we focus on the former rather than the latter.

At each time step t, each agent a_i (in random order) selects a category c_j (again, in random order, with the constraint that the agent is not an expert on the category) and searches for recommendations on the network. In informal notation, the protocol for the agent's search proceeds as follows:

1. Agent a_i prepares a query(a_i, c_j) for category c_j and then transmits it to its neighbours.
2. Each neighbour a_k receives query(a_i, c_j) and either
   a) returns a response$(a_i, a_k, (o_j, r_j), T_{a_i,\ldots,a_k})$, if it knows a rating r_j for a particular object o_j in c_j that it can recommend, i.e. if $p_k(o_j) = r_j > 0$:
   b) or, passes query(a_i, c_j) on to its own neighbours if it does not know a rating r_j for the particular category c_j.

It is assumed that agents keep track of the queries they have seen. Now there are two strategies to guarantee that the algorithm terminates: either,

- agents do not process queries that they have already seen again ("incomplete search", IS); or,
- agents pass on queries only once, but, if they have an appropriate recommendation, can return responses more than once ("complete search", CS).

In essence, both are a form of *breadth-first search* on the social network of agents, but with different properties: the former returns, for each possible recommendation, only one possible path in the network from the querying to the responding agent; the latter, however, returns, for each possible recommendation, each of the possible paths in the network from the querying to the responding agent.

As we will see later, this is a crucial difference for the decision making of agents. For a given recommendation, there might be several paths between the querying and the responding agent. The IS returns a recommendation along one of these paths, while the CS returns a set of recommendations along all possible paths. Some paths between two agents have high trust, some have low trust. The IS may return a recommendation along a low-trust path even though there exists a high-trust path, thus providing an agent with insufficient information for proper decision making. Of course, there is also a pitfall with the CS – it is computationally much more expensive.

4.4 Decision Making

As a result of a query, each agent a_i possesses a set of responses from other agents a_k. It now faces the issue of making a decision on the ratings provided. The agent needs to decide, based on the recommendations in the response, what would be the appropriate choice of all the objects recommended. We denote query$(a_i, o_j) = Q$ and a response$(a_i, a_k, (o_j, r_j), T_{a_i,...,a_k}) \in R$ where R is the set of all responses. The values of trust along the path provide a ranking of the recommendations. There are many ways of choosing based on such rankings; we would like to introduce an exploratory behaviour of agents and an established way of doing so consists in choosing randomly among all recommendations with probabilities assigned by a logit function [30]. For this purpose, it is convenient to first map trust into an intermediate variable $\hat{T}$, ranging in $[-\infty, \infty]$:

$$\hat{T}_{a_i,...,a_k} = \frac{1}{2} \ln \left(\frac{1 + 2(T_{a_i,...,a_k} - 0.5)}{1 - 2(T_{a_i,...,a_k} - 0.5)} \right) \in [-\infty, \infty] \quad (2)$$

$$P(\text{response}(a_i, a_k, (o_j, r_j), T_{a_i,...,a_k})) = \frac{\exp(\beta \hat{T}_{a_i,...,a_k})}{\sum_R \exp(\beta \hat{T}_{a_i,...,a_l})} \in [0, 1] \quad (3)$$

where β is a parameter controlling the exploratory behaviour of agents. For $\beta = 0$, the probability of choosing each response will be the same (i.e. this is equivalent to a random choice), but for $\beta > 0$, responses with higher associated values of $Ta_i, ..., a_k$ have higher probabilities. To decide for one of the objects, the agent chooses randomly between all recommendations according to these probabilities. This process is illustrated in Figure 4.

For benchmarking the trust-based approach of selecting recommendations, we consider an alternative decision making strategy, namely a *frequency-based approach* without any trust relationships being considered at all. In this approach, an agent chooses randomly among each of the recommendations with equal probability for each of the recommendations.

4.5 Trust Dynamics

In order to enable the agents to learn from their experience with other agents, it is necessary to feedback the experience of following a particular recommendation into the trust relationship. This is done as follows: subsequent to an

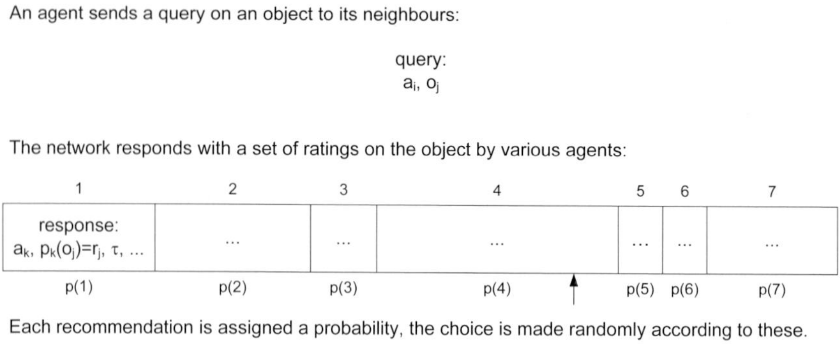

Fig. 4. Search for Recommendations and Decision Making: agents send queries, they receive responses, and then decide for one randomly according to probabilities they have assigned to each recommendation [1]

interaction, agent a_i who has acted on a rating through its neighbour, agent a_j, updates the value of trust to this neighbour, based on the experience that he made. Let o_k be the chosen object. Then, assuming agent a_i having profile p_i, $p_i(o_k) = r_k$ is the experience that a_i has made by following the recommendation transmitted through a_j. It is convenient to define the update of $T(t+1)$ in terms of an intermediate variable $\tilde{T}(t+1)$:

$$\tilde{T}_{a_i,a_j}(t+1) = \begin{cases} \gamma\tilde{T}_{a_i,a_j}(t) + (1-\gamma)r_k & \text{for } r_k \geq 0 \\ (1-\gamma)\tilde{T}_{a_i,a_j}(t) + \gamma r_k & \text{for } r_k < 0 \end{cases} \tag{4}$$

where $\tilde{T}_{a_i,a_j}(0) = 0$ and $\gamma \in [0,1]$. Because $\tilde{T}_{a_i,a_j} \in [-1,1]$, we have to map it back to the interval $[0,1]$:

$$T_{a_i,a_j}(t+1) = \frac{1 + \tilde{T}_{a_i,a_j}(t+1)}{2} \in [0,1] \tag{5}$$

The distinction between $r_k \geq 0$ and $r_k < 0$ creates, for values of $\gamma > 0.5$, a slow-positive and a fast-negative effect which usually is a desired property for the dynamics of trust: trust is supposed to build up slowly, but to be torn down quickly. The trust update is only applied between neighbouring agents – the trust along a pathway between two non-neighbour agents $T_{a_i,\ldots,a_j}$ changes as a result of changes on the links of the path. The performance of the system results from the development of pathways of high trust and thus is a emergent property of local interactions between neighbouring agents.

It is important to note that – in the current version of the model – trust turns out to reflect the similarity of agents. In further extensions of the model, it should reflect other notions such as "agent a_j cooperated with agent a_i", "agent a_j gave faithful information to agent a_i", or "agent a_j joined a coalition with agent a_i". In other words, the metric should be an aggregate of different dimensions of trust, possibly measuring the faithfulness, reliability, availability, and quality of advice from a particular agent.

4.6 Utility of Agents, Performance of the System

In order to quantitatively measure the difference of the trust-based approach of selecting recommendations as compared to the frequency-based approach, it is necessary to define measures for the utility of agents as well as for the performance of the system.

We define an instantaneous utility function for an agent a_i following a recommendation from agent a_j on object o_k at time t as follows:

$$u(a_i, t) = r_i \tag{6}$$

where agent a_i's profile determines $p_i(o_k) = r_i$. We consider the performance of the system to be the average of the utilities of the agents in the system:

$$\Phi(t) = \frac{1}{N_A} \sum_{a_i \in S_A} u(a_i, t) \tag{7}$$

This gives us a measure for quantitatively comparing the difference that the trust-based approach makes towards the frequency-based approach, both on the micro-level of an agent and the macro-level of the system. In the following, we will use the instantaneous measures for utilities and performance rather than the cumulative ones (if not indicated otherwise).

5 Results and Interpretation

One of the most important results of the model is that the system self-organises in a state with performance near to the optimum. Despite the fact that agents only consider their own utility function and that they do not try to coordinate, long paths of high trust develop in the network. This allows agents to rely on recommendations from agents with similar preferences, even when these are far away in the network. Therefore, the good performance of the system is an emergent property, achieved without explicit coordination.

5.1 Key Quantities

Three quantities are particularly important for the performance of the system: the network density, the preference heterogeneity among the agents, and the sparseness of knowledge. The core result is that recommendation systems in trust-based networks outperform frequency-based recommendation systems within a wide range of these three quantities:

- *Network density*: if the network is very sparse, agents receive useful recommendations on only a fraction of the items that they send queries about; the denser the network, the better the performance, but above a critical threshold for the density, the performance stabilises. The proximity of this value to the optimum depends on the other two quantities.
- *Preference heterogeneity*: if the preferences of agents are homogeneous, there is no advantage for filtering the recommendations; however, if the preferences of agents are all different, agents cannot find other agents to act as suitable filters for them. In between, when preferences are heterogeneous, but 'not too much', the system performance can be near to the optimum.
- *Knowledge sparseness*: when knowledge is dense (N_c and/or N_p small), it is easy for an agent to receive recommendations from agents with similar preferences. In the extreme situation in which, for each category there is only one expert with any given preference profile, agents can receive useful recommendations on all categories only if there exists a high-trust path connecting any two agents with the same profile. This is, of course, related to the density of links in the network.

The performance of the system thus depends, non-linearly, on a combination of these three key quantities. Under certain assumptions, the model can be investigated analytically and in a mean-field approach it is possible to make quantitative predictions on how these factors impact the performance. These results are presented in [1]. Here, we illustrate the properties of our recommendation system by describing the results of multi-agent simulations of the model. As a benchmark, we compare the trust-based recommendation system to a frequency-based recommendation system.

5.2 Simulation Parameters

For the simulations we have used the following parameters to the model: we consider $N_a = 100$ agents, and the simulations are averaged over $N_r = 100$ runs. The size of each category is the same and we vary $N_c \in \{10, ..., 50\}$ and $N_p \in \{2, 4, 6\}$; N_o is usually adjusted such that there are at least 2 objects in each category. Profiles are distributed such that the sum over a profile is 0 on average – across the profile, categories, and agents. Each agent is an expert on one category. Further, for the social network we assume a random directed graph with a given number of agents, N_a, and a given total number of links, ℓ. The *network density* is then defined as $p = \ell / N_a(N_a - 1)$. Agents are connected randomly with respect to their profile.

5.3 Trust and Decision-Making Dynamics

Figure 5 (left) shows that the update rule of trust as described by eq. 4 and eq. 5 produces a slow-positive fast-negative dynamics. Trust between two agents of the same profile evolves to 1 (red line, partially covered by the green one). Trust between two agents of opposite profiles evolves to 0 (blue line). In case that an agent recommends an object that is rated negatively, trust drops quickly and recovers slowly (green line). The probability of choosing a recommendation depends critically on the parameter β, which controls the exploratory behaviour of agents, as shown in Figure 5 (right).

5.4 Performance over Time and Role of Learning

Over time, each agent develops a value of trust towards it neighbours which reflects the similarity of their respective profiles. After some time, paths of high trust develop, connecting agents with similar profiles. As a result, the performance of the system, as defined in eq. 7 increases over time and reaches a stationary value which approaches the optimum. This is shown in Figure 6, where coloured curves correspond to different values of γ.

We have also simulated a situation in which, prior to the start of the dynamics, there is a learning phase in which the agents explore only the recommendations of their direct neighbours on the categories that these claim to be expert on. This way, the trust dynamics already start from a value deviating from the neutral point of 0.5 and closer to one of the fix points (see eq. 4). In this case, the performance is optimal from the beginning on (black curve). Interestingly, the system evolves, even in the normal dynamics, to the same value that is reached with the learning phase, supporting the idea that the optimal performance is an emergent behaviour of the system.

5.5 Impact of Network Density and Search Type

In the model description, we have described two types of search. Figure 7 – the performance of the system plotted against increasing values of density

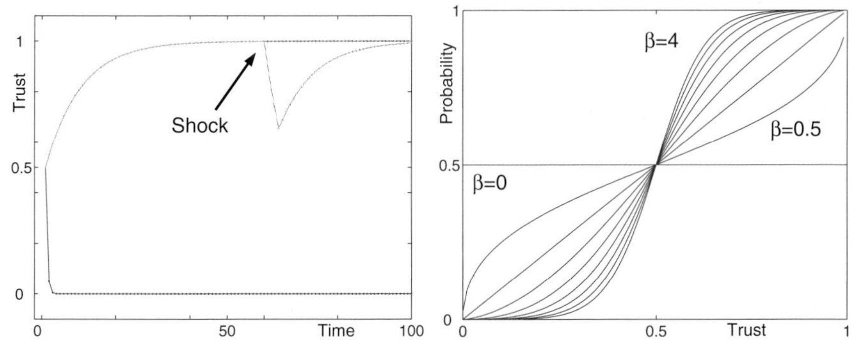

Fig. 5. Trust and Decision-Making Dynamics. The left illustrates the slow-positive fast-negative dynamics of trust and the right the impact of the choice of the exploration parameter β on the decision making [1]

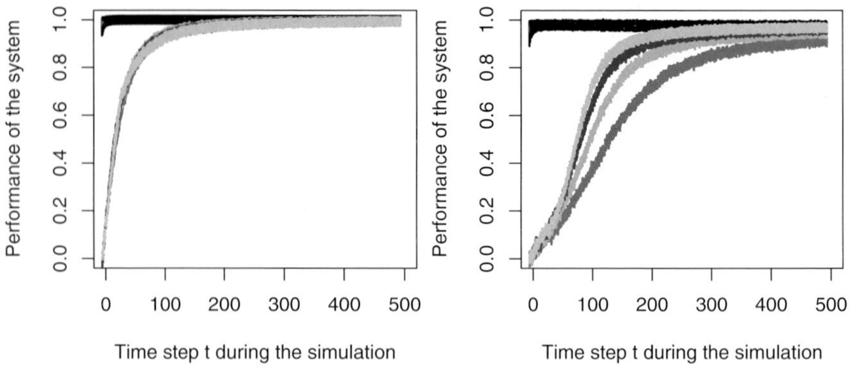

Fig. 6. Performance Φ vs. Time for $N_c = 10$ (**left**) and $N_c = 50$ (**right**). Over time, performance approaches the optimum – with learning (black line), this process is accelerated. Different colours represent different values of γ [1]

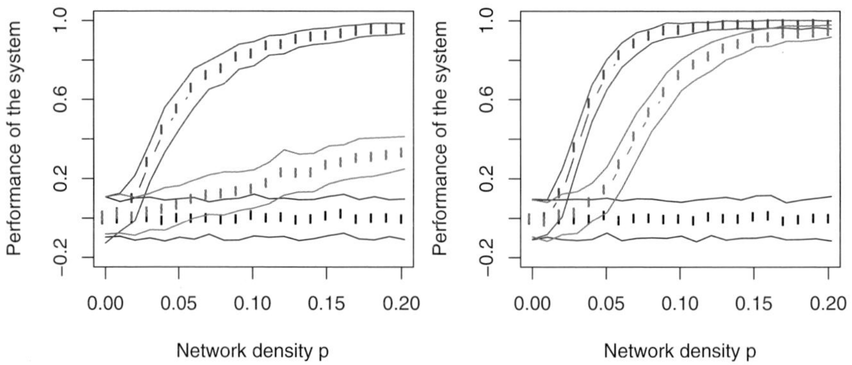

Fig. 7. Performance Φ vs. density for different N_c. Incomplete search (**left**) and complete search (**right**). For sparse knowledge, the complete search performs much better than the incomplete search [1]

in the network – shows that the search type becomes important when the knowledge is sparse. We notice a sigmoid shape which would become steeper for systems with larger numbers of agents. We consider different N_c, corresponding to levels of sparseness of knowledge (in blue and red, 10 and 50 categories, respectively, $N_p = 2$). With the incomplete search algorithm, the performance deteriorates. With the complete search algorithm, the system reaches the optimal performance even in the case of maximally sparse knowledge (50 categories means that there is only 1 expert from each profile in each category). In both (left) and (right) the black curves correspond to the frequency-based recommendation system used as benchmark. In fact, without trust, the performance is 0 on average, because random choices lead to an equal distribution of "good" and "bad" objects (with respect to profiles).

5.6 Preference Heterogeneity and Knowledge Sparseness

We now illustrate the role of preference heterogeneity. We consider first the case in which there are two possible, opposite, profiles in the population, say p_1 and p_2. We define the fraction of agents characterised by the first profile as n_1. In Figure 8 (left), we plot the performance of the system with and without trust (yellow and black, respectively) against increasing values of n_1. When $n_1 = 0.5$ there is an equal frequency of both profiles, while when $n_1 = 1$ all agents have the first profile. For the system without trust, the performance increases for increasing n_1. In fact, despite that choices are random, agents receive recommendations which are more and more likely to match the preferences of the majority. On the other hand, the minority of agents with the profile p_2 are more and more likely to choose wrong recommendations, but their contribution to the performance of the system decreases. The simulation results are in good agreement with the predictions obtained in an analytical approximation (red and blue), see [1].

For the system with trust the performance is almost unchanged by the frequency. This very strong result has the following explanation: The social network is a random graph in which agents have randomly assigned profiles. Agents assigned to p_2 decrease in number, but, as long as the minority, as a whole, remains connected (there is a path connecting any two such agents) they are able to filter the correct recommendations. At some point the further assignment of an agent to p_1 causes the minority to become disconnected and to make worse choices. In the simulations, this happens when $n_1 = 0.9$ and $n_2 = 0.1$. Another way of investigating the role of heterogeneity of preferences is to consider an increasing number of profiles in the population, each with the same frequency. In the extreme case in which, for each category there is only

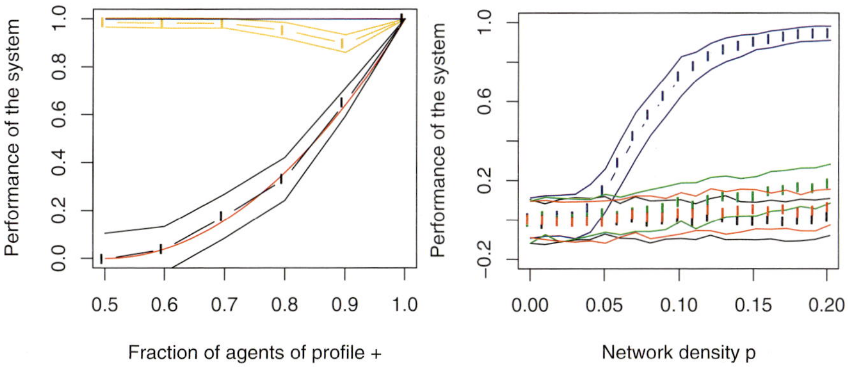

Fig. 8. Effect of heterogeneity on performance. Performance as a function of the heterogeneity of preferences (**left**) and with different N_p (**right**). The trust-based approach performs well also in very homogeneous systems; in the extreme case of very heterogeneous systems, performance drops [1]

one expert with any given preference profile, the performance, at constant values of network density, drops dramatically, as shown in Figure 8 (right).

6 Application Scenario

We consider a portal on which users can register with their *name* and a *brief profile* containing personal or contact information. Similar to many other social networking services, each user maintains a list of other users which he knows or is a friend of – a list commonly known as the "buddy list" of a user. The system provides a *search facility* in which a user can search for people on the platform by their name, address, or other details, as to make it straightforward for people to find other people they know in the real world. The set of users in the system and the connections between them constitutes a *social network*.

Furthermore, the system maintains a *list of objects*, an object being a unique representation for a user, product, buyer, seller, etc. Each object has a *name* but also several *keywords* and a brief *description* as to enable users to search for objects not only by their name. Each user now maintains a list of objects that it has an opinion on and associates a rating with each of these objects. A rating consists of

- A value of *'like'*, *'neutral'*, or *'dislike'*, corresponding to numeric values in the set $R = \{1, 0, -1\}$.
- Optionally, a *brief textual description* with an explanation of the rating in human-understandable format.

This scheme (please note that it is similar to the one used by `ebay.com`) has several benefits, the main ones being the following:

- It is *simple*. Complicated schemes – for examples, ones requiring users to make more fine-grained ratings – suffer from the fact that no two users will interpret the metrics used in exactly the same manner, thus leading to inaccuracies being amplified by the finer granularity of ratings. Additionally, such schemes tend to be too confusing to use.
- It is *two-fold* in the sense that the numeric value can be processed automatically by algorithmic means but the textual string can still be used by humans in case that they would like to obtain more information on a particular rating.

Consider a university setting and let the users of the system be students, researchers, and professors. The social network is built by acquaintance and thus spans among the people in different groups of a university, but also across groups and universities between people that know each other through collaboration, projects, or conferences. Furthermore, let the objects in the system be publications. Each publication has a unique identifier, information about the title, authors, and similar information, as well as a set of keywords and possibly an abstract. Each user maintains a list of publications he knows – a subset of all publications known to the system – and with each of these, associates a rating and possibly a brief textual description with more information.

The purpose of the recommendation system is to provide users with a unique gateway to more information on the objects listed in the system. Users can search for objects based on the name, description, or keywords. They then see a ranking of objects matching their search and upon selecting a particular object, they are displayed

- information on that object,
- an *aggregate rating* derived from the ratings of users in their social network and weighted by the trust relationships to these users
  and possibly
- a *representative subset of the ratings* (numeric values as well as textual descriptions) used in construction of the aggregate rating.

Based on the recommendation provided by the system, users can then decide to use a particular object. When they do so, they experience this object and thus are able to provide a rating themselves. The system detects and records such ratings and uses them as *feedback* on the trustworthiness of ratings by other users. Over time, the system learns which users provide particularly useful/useless recommendations for which other users, and uses this knowledge to adjust the computation of aggregate ratings for individual users. Along these lines, returning to the example scenario in the university setting, the system allows users to

- Search for publications based on title, authors, keywords, and so on.
- Obtain a recommendation for any such publication; the recommendation is based on the ratings of other users in a user's social network and the trust relationships existing to these other users.
- Obtain a ranking of publications for a particular set of tags, e.g. publications in a particular field or by a particular author.

The benefit of the system is that users are able to select, from a possibly huge set of publications, those that may be relevant for them based on what the people they trust find relevant. Thus, the system provides a filtering technique for people to cope with information overload.

Any such recommendation system can implemented in an according way that it can be accessed through a web interface but also through a web service which seamlessly and transparently makes it available to all sorts of mobile devices such as notebook computers, handheld devices, or mobile phones. This might be more suitable for scenarios different from the example in a university setting, such as recommendation systems for restaurants and bars, products in supermarkets, and so on – returning to the example of Swiss wine and cheese, imagine the scenario of a person querying for recommendations in a supermarket and receiving responses with ratings through a PDA in a matter of seconds, while standing between two aisles in the supermarket.

7 Extensions to the Model

At this point, let us return to the model itself. So far, we have made the assumptions that

- agents are self-interested in the sense of bounded rationality, but do not act randomly, selfishly, or maliciously and that
- the social network of agents is fixed and does not change over time – no agents join or leave the networks and no links are rewired, added, or dropped.

In reality, both of these assumptions need to be relaxed, so in further work, we plan to investigate the behaviour of the system in an evolving network as well as its robustness in the presence of agents which act randomly, selfishly, and maliciously.

7.1 Evolving Social Network

Considering a static social network between agents does not appropriately depict reality; usually, *social networks evolve over time* with links being created and deleted at each time step. People tend to establish contacts to new people and lose contact to old acquaintances. Both of these actions lead to an evolution of the underlying social network.

Thus, to stay close to reality, we have to analyse the model having an underlying dynamic social network with the possibility of the

- *Creation of links* between agents which have mutually benefited from each other's recommendations for a particular number of times.
- *Deletion of links* unilaterally or bilaterally between agents that believe the other agent to give useless recommendations.

It is conceivable that the evolution of the social network has a crucial impact on the performance of the system: over time, agents learn which other agents are trustworthy as well as which are not and adjust their links accordingly. A priori, it is not clear whether this leads to better performance (possibly because agents have similar agents as their immediate neighbours that they can rely on) or worse performance (possibly because agents focus too much on their immediate neighbours to see that there are more opinions than these). It might also be interesting to analyse to what extent the global and local properties of the underlying social network change.

It is reasonable to assume that a person is more likely to keep a link towards a neighbour the more he/she trusts the neighbour and vice versa.

We do not model the decision-making process explicitly but we capture this tendency with a stochastic rule as follows:

$$P(\text{rewire}) = 1 - T_{a_i,a_j}, \quad P(\text{keep}) = T_{a_i,a_j} \tag{8}$$

i.e. $P(\text{rewire}) + P(\text{keep}) = 1$. Equation 8 implies that the probability to randomly rewire the link from agent a_i to a_j is high if the trust from a_i to a_j is low. Thus, trustworthy links are kept while untrustworthy links are replaced.

Figure 9 shows how snapshots of the evolution of a sample network of agents at different stages for different values of β (more and less explorative agents) look like when applying this mechanism. This illustrates the dilemma between exploration and exploitation faced by the agents. For $\beta = 0$, agents choose randomly, thus performing worse, but they explore many the other agents repetitively and their trust relationships converge to the steady state of the trust dynamics. Then, over time, links with low trust are rewired and links with high trust are kept. This leads to the emergence of two disconnected clusters. Eventually, subsequent to the formation of clusters, such agents will

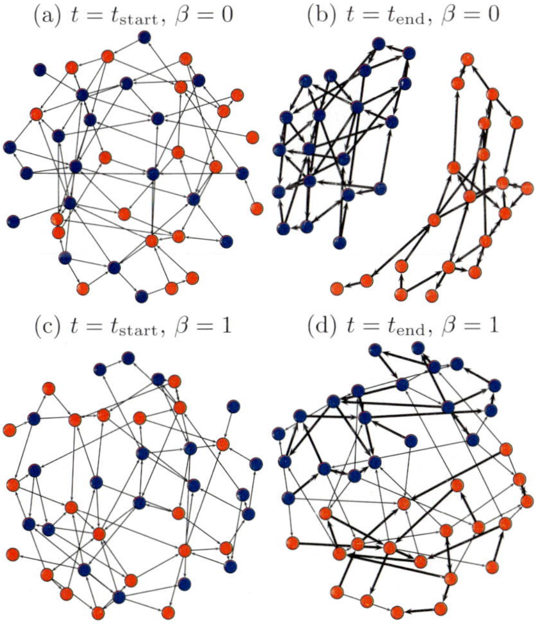

Fig. 9. Snapshots of the evolution of a network of 40 agents in 2 profiles and 80 links at time $t = t_{\text{start}}$ and $t = t_{\text{end}}$ for $\beta = 0$ and $\beta = 1$, respectively. When $\beta = 0$ (very explorative agents, see eq. 3), disconnected clusters of agents with the same profile form, when $\beta = 1$ (less explorative agents, see eq. 3), interconnected clusters of agents with the same profiles form. For $\beta > 0$, agents develop stronger ties to agents of the same profile than to agents of different profiles [1]

perform well, as any recommendation will come from an agent of the same profile. For $\beta = 1$, agents choose according to the strength of trust relationships, thus performing better, and they are able to exploit their knowledge. However, they exploit stronger links while not even exploring weaker ones. This results in clustering, but with interconnections between clusters. As networks in reality are evolving, it is important to study the impact of such behaviour on the system in more detail.

7.2 Robustness against Attacks

The model also allows us to focus on the robustness of the recommendation system against attacks. This is a very important aspect because of the fact that in real-life systems there will be users that try to cheat the system as soon as money is involved – which would be the case even in the illustrative example of Swiss wine and cheese. The financial incentive for some of the agents in the system may have a level high enough to, for example, lead to the following: wine and cheese manufacturers may be tempted to improve recommendations for their products so as to increase their revenues, upset customers may try to defame products that they made bad experiences with as an act of retaliation, and so on.

To illustrate and further stress this point, consider a similar example from the field of search engines: Google, currently the most widely used search engine builds its search engine rankings according to the page rank algorithm. The basic idea is that the more links point to a page, the higher up in the search ranking this page will be placed. Of course, as Google has a vast market share in the search engine domain, it is of utmost importance for the manufacturers of a certain product or the providers of a certain service that they rank among the top 5 of the search engine results for certain keywords. There is a strong incentive for manipulation of the search engine results by means of increasing the number of links to particular pages in the context of certain keywords. This can be done, for example, by setting up large numbers of artificial web pages with hardly any content except a number of keywords which all cross-link to a desired web page and thus increase the number of links to this page with respect to the keywords. This has become known as "Google bombing" and is an ongoing issue that all search engines have to deal with.

Thus, in the construction of a real-life recommendation system, cheaters and attackers have to be considered. For example, it would be possible to consider three different additional types of agents:

- *Random agents* are agents that, instead of giving correct recommendations, give a random recommendation. This is not necessarily due to selfish or malicious intentions, but may just as well result from a pure lack of knowledge. In a sense, having such agents in the system mimics the effect of noise on communication channels.

- *Selfish agents* are agents that do not return recommendations except in the case that they have already received responses through the agent that initiated the query. Obviously, if all the agents in a system are selfish, the system is in a deadlock state where no one gives anyone else recommendations.
- *Malicious agents* are agents that intentionally give recommendations that do not correspond to their own beliefs – i.e., they recommend what they would not use themselves, and vice versa. An ideal recommendation system should be able to cope with such agents.

In each of the cases, we are interested in the performance of the recommendation system with respect to differing fractions of such random, selfish, and malicious agents in the system: Does the presence of random/selfish/malicious agents impact the performance of the recommendation system? Is there a critical value of the fraction of random/selfish/malicious agents for which the recommendation system becomes unusable/usable?

It may also be interesting to look at more sophisticated agents, e.g. ones that alternate between these types of behaviour, or agents which form networks with other agents to influence the system in a particular way. Understanding the aspect of the robustness against attacks is crucial for real-life systems.

8 Conclusions

In this chapter, we have presented trust-based networks as an application of complex systems theory to cope with information overload on the Internet. By combining recommendation systems, trust, and social networks, it is possible to build a system in which agents use their trust relationships to filter the information that they have to process, and their social network to reach knowledge that is located far from them. The emergent property of the system is that it self-organises in a state with performance near to the optimum without explicit coordination of the agents. In this chapter, we have given one example of a real-world application, but we believe that the system is applicable to a vast variety of domains ranging from low-involvement products such as books or groceries to high-involvement services such as insurance or health-care.

Acknowledgements

We thank Markus M. Geipel, Patrick Groeber and João F. M. Rodrigues for their valuable comments. The network graphs were layouted with an elastic-band-layouter developed by Markus M. Geipel. For details, please refer to `www.sg.ethz.ch/research/graphlayout`.

References

1. Walter, F.E., Battiston, S. and Schweitzer, F.: A Model of a Trust-based Recommendation System on a Social Network. Journal of Autonomous Agents and Multi-Agent Systems (forthcoming, 2007). Available on Arxiv.org, http://arxiv.org/abs/nlin.AO/0611054 (2006)
2. Wellmann, B.: Computer networks as social networks. Science **293** (2001) 2031–2034
3. Huberman, B.A., and Adamic, L.A.: Growth dynamics of the World-Wide Web. Nature **401** (1999) 131
4. Marsh, S.: Formalising Trust as a Computational Concept. Phd. Thesis - University of Stirling (1994)
5. Mui, L., Mohtashemi, M. and Halberstadt, A.: A Computational Model of Trust and Reputation for E-Businesses. In: Proceedings of the 35th Annual Hawaii International Conference on System Sciences IEEE Press (2002)
6. Castelfranchi, C. and Falcone, R.: Social trust: a cognitive approach. In: Castelfranchi, C., and Tan, H.-J., eds.: Trust and deception in virtual societies. Kluwer Academic Publishers (2001) 55–90
7. Gray, E., Seigneur, J.-M., Chen, Y. and Jensen, C.D.: Trust Propagation in Small Worlds. In: Nixon, P. and Terzis, S., eds.: Proceedings of the First International Conference on Trust Management, Lecture Notes in Computer Science Springer **2692** (2003) 239–254
8. Guha, R., Kumar, R., Raghavan, p. and Tomkins, A.: Propagation of Trust and Distrust. In WWW '04: Proceedings of the 13th International Conference on the World Wide Web ACM Press (2004) 403–412
9. Sabater, J. and Sierra, C.: Review on Computational Trust and Reputation Models. Artificial Intelligence Review **24** (2005) 33–60
10. Brin, S. and Page, L.: The anatomy of a large-scale hypertextual Web search engine. Computer Networks and ISDN Systems **30** (1998) 107–117
11. Montaner, M., López B., De La Rosa, J.-L. A Taxonomy of Recommender Agents on the Internet. Artificial Intelligence Review **19** (2003) 285–330
12. Massa, P. and Bhattacharjee, B.: Using Trust in Recommender Systems: An Experimental Analysis In: Jensen, C.D., Poslad, S. and Dimitrakos, T., eds.: Proceedings of the Second International Conference on Trust Management (ITRUST 2004), Lecture Notes in Computer Science Springer **2692** (2004) 221–235
13. Massa, P. and Bhattacharjee, B.: Using Trust in Recommender Systems: An Experimental Analysis In: Gini, M., Ishida, T., Castelfranchi, C., and Lewis Johnson, W., eds.: Proceedings of the First International Joint Conference on Autonomous Agents and Multiagent Systems (AAMAS'02) ACM Press (2002) 304–305
14. Erdos, P. and Rényi, A.: On random graphs. Publicationes Mathematicae Debrecen **6** (1959) 290–291
15. Watts, D.J. and Strogatz, S.H.: Collective dynamics of 'small-world' networks. Nature **393** (1998) 440–442
16. Barabasi, A.-L. and Albert, R.: Emergence of Scaling in Random Networks. Science **286** (1999) 509–512
17. Newman, M.: The Structure and Function of Complex Networks. SIAM Review **45** (2003)

18. Kleinberg, J. and Lawrence, S.: The structure of the Web. Science **294** (2001) 1849–1850
19. Amaral, L.A.N., Scala, A., Barthélémy, M. and Stanley, H.E.: Classes of small-world networks. Proceedings of the National Academy of Sciences **97** (2000) 11149–11152
20. Golder, S. and Huberman, B.A.: The Structure of Collaborative Tagging Systems. http://arxiv.org/cs/0508082 (2005)
21. Gruber, T.R.: A Translation Approach to Portable Ontology Specifications. Knowledge Acquisition **5** (1993) 199–220
22. Battiston, S., Bonabeau, E. and Weisbuch, G.: Decision making dynamics in corporate boards. Physica A **322** (2003) 567
23. Battiston, S., Weisbuch, G. and Bonabeau, E.: Spread of decisions in the corporate board network. Advances in Complex Systems **6** (2003)
24. Battiston, S., Weisbuch, G. and Bonabeau, E.: Statistical properties of board and director networks. European Journal of Physics B **38** (2004)
25. Laureti, P. and Moret, L. and Zhang, Y. -C. and Yu, Y. -K.: Information Filtering via Iterative Refinement. Europhysics Letters **75** (2006) 1006
26. Laureti, P., Slanina, F.,Yu, Y.-K. and Zhang, Y.-C. Buyer Feedback as a Filtering Mechanism for Reputable Sellers. Physica A **316** (2002) 413
27. Dorigo, M., Maniezzo, V. and Colorni, A.: The Ant System: Optimization by a colony of cooperating agents. IEEE Transactions on Systems, Man, and Cybernetics Part B: Cybernetics **26** (1996) 29–41
28. Ziegler, C.-N. and Golbeck, J.: Investigating Correlations of Trust and Interest Similarity. Decision Support Systems (2006)
29. Schweitzer, F. and Lao, K. and Family, F.: Active random walkers simulate trunk trail formation by ants. BioSystems **41** (1997) 153–166
30. Weisbuch, G., Kirman, A. and Herreiner, D.: Market Organisation and Trading Relationships. The Economic Journal **110** (1998) 411–436
31. Palau, J., Montaner, M., López, B. and de la Rosa, J.-L: Collaboration Analysis in Recommender Systems Using Social Networks. In: Klusch, M., Ossowski, S. Kashyap, V. and Unland R., eds.: Proceedings of the 8th International Workshop on Cooperative Information Agents (CIA 2004), Lecture Notes in Computer Science Springer **3191** (2004) 137–151
32. Battiston, S., Walter, F.E. and Schweitzer, F.: Impact of Trust on the Performance of a Recommendation System in a Social Network In: Proceedings of the Workshop on Trust at the Fifth International Joint Conference on Autonomous Agents and Multi-Agent Systems (AAMAS'06) (2006)
33. Abdul-Rahman, A. and Hailes, S.: Supporting Trust in Virtual Communities. In: Proceedings of the 33th Annual Hawaii International Conference on System Sciences (2000)
34. Bollobas, B.: Random Graphs. Academic Press (1985)
35. Luhmann, N.: Trust and Power John Wiley & Sons (1979)
36. Sztompka, P.: Trust: A Socialogical Theory Cambridge University Press (1999)
37. Gambetta, D.: Trust: Making and Breaking Cooperative Relations Electronic Edition, Oxford University (2000)
38. Sarwar, B., Karypis, G., Konstan, J., and Riedl, J.: Analysis of Recommendation Algorithms for E-Commerce. In: EC '00: Proceedings of the 2nd ACM conference on Electronic Commerce ACM Press (2000) 158–167

39. Golbeck, J.: Generating Predictive Movie Recommendations from Trust in Social Networks. In: Proceedings of the 4th International Conference on Trust Management (2006) 93–104

Part VI

Conflict and Consensus

Complexity in Human Conflict

Neil F. Johnson

Physics Department, University of Miami, Coral Gables, Florida FL 33124, U.S.A
njohnson@physics.miami.edu

1 Introduction

Nowadays, media headlines invariably contain a report of fresh casualties from an ongoing conflict somewhere in the world: 'twenty wounded in Iraq', 'three killed in Afghanistan', 'a terrorist bomb in Israel', 'three guerilla attacks in Colombia'. We are also bombarded by analyses from experts in the various geographical regions, offering explanations which might rationalize these seemingly random strings of casualty figures. Yet despite the numerous insights from either historical, geographical, social or economic perspectives, the casualty numbers themselves continue to sound surprising, irrational, and ultimately random.

One might think this is to be expected, since there is arguably nothing more disordered, chaotic, and unpredictable, than a war. But could there in fact be some general, common characteristics which connect together all wars, despite their very different origins, locations, ferocities, and durations? Indeed, just like scientists from the field of Complexity have recently shown for financial markets – where markets as diverse as Shanghai and New York have been shown to have similar statistical properties in terms of their price movements [1, 2] – might the same thing hold true for wars? And even for a global war such as terrorism?

This Chapter discusses recent research which suggests that there is indeed a universality in human conflict [3]. Those seemingly random casualty figures – while of course senseless from a humanitarian perspective – are actually telling us something important about the character of wars, and more generally about the character of all human conflict. Indeed, we are able to make quantitative sense out of these violent events and can also interpret what such patterns mean. Using the mathematical concepts of power laws, networks and multi-agent dynamics from Complexity Science, we are able to show that modern insurgent conflicts, terrorism and violent crime exhibit remarkably universal dynamics. Furthermore, we can provide simple yet realistic models which are able to reproduce this quantitative behaviour.

These findings are not simply statistical exercises of purely academic interest. Rather, they suggest that the way in which organized violence evolves has less to do with geography or ideology and more to do with the day-to-day mechanics of human insurgency and violence. It is simply the way in which groups of human beings operate when faced with a generally stronger, but more rigid, opponent. It is therefore a parallel finding to the one for financial markets: The way in which the system (e.g. financial market, or human conflict) evolves has less to do with the specific details of where and when it occurs, and more to do with the generic behaviour of collections of individuals. For the specific case of insurgent conflicts where the forces involved are highly asymmetric in terms of their total strength, we can conclude that the insurgent groups are operating in essentially the same way regardless of the origins and locations of these conflicts – going further, we might justifiably claim that the enemy on all fronts is effectively the same.

2 Conflict and Complexity

For as long as there have been humans, there have been human conflicts – and for as long as there has been human conflicts, observers have been trying to understand and explain their evolution. There are countless books and academic papers which have been written about human conflicts, both past and present. Each conflict is complicated, and it will take many such works to get to the bottom of what is really happening in each – if indeed this will ever happen. For example, there are a plethora of books on the current wars in Iraq and Afghanistan, and the ongoing global 'war' faced by all of us in terms of global terrorism. Indeed it would seem fair to claim that the global nature of terrorism and modern wars represents a major threat to the future of our civilization as a whole.

Although human conflict has traditionally been the domain of historians, military strategists, sociologists and political scientists, there are a number of reasons why conflict should be of interest to Complexity scientists. First, the increased availability of computerized datasets means that there is a data revolution underway across the social sciences – just as the field of astronomy recently caught alight as a result of improved data collection. Human conflict is as old as mankind itself – however a lack of reliable time-series data in the past has kept it out of reach of the quantitative sciences. This has now changed with the media, governments and non-governmental organizations all now regularly collecting data on ongoing conflicts. Admittedly the analysis of their datasets is not always straightforward – not only do the individual agencies differ in their numbers, but the way in which the figures are reported can differ quite markedly. Extensive cross-checks from the various sources must therefore be carried out, prior to any data analysis.

The second reason touches the fascinating aspect of Complexity Science itself. In particular, modern wars seem to exhibit all the common characteristics of Complex Systems: (1) There is feedback, both at the microscopic and macroscopic scale, yielding a system with memory and hence so-called non-Markovian dynamics. (2) The time-series of events is non-stationary, meaning that the character of the distribution may change over time. (3) There are many types of 'particle', according to the various armed actors, and they interact in possibly time-dependent ways. A conflict's evolution is then driven by this ecology of agents. (4) The agents can adapt their behaviour and decisions based on past outcomes. The system is far from equilibrium and can exhibit extreme behavior – for example, if the strategies of several groups of agents suddenly coincide. (5) The observed conflict constitutes a single realization of the system's possible trajectories. (6) The system is open, with this coupling to the environment making it hard to distinguish between exogenous (i.e. outside) and endogenous (i.e. internal, self-generated) effects. Point (3) in particular, deserves some extra discussion. Wars involving three or more actors – be they insurgents, guerillas, paramilitaries or national armies – are far more complicated than those involving just two. If A hates B, and B hates C, does that mean that A must therefore like C? Not necessarily. Hate is many-sided, just as love can be. Again we only need to think about the ongoing insurgencies in places such as Colombia, where there are many armed groups, to see the potential complications. A sides with B, B sides with C, but A hates C. Therefore A starts to fight B so as not to favour C – and the whole process becomes a self-driven perpetual conflict. Indeed, such frustration may be why many modern conflicts seem to go on and on without reaching any definite conclusion.

Third, there is a wider context which relates to the important question of group formation among collections of semi-autonomous agents, whether these agents be individually alive or robotic in nature. Throughout the human, animal, insect and fish kingdoms, groups form and break up at different moments in time and at different points in space [4, 5]. Finding out what drives such group dynamics, and how it might be controlled, has become a fundamental research problem across the biological and social sciences. Furthermore, understanding the process governing the formation of human groups in relation to violence – from street-gangs and organized crime through to Mafia, guerilla groups and terrorist networks – is of crucial importance to the well-being of Society as a whole. Even at the level of cellular behaviour, the dynamics of such group formation, and in particular the underlying cost-benefit interplay which governs this dynamics from the perspective of the individuals involved, is of deep scientific interest [6, 7]. Therefore any insights, parallels, or differences, between the various domains is likely to be of broad interest. In particular, the goal of understanding how the various species fight battles is a fascinating new research area [8] and may eventually offer practical insights into how to best tackle street-gangs and organized crime networks.

3 Data Distributions

Before we discuss casualty distributions, we need to briefly summarize some key points about data distributions as a whole. Imagine we know the heights of all the adults in our street, city, or country. If we then make a graph of the distribution of these heights, we will get a graph like Figure 1. Since noone is 12 feet tall, and noone is less than a foot tall, it makes sense that the curve will rise up and then drop back down again. This also means that there will be a peak – like the top of a mountain is a peak. This peak occurs at the height which describes the largest number of people – it therefore represents the typical, or average, height of humans. Everyone has a height close to this value. An important point for our story is that there is therefore such a thing as a typical height – so imagine that we subsequently had to guess someone's height without ever having seen them. If the peak of the curve in Figure 1 occurs at 5 feet 9 inches, and the spread around this value is 6 inches, then we would be pretty safe in suggesting that this unknown person's height is 5 feet 9 inches give or take about 6 inches.

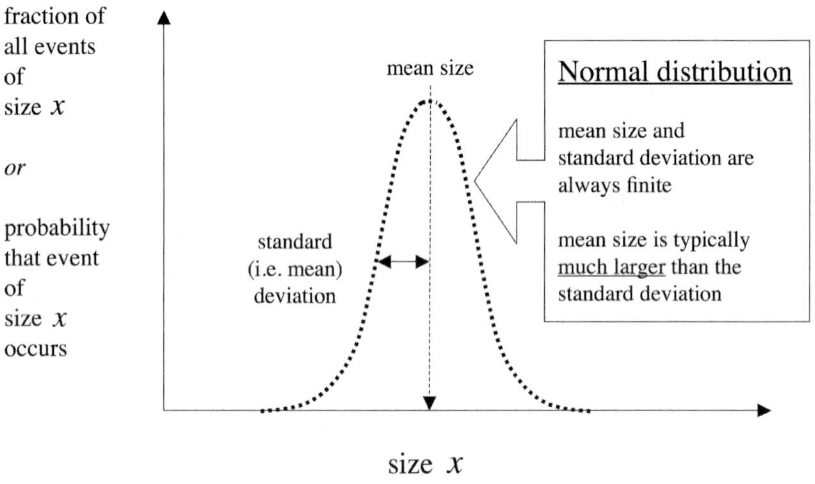

Fig. 1. Schematic graph showing a Normal distribution of data values measured in some system or population. Possible everyday examples include the height of adults in a population, or the speed of cars on a road

Many other curves produced by everyday data will look like this one – for example the speed of cars on a road. In all these cases, the resulting bell-shape of Figure 1 emerges and is referred to as a Normal distribution. There is a good reason why many systems produce a Normal distribution – in each case, the average value of the quantity being measured is dictated by something intrinsic to the individuals themselves, and hence something structural and pre-determined, while the spread (or so-called, fluctuation) in values around this average is usually due to environmental ad hoc reasons. As far as adult heights go, a person's body has an implicit reason based on genes and inheritance to grow to a certain approximate height. Then if the person has an extreme oversupply or undersupply of nutrition, they will probably end up somewhere just above or below this value. The same idea holds for traffic in that there is some pre-existing speed limit on a given road which tends to control the average speed of the cars. Then on top of that average speed, there are everyday environmental and behavioural reasons why individual drivers may drive slightly above or below this value.

By contrast, there are many social, economic and biological systems which do *not* follow a peaked distribution as in Figure 1, but instead resemble far more closely the distribution shown in Figure 2 which is called a 'power law'. The term 'power law' describes the fact that the plot of the fraction of events of size x as a function of x has the form $x^{-\alpha}$ where α is a number. For more details on such power laws, please see Ref. [9]. Although power laws have many interesting mathematical properties, the following discussion summarizes the key elements needed for understanding our subsequent conflict story.

A power law distribution tends to arise in systems containing a collection of objects which are interacting and in which there is *no* central control or 'invisible hand' – in other words, Complex Systems. In contrast to the Normal distribution where the values being measured have a typical or average size which occurs with a very high probability, accompanied by occasional small fluctuations around this value, a power law exhibits a wide spread of likely values. The underlying reason why power law-like distributions are so ubiquitous can be understood as follows: The lack of any central control in such systems implies that at any moment in time, arbitrarily-sized subsets of the objects may be acting in a strongly correlated way. Groups of any size can form, act and breakup, at any time – hence events of any size can occur with relatively high probability. The members of the individual groups may be connected by real physical links or through shared strategies, and the fact that they act as a single unit (even if the group membership and size subsequently change) means that this common action will register some kind of macroscopic event – for example, an individual earthquake in a physical population of interacting tectonic plates.

Given that power laws arise from aggregate collective behaviour of constituent objects in a Complex System, and that this in turn leads to a sizeable event, it is not surprising that power laws tend to emerge at higher values of x as opposed to lower values [9]. In short, it is at high x values that such

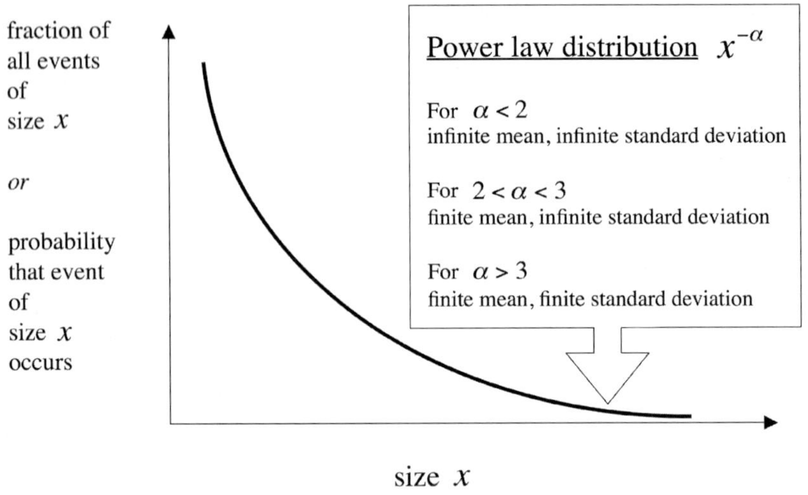

Fig. 2. A power law distribution. The term 'power law' has nothing to do with physical strength. Instead it describes the fact that the plot of the fraction of events of size x (*vertical axis*) as a function of x (*horizontal axis*) has the form of 'x to some power' which in more formal mathematical terms is written as $x^{-\alpha}$ where α is a number

aggregate collective behaviour tends to show itself. What happens in practice, therefore, is that a power law is observed to a good approximation above some particular value of x which we refer to as x_{min}.

But instead of plotting this on linear graph paper as in Figure 2, let us now imagine that we use log-log paper. In other words, each successive equal division represents an increment in the *logarithm* of the number. Consequently, successive equal increments in the underlying number do not correspond to successive equal increments on the axis shown on the log-log paper. For example, choosing logarithms in Base 10 implies that an increment $1 \rightarrow 2$ on the graph represents $10^1 \rightarrow 10^2$ which is $10 \rightarrow 100$, while $2 \rightarrow 3$ represents $10^2 \rightarrow 10^3$ which is $100 \rightarrow 1000$. Figure 3 shows the power law of Figure 2 plotted on log-log paper. Since the logarithm of $x^{-\alpha}$ is simply $-\alpha$, the power law above x_{min} simply appears as a straight line whose slope can be easily measured. Hence the number α which is the only parameter characterizing the power law, can be essentially just read off from the plot.

In practice, the empirical values obtained from any real-world Complex System, will typically be discrete – for example for the case of heights, there

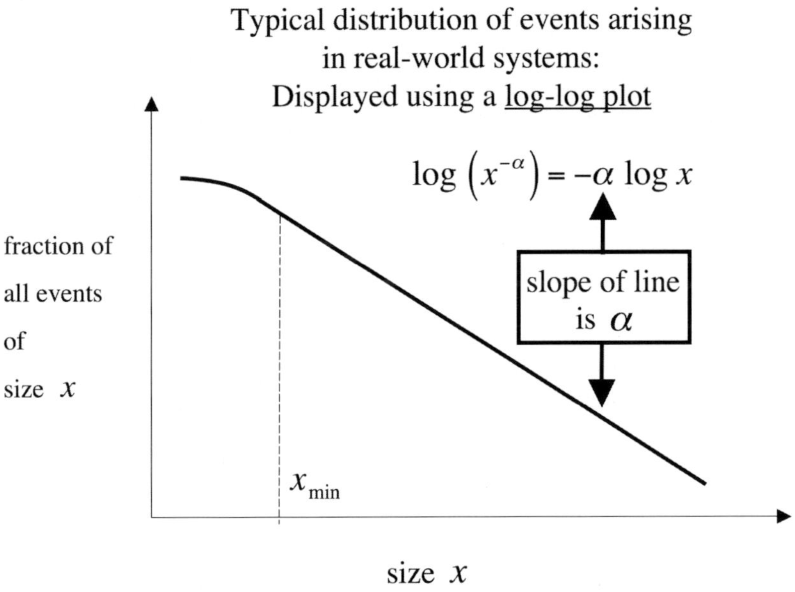

Fig. 3. The power law distribution of Figure 2, becomes a straight line when plotted on log-log graph paper. To reflect the usual situation of real-world Complex Systems, the power law is shown to arise for x values larger than $x_{\min}$

are only an integer number of people of a given height and this will depend slightly on the specific finite sample which was chosen. Likewise in a conflict, there are an integer number of casualties and a finite number of events. This means that the data in Figure 3 will appear like a rugged landscape at large x where there are fewer events. Hence it is preferable to produce a cumulative version of Figure 3, such that each point represents the fraction of events having *size x or greater*. If the original distribution has the form $x^{-\alpha}$ as in Figure 3, and the cumulative version is effectively just the integral of this mathematical function, the form of this cumulative version above $x_{\min}$ will be $x^{-[\alpha-1]}$. Hence the corresponding slope on a log-log plot will be simply $-[\alpha - 1]$. This cumulative version, which is shown in Figure 4, is called the complementary cumulative distribution of the original power law distribution.

4 Richardson's Search for a Law of War

The quantitative analysis of conflict casualties was started in earnest around the time of World War II, by Lewis Fry Richardson. Richardson had been an ambulance driver in World War I, and decided to collect together the total casualty figures from every war that had taken place between 1820 and 1945. For full details of Richardson's work, see Refs. [10, 11] – also see the work of Lars Cederman who has given a possible interpretation of Richardson's findings using a model based on the spread of forest fires [12]. When he plotted the number of wars having a given number of casualties, Richardson found that the distribution was very different from that in Figure 1. This finding is very surprising in itself, since one might have expected that there would be a typical size of a war with a typical number of casualties – and a spread around this value according to each war's particular circumstances, like in Figure 1. But even more remarkable was Richardson's next realization that the actual distribution closely resembled a power law as in Figures 2-4. This finding is quite unexpected: wars have different causes, are fought by different people in different parts of the globe, and seem so horrifyingly unique that it would appear impossible that any significant inter-relationships would arise. Yet what Richardson found was not just a statement of similarity in words

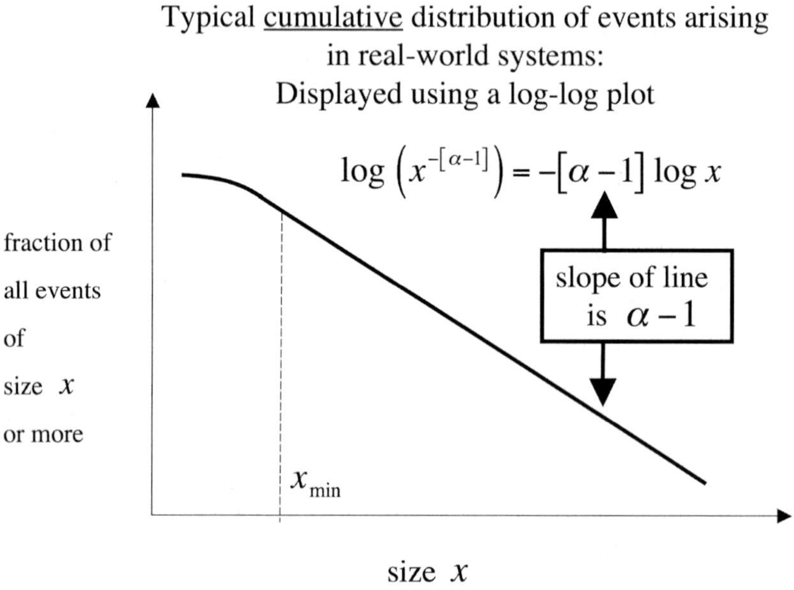

Fig. 4. The complementary cumulative distribution of the original power law distribution from Figure 3. This cumulative version becomes a straight line for x values larger than $x_{\min}$, when plotted on log-log graph paper. The slope is now $-[\alpha - 1]$, as opposed to $-\alpha$ for the power law distribution itself in Figure 3

– instead it suggested that all wars from 1820 to 1945 were connected by a single mathematical power law relationship.

The fact that these wars seem to follow a power law has some important consequences as compared to the bell-curve distributions typified by Figure 1. First the good news: The most frequent wars will be the ones with fewest casualties, unlike the case of people's heights. Now the bad news: Very deadly wars and attacks with many casualties will occur – rarely, but they will occur. This is unlike the case of heights, where the chances are zero that someone will be taller than 12 feet. For this reason, planning for wars is inherently a complex task. House designers can happily put the height of an entrance at something less than 12 feet knowing that such a tall housebuyer will never appear. They can also put step heights above 6 inches, knowing that such a small person will also never appear. However, the presence of a power-law means that this type of assumption will not work for wars. Unlike the bell-curve, the distribution of wars predicts that future conflicts can have an extremely wide range of casualties. Figure 2 indicates this by listing the behaviour of the average of a power law distribution and the fluctuations (i.e. standard deviation) as a function of the power-law parameter α. The large possible fluctuations suggest that instead of planning for some typical future war, planners should indeed plan for the worst case.

5 Global Conflicts and Global Terrorism

Two University of New Mexico researchers – Aaron Clauset and Maxwell Young – recently took another look at the work of Richardson, but this time in the context of terrorism [13]. They repeated what Richardson did, but used instead the number of casualties per terrorist attack rather than the number per war. What they found was equally remarkable to Richardson's original results.

Despite the fact that terrorist attacks are typically well spread out in time, and in space – in other words, they occur quite rarely, and are spread out over the entire globe – Clauset and Young found that when they made a log-log plot of the number of events with a total of x casualties versus the number of casualties (i.e. event size) x, they also saw evidence of a power law. In other words, the number of terrorist attacks with a total of x casualties varies according to $x^{-\alpha}$. When they restricted themselves to terrorist attacks occurring in non-G7 countries, they found the value of α to be 2.5. For terrorist attacks occurring in G7 countries, they also found a power-law, but with α now equal to 1.7. However, like Richardson, they were unable to offer a plausible explanation for these numerical results in terms of the behaviour of the underlying violent groups.

6 Measuring the Character of Ongoing Wars

We recently built and analyzed a very detailed dataset for the individual conflict events in the twenty-plus year, ongoing war in Colombia [3]. Then we did the same for the war in Iraq [3]. In principle, we could have added together all the casualties for each of these 'new' wars, and then added these two new datapoints to Richardson's dataset of total casualties for individual wars. However, we instead pursued the following line of thinking: Wars follow a power-law, and wars are a human activity. But given that a war is generally made up of lots of smaller battles or clashes, like 'wars within wars', would we also see a similar power law pattern emerging *within* a single war? In other words, can a single war be seen as a set of wars-within-wars?

This is indeed exactly what we found when we plotted the histogram of the number of events within a given war with x or more casualties, versus x, on a log-log plot as in Figure 4. The datapoints for each war fell neatly on to a straight line, as shown in Figures 5 and 6 for Iraq and Colombia respectively. Despite the very different origins, motivations, locations and durations

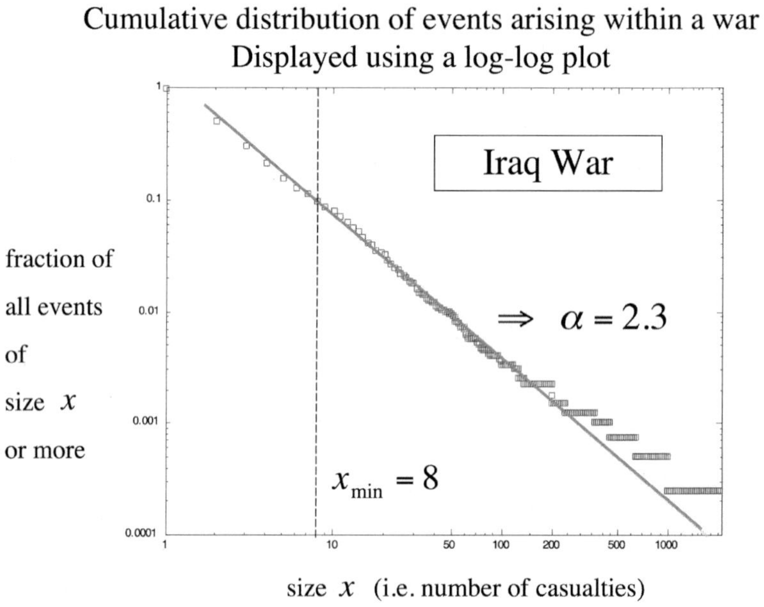

Fig. 5. Pattern of casualties from the events within the Iraq War. The log-log plot shows the fraction of events with x casualties within a given war, plotted against the number of casualties x

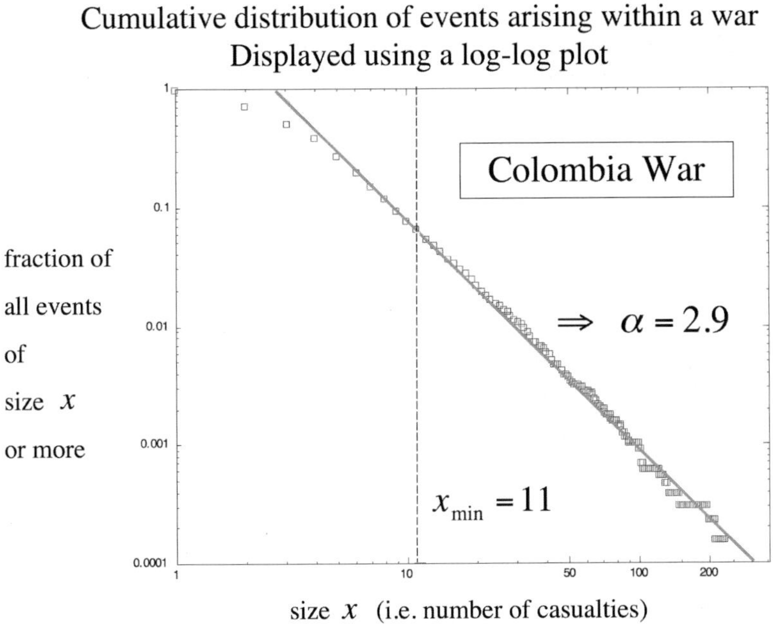

Fig. 6. Pattern of casualties from the events within the Colombia War. The log-log plot shows the fraction of events with x casualties within a given war, plotted against the number of casualties x

of the wars in Iraq and Colombia, we found similar power law patterns in the casualty figures for the events within each war. This finding is surprising not just because of the different conditions of the wars, and their very different locations, but also their different durations. The Iraq war is basically being fought in desert conditions and, at the time of writing, has only been going on for a few years – meanwhile the guerilla war in Colombia is mainly fought in mountainous jungle regions, and has been ongoing for more than twenty years against a fairly unique back-drop of drug-trafficking and Mafia activity. We then repeated this exercise for wars as diverse as Israel, Senegal, Peru and Afghanistan – in each case, we obtained a power law. What these findings suggest, therefore, is that these modern wars' character has less to do with geography or ideology and much more to do with the day-to-day mechanics of human insurgency – in other words, it has to do with the way in which groups of human beings fight each other. As mentioned earlier, this is exactly the same kind of common feature that Complex Systems researchers have found in financial markets [1]. When left to their own devices, without any 'invisible hand' or central controller, human groups interact in such a way as to produce markets with similar characteristics, and wars with similar characteristics.

Not only did we obtain straight-line slopes in each war, but these slopes all produced a power law exponent α near 2.5. Furthermore, this is the same value found by Clauset and Young for non-G7 terrorism. By contrast when we look at data from within older wars – such as the civil wars in the US, Spain and Russia – we find no statistical evidence for a power law. In all this empirical data analysis, there is an important practical point: the power law exponent α is insensitive to any systematic over- or under-reporting of casualties because the overall number of casualties is essentially just a normalizing factor for the overall power law distribution. Hence the power law signature successfully focuses on the war's internal pattern of events and hence casualties, as opposed to simply monitoring the aggregate number of casualties.

7 A Model War

Why should 2.5 emerge as an apparent magic number, connecting together all modern wars and global terrorism? To answer this, we developed a model of dynamical group-formation to describe an insurgent force. Our cue came from the fact that most modern wars, including terrorism, can be characterized by an asymmetric 'David-and-Goliath' structure in which a small, but agile, insurgent force faces a much stronger, but more rigid, institutional force such as a state's army. Because of its less rigid structure, the insurgent force is able to self-organize itself into a loosely connected soup of attack units which combine and dissociate over time in response to their own ad hoc operations, and in response to the state army's operations. These attack units are shown schematically in Figure 7. The number of dark shadows in each unit is proportional to the number of casualties that that unit will inflict in a typical conflict event. In other words, each attack unit has a particular 'attack strength' which indicates the average number of casualties which will arise in an event in which this attack unit is involved.

Each attack unit comprises a group of people, weapons, explosives, machines, or even information, which temporarily organizes itself to act as a single unit. In the case of people, this means that they are probably connected by a common location, or by some common communication system. However, an attack unit may also consist of a combination of people and objects for example, explosives plus a few people, such as in the case of suicide bombers. Such an attack unit, while only containing a few people, could therefore have a high attack strength. Information could also be a valuable part of an attack unit. A lone suicide bomber who knows when a certain place will be densely populated – for example a military canteen at lunchtimes – and who knows how to get into such a place unnoticed, will also represent an attack unit with a high attack strength. When a given attack unit undertakes an attack, it creates a number of casualties proportional to its strength – hence the distribution of

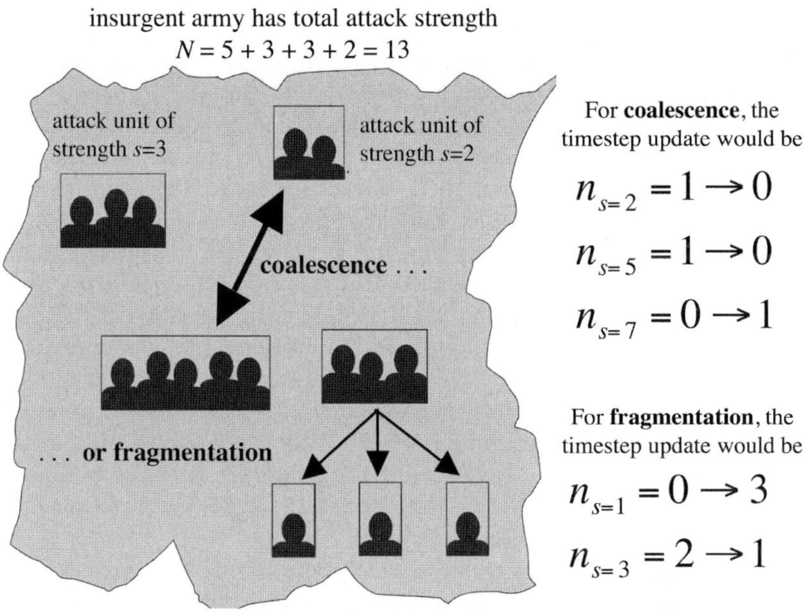

Fig. 7. A model which successfully reproduces the power law pattern observed in a wide range of modern wars including those in Iraq, Colombia and global terrorism in non-G7 countries. Here $n_{s=1}$ represents the number of attack units of attack strength $s = 1$, and similarly for $n_{s=2}$, $n_{s=3}$ etc

attack-unit strengths will reflect the distribution of casualties which arise in the war.

As the war unfolds in time, the attack units either join forces with other attack units (i.e. coalescence) or break up (i.e. fragmentation). In the real war, joining forces or breaking up would probably involve a decision process – so a model of whether to choose option 1 (i.e. combine with another attack unit) or option 0 (i.e. break up) might make sense. However, such a model would be difficult to analyze mathematically. Instead, we have found that we can use a much simpler description of insurgent decision-making, and yet still explain the observed data. In particular, we assumed that the attack units effectively use a coin to make decisions about whether to combine or break up. In particular, we assumed that attack units join together with a given probability $1 - p$, and break apart with a probability p. We then allowed this process whereby attack units break up or combine, to carry on indefinitely. To our surprise, we found that the insurgent force reaches a kind of status quo in which the distribution in the number of attack units with a given attack strength, follows a power law. Since each attack unit will produce an average number of casualties equal to its attack strength in any given event, this distribution also represents the distribution of casualties per event. So this agrees nicely with the real casualty data – but the surprises don't stop there. Remarkably, the value of the power-law slope which emerges from the model is 2.5, which is the same value of α as that obtained for the real wars and non-G7 terrorism. If we then make the group formation-dissociation probabilities depend on the existing group sizes, this α value can be moved toward 2.0 or 3.0, thereby incorporating all the results which we obtained for modern wars. Generalizing the model further to include multiple insurgent groups, yields a near-perfect fit with the real data over the entire range of α, including the nonlinear deviations at high and low x. Hence we can explain the entire range of casualty events in all modern wars and terrorism, just by using slight variations of the same basic model. This suggests that the dynamics of insurgent group formation are the same across all arenas – and as a consequence of this, it would seem that unless the stronger, but more rigid, opponent can change its tactics, the same statistical patterns of casualties will be repeated indefinitely into the future.

The reason that a power law arises from our model, is itself interesting from the perspective of Complex Systems. The precise distribution of attack units that are available for breaking up or combining at a given moment in the war, will depend on what has been happening to the soup of attack units leading up to this moment – and this, by definition, means that there is feedback from the past. As a result of this feedback, the distribution of attack units, and hence casualties, will neither be completely disordered nor completely ordered. Instead, it gives us something which is complex.

8 The 'When' of War

We have now taken this analysis further, looking at how the war evolves in time. In particular, we chopped up the length of each war since its beginning into little sections, and found that the data in each piece also followed a power-law. We then deduced the slope of the power-law for each piece. The result is shown in Figure 8. Remarkably, the size of the power-law slope in each war has crept even closer to 2.5 over time. This suggests that both these wars and global non-G7 terrorism currently show the same underlying patterns and hence character. This in turn suggests that the insurgent forces underlying these modern wars and terrorism, are now effectively identical in terms of how they are operating. These results also let us re-interpret the history of a given war. In particular, the Iraq war began as a conventional confrontation between large armies, but continuous pressure applied to the Iraqis by coalition forces broke up the insurgency into a collection of attack units. In Colombia, on the other hand, the same end result has been arrived at in the opposite way. In particular, the guerrillas in the early 1990 were unable to join up into

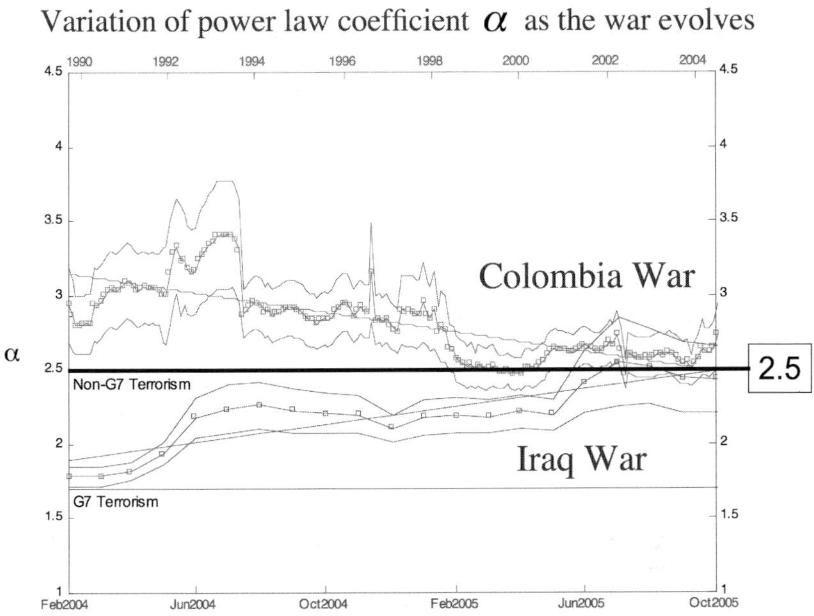

Fig. 8. Two modern, but very different, wars in Iraq and Colombia seem to have evolved in such a way that they currently have the same power law exponent α. This power law exponent matches that obtained from the distribution of terrorist attacks in non-G7 countries

high-impact units – hence the attack units were all very small. But since then, they have gradually been acquiring comparable capabilities, and now have a distribution of attack units which is as wide as that of the insurgents in Iraq. Furthermore, the fact that both Iraq and Colombia currently have the same power-law slope as non-G7 terrorism, suggests that the attack unit structures in all three arenas are currently the same. But what if someone has artificially inflated or deflated the casualty numbers for Iraq or Colombia? Fortunately it turns out that it doesn't matter too much, since any systematic multiplication of the raw numbers by some constant factor has no affect on the slope and hence the value of α.

These results provide us with significant new insight into the character of wars. However as anyone operating in a conflict on the ground knows, it would be even more useful to know something about the pattern of attacks in time – in particular, on a daily scale. Judging from the news from Iraq that we hear, it certainly doesn't seem like there is any pattern. Tuesday there might be three attacks in Baghdad, with eight casualties in one attack and twenty in the other two. Wednesday, there might be one attack in Tikrit with fifteen casualties. And so it goes on. So is there any method at all underlying this madness? To answer this question, we took the output time-series from this Complex System – in particular, we took the list of the number of attacks per day in Iraq – and started looking for patterns. Unfortunately most statistical tests require lots of data – and the Iraq war is a one-off event, so it only has one set of data. We were therefore faced with a problem which is analogous to the following situation. Imagine someone has told you that they have shuffled a deck of cards. You don't believe them, and so you want to check. If they have indeed shuffled them, then the sequence in which the cards appear should look random. But what does this mean? It means that the actual sequence of cards should look similar to a deck which has been thoroughly shuffled. Now let's suppose that the sequence of cards in the deck represents the sequence of attacks-per-day in the Iraq war. In particular, each card represents a day, and the number of points on each card represents the number of attacks on that day. For example, the three of clubs, hearts, diamonds or spades would correspond to a day with three attacks. Hence the total number of attacks that the insurgent force can produce over the length of the war, is equal to the total number of points in the deck. What we wanted to find out is if there is any specific order in which the insurgent force is performing these daily attacks – in other words, if there is any specific order in which the cards are arranged?

This card analogy gave us the clue as to how to proceed with the real Iraq data. We took the deck of cards – or equivalently the set of attacks-per-day – and shuffled them thoroughly. In doing so, we produced a 'random Iraq war' in which the numbers of attacks on consecutive days are unrelated. We then repeated this process in order to obtain a large set of such random Iraq wars. Since this analysis of the number of events doesn't involve the size of each event, each of these random wars has exactly the same distribution of casualties as the actual Iraq war, i.e. it would produce exactly the same power-law as in Figures 5 and 6 and with the same slope. However, the order in which the attacks-per-day occurred would be different in each version. By repeating this procedure many times, we were able to get a picture of what the war in Iraq would be like if the sequence of daily attacks was random.

We found that the actual sequence of daily attacks in Iraq shows more order than for a random war. In other words, there does indeed seem to be some systematic timing in the attacks and hence some forward planning by the insurgent groups – just as we would expect from a Complex System containing a collection of competitive, decision-making agents. Going further, we have been able to deduce the particular sequences of daily attacks which occur more often than expected, and those which occur less often than expected. What is even more surprising is that we find similar results for the case of Colombia. Needless to say, we are currently hard at work on further tests to uncover the full extent of the temporal patterns underlying such attacks.

9 Future Work and Perspectives

Having looked at the sizes and timing of events in modern wars, we are extending this to organized crime activity including homicides, kidnappings and extorsion. In particular, we have successfully created multi-agent models which mimic the decision-making dynamics of insurgent groups, just as had been done earlier for groups of financial traders [1]. By analyzing the size, timing and spatial coordinates of a given event, as well as the groups involved, we are now able to reconstruct the possible trails which a particular insurgent group might have followed. Just as in a multi-species ecological setting within the natural world, we are interested in determining the behaviours and possible protocols which arise when a particular group from insurgent army A happens to cross the path of a particular group from insurgent army B. In particular, we are trying to deduce whether they decide to fight each other, collaborate, ignore each other – or even consciously avoid each other. Going further, we know that wars like the ones in Colombia and Afghanistan have taken place against the backdrop of an illicit trade such as drug trafficking. This activity provides an effective nutrient supply in the form of money for buying supplies and weapons, and thereby helps feed the war as a whole. So just like a fungus will thrive in a forest, or a cancer tumour will thrive in a host, these armed groups are fed by a rich source of nutrients which allows them to self-organize into a robust structure.

Finally we note that far from threatening traditional interpretations of individual wars by historians, sociologists and political scientists, this work provides a complementary cross-conflict analysis which in turn offers a framework for comparing wars – not in terms of their origins, geography or duration, but in terms of the 'way' in which the war is fought. In other words, we are looking at the character of human conflict.

References

1. Johnson, N.F., Jefferies, P. and Hui, P.M.: Financial Market Complexity. Oxford University Press (2003)
2. Johnson, N.F.: Two's Company, Three is Complexity. Oneworld Publishing (2007)
3. Johnson, N.F., Spagat, M., Restrepo, J.A., Becerra, O., Bohorquez, J.C., Suarez, N., Restrepo, E.M., Zarama, R.: For full details of this work, and the underlying model, see the e-print titled 'Universal patterns underlying ongoing wars and terrorism'. available at http://xxx.lanl.gov/abs/physics/0605035
4. Couzin, I.D. and Krause, J.: Self-organization and collective behavior of vertebrates. Advances in the Study of Behavior **32** (2003) 1
5. Couzin, I.D., Krause, J., Franks, N.R. and Levin, S.A.: Effective leadership and decision making in animal groups on the move. Nature **433** (2005) 513
6. Rainey, P.B.: Unity from Conflict, Nature **446** (2007) 616
7. Nowak, M.A.: Five Rules for the Evolution of Cooperation. Science **314** (2006) 1560

8. Plowes, N.J.R. and Adams, E.S.: An empirical test of Lanchester's square law: mortality during battles of the fire ant Solenopsis invicta. Proceedings of the Royal Society of London, Series B **272** (2005) 1809
9. Newman, M.E.J.: Power laws, Pareto distributions and Zipf's law Contemporary Physics **46** (2005) 323
10. Richardson, L.F.: Variation of the Frequency of Fatal Quarrels with Magnitude. Amer. Stat. Assoc. **43** (1948) 523
11. Richardson, L.F.: In: Wright, Q. and Lienau, C.C.: (eds.) Statistics of Deadly Quarrels Boxwood Press, Pittsburgh (1960)
12. Cederman, L.: Modeling the Size of Wars: From Billiard Balls to Sandpiles. Amer. Pol. Sci. Rev. **97** (2003) 135
13. Clauset, A. and Young, M.: Scale invariance in Global Terrorism. e-print available at http://xxx.lanl.gov/abs/physics/0502014

Fostering Consensus in Multidimensional Continuous Opinion Dynamics under Bounded Confidence

Jan Lorenz

Universität Bremen, Bibliothekstr. 1, 28359 Bremen, Germany `mathjanlo.de`

1 Introduction

Consider a group of agents which is to find a common agreement about some issues which can be regarded and communicated as real numbers. Each agent has an opinion about each issue which he may change when he gets aware of the opinions of others. This process of changing opinions is a *process of continuous opinion dynamics*. Examples for discussing groups are parliaments, commissions of experts or citizens in a participation process. The opinion issues in parliaments can be tax rates or items of the budget plan, in commissions of experts predictions about macroeconomic factors and for citizens the willingness to pay taxes or the commitment to a constitution.

In many processes of opinion dynamics it is desirable that the agents reach *consensus*, either for reaching a good approximation to the truth or for the reason, that reaching consensus is a good in itself (e.g. in the commitment to the constitution). Often, all relevant information about a societal issue has been collected and published but it is not reliable enough to bring a collective opinion or 'the truth' without opinion dynamics where agents judge, communicate and negotiate about the 'right' opinion. In the need of a collective decision it is the best for the group to achieve consensus because it does not need a decision by voting or other mechanisms with potential to conflict. In this study we make simple but reasonable assumptions on humans in opinion dynamics. The models reproduce the formation of parties and interest groups and some other reasonable facts in real opinion dynamics. But there remain many reasonable free parameters of opinion dynamics, where we check a few with the aim to find structural conditions which might foster the achievement of consensus in the group.

We define the models based on two facts from social psychology. First, people adjust their opinions towards the opinions of others. This may be for normative or for informational reasons. So either because they feel conformational pressure and want to assimilate or because they appreciate the information of others to be relevant. Further on, people perceive themselves as members of a subgroup, according to the theory of self categorization. In our setting one feels as in a group with the people who have similar opinions. We put these descriptions of peoples behavior regarding opinion dynamics into rules for agents behavior: Agents find new opinions as averages of opinions of others and they will do this only with respect to agents which lie within their area of confidence.

Repeated averaging and bounded confidence lead to clustering dynamics. If the agents in our model have big enough areas of confidence they are able to find a consensus. If they are small they will fail and form several clusters. Are their structural properties of the opinion dynamics environment that have a positive effect on the chances of finding a consensus? Here, we will ask how structural properties of the opinion dynamics process as the communication regime, the number of opinion issues, their interdependence and the mode how agents form their area of confidence affect the chances for consensus?

With the question about conditions for consensus we grab an old research line of DeGroot [3] and Lehrer and Wagner [7] about the problem how to aggregate opinions to a rational consensus in science or society. They model aggregation by averaging with powers of reputation matrices . The work in [7] was in the flavor of the social choice problem . In recent times Hegselmann and Krause [4] grabbed on this with the idea of bounded confidence and formulated a model (now nonlinear) of opinion dynamics which can be seen as repeated meetings of agents with bounded confidence. Independently, Weisbuch, Deffuant and others [2, 11] formulated a similar bounded confidence model with random pairwise interaction, what we call gossip communication. They came with the background of social simulation, sociophysics and complexity science.

In section 2 we will outline and discuss the parameter space and define the two opinion dynamic processes. Section 3 shows the basic dynamics which are universal in these models: cluster formation in the time evolution and the bifurcation of cluster patterns in the evolution of the bound of confidence. We will set up on them in section 4 where we present and discuss the simulation results with a focus on the consensus transition. We show e.g. that raising the number of opinion issues fosters consensus if the issues are under budget constraints, but diminishes consensus if they are not. We conclude by giving a colloquial summary and pointing out further research directions.

2 Continuous Opinion Dynamics and Bounded Confidence

Here, we define the basic models of [4] and [11] such that they extend to more dimensional opinions and to different areas of confidence. We briefly discuss real world interpretations.

The agents

Often, analytical results are either possible for very low numbers of agents or in the limit for a large number of agents. Complexity arises with finite but huge numbers of agents. The fuzzy thing is that some macro level dynamics work, while at critical points changes appear very sensitive due to specific finite size effects . In the simulation studies we chose $n = 200$ because we regard this as applicable to a wide range of real groups of agents. We also checked $n = 50, 500$ to ensure that the results hold also in this range, which they do. This range of group sizes coincides with the social brain hypotheses [6] that humans can only hold about 150 relationships on average.

The opinion space and the initial profile

The opinion space is the set of all possible opinions an agent may have. In continuous opinion dynamics about d issues this is $\mathbb{R}^d$. So, we call $x^i(t) \in \mathbb{R}^d$ the opinion of agent i and $x(t) \in (\mathbb{R}^d)^n$ the opinion profile at time $t \in \mathbb{N}$. The evolution of an opinion profile is the *process of continuous opinion dynamics.* Dynamics depend heavily on the initial opinion profile. If we model dynamics by repeated averaging, then dynamics take place in the convex region spanned by the initial opinion profile $x(0)$, we call this the *relevant opinion space.* For $d = 1$ this is always an interval. For higher d there are many shapes. In this study we will restrict us to $d = 1, 2, 3$ and two shapes of the initial relevant opinion space: the *cube* $\square^d := [0, 1]^d$ and the *simplex* $\triangle^d := \{y \in \mathbb{R}^{d+1}_{\geq 0} \mid \sum_{i=1}^{n} y_i = 1\}$ (see Figure 1). Notice that $\triangle^d$ is a subset of $\square^{d+1}$. These two shapes stand for two different kinds of multidimensional problems. The cube represents opinions about d issues which can be changed independently. The simplex represents opinions about issues where the magnitude of one can only be changed by changing others in the other direction. The main example is a budget plan with a fixed amount of money to allocate. Further on, we restrict us to random initial opinions which are equally distributed in the relevant opinion space. (It is not trivial to produce an equal distribution on a simplex! Normalization to sum-one of a $d + 1$-dimensional cube would be skewed. We produce it by taking a d-dimensional cube and throwing away all opinions with sum bigger than 1. Then we compute the missing least component for each opinion.)

The area of confidence

The *area of confidence* is a region in the opinion space around an agent's opinion. He regards all opinions in this region as relevant and all others as irrelevant. This region moves when the agent changes his opinion. Formally, it is a compact and convex subset of the opinion space including the origin. The origin is mapped to the opinion of the agent. In a one dimensional opinion space the only relevant areas are intervals. In more dimensions several areas seem appropriate. We restrict this study to the unit balls of the 1- and the

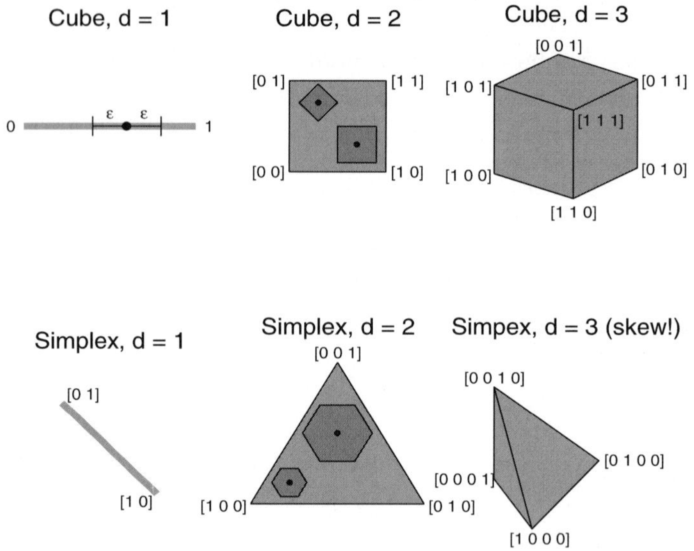

Fig. 1. □ and △ opinion spaces with example areas of confidence for $p = 1, \infty$

∞-norm (see Figure 1) centered on the opinion and scaled by a *bound of confidence* $\varepsilon > 0$. Thus, agents measure the distance of opinions $x^1, x^2 \in \mathbb{R}^d$ as $\left\|x^1 - x^2\right\|_1 = \sum_i \left|x_i^1 - x_i^2\right|$ or as $\left\|x^1 - x^2\right\|_\infty = \max_i \left|x_i^1 - x_i^2\right|$ and judge their relevance by the threshold ε. We use these norms because they are close to how humans may judge differences in opinion. Agents using the 1-norm are willing to compensate between the opinion issues. If the other agent's opinion differs a lot in one issue this can be *compensated* by differing low in another issue. Agents using the ∞-norm are *noncompensators.* Their distance in each opinion issue should be below ε to accept another's opinion. For $d = 1$ the area is always an interval. For the cube and $d = 2, (3)$ the ∞-ball is a square (cube); for the 1-ball it is a diamond (octahedron). The intersection of the 2-dimensional simplex and the 3-dimensional area of confidence is a hexagon with edge length ε for the ∞-norm and with edge length $\varepsilon/2$ for the 1-norm. The intersection of the 3-dimensional simplex and the 4-dimensional area of confidence is an octahedron for the ∞-norm and a cuboctahedron for the 1-norm. Things get more fuzzy when going to more dimensions.

The communication regime

The models of [4, 11] can both be extended naturally to the different opinion spaces and the areas of confidence outlined above. They differ in their communication regime. In the model of Hegselmann and Krause [4] each agent chooses his new opinion as the arithmetic mean of all opinions in his area of confidence. All agents do this at the same time. To do this, they need to know the opinions of all agents. We call it communication by *repeated meetings.* In the basic model of Deffuant, Weisbuch and others [11] two agents were chosen at random. They compromise in the middle if their opinions lie in the area of confidence of each other. We call this communication regime *gossip*.

Now we are ready for the mathematical definition of the two processes of continuous opinion dynamics.

Given an initial profile $x(0) \in \mathbb{R}^n$, a bound of confidence $\varepsilon > 0$ and a norm parameter $p \in \{1, \infty\}$ we define the *repeated meeting process* $(x(t))_{t \in \mathbb{N}}$ recursively through

$$x(t+1) = A(x(t), \varepsilon)x(t), \tag{1}$$

with $A(x, \varepsilon)$ being the *confidence matrix* defined

$$a_{ij}(x, \varepsilon) := \begin{cases} \frac{1}{\#I(i,x)} & \text{if } j \in I(i, x) \\ 0 & \text{otherwise,} \end{cases}$$

with $I(i, x) := \{j \mid \left\|x^i - x^j\right\|_p \leq \varepsilon\}$. ("#" stands for the number of elements.)

We define the *gossip process* as the random process $(x(t))_{t \in \mathbb{N}}$ that chooses in each time step $t \in \mathbb{N}$ two random agents i, j which perform the action

$$x^i(t+1) = \begin{cases} x^i(t) + \frac{1}{2}(x^j(t) - x^i(t)) & \text{if } \left\|x^i(t) - x^j(t)\right\|_p \leq \varepsilon \\ x^i(t) & \text{otherwise.} \end{cases}$$

The same for $x^j(t+1)$ with i and j interchanged.

Figure 2 demonstrates one time step in each process.

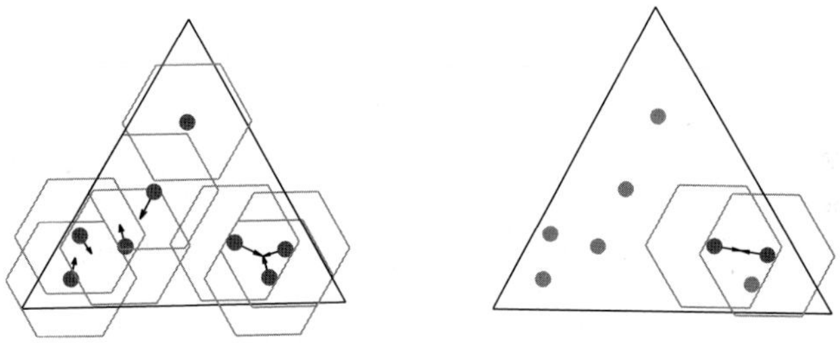

Fig. 2. Examples of one step in meeting (*left hand*) and gossip (*right hand*) dynamics in the opinion space $\triangle^2$

3 General Dynamics

Clustering dynamics in the time evolution

Every gossip and meeting process converges to a fixed configuration of opinion clusters [8, 10]. We call this fixed configuration the *stabilized profile.* A general dynamic is that opinion regions with high agent density attract agents from around. This attraction comes due to a higher probability to meet an agent in this region in gossip communication and due to the fact that the barycenter of opinions in an area of confidence is often close to a high density region.

If we consider an initial profile with uniformly distributed opinions on a certain relevant opinion space ($\square$ or $\triangle$) than the density distribution of opinions evolves over time as follows. (The following dynamic description can be traced in Figure 3(a) for $\square^1$ and for $\triangle^2$ in figure 4.) Agents at the border of the relevant opinion space move closer to the center because opinions in their area of confidence are not equally distributed. Density in the center changes only due to random fluctuations in the initial conditions. So the relevant opinion space contracts but holds mainly the same shape but with a higher agent density at the border. If a more dimensional opinion space had some vertices (as $\square$ and $\triangle$ have) the density in the evolving high density regions is even higher at the vertices due to opinions coming from more sides.

These high density regions at the vertices of the relevant opinion space attract agents from the center and may get disconnected from the center and from the other vertices at some time, due to absorbtion of the connecting agents, and form a cluster. The dynamics goes on similar in the remaining cloud of connected opinions.

If some of these high density regions lie as close to each other that a small group of agents holds contact to both, then it may happen that they attract the agents in these high density regions and both join to form a bigger cluster. This may also happen to more clusters at the same time or with some delays (see Figure 4 for an example). The fuzzy thing in more dimensions is that this contracting process happens on all face levels (e.g. faces and edges) of the shape of the opinion space on overlapping time scales. Further on, some clustering in the center may also occur due to slow deviations of uniformity. The time when some high density regions have formed but have not completely disconnected from the rest is thus the critical time phase. In more than one dimension it is unpredictable which of the intermediate clusters joins with which others. Changes may happen due to very low fluctuations in the initial profile or the communication order.

$\triangle^d$ has $d+1$ vertices and thus the same number of intermediate high density regions. The number of possible final cluster configurations that may evolve by disconnecting or joining of these high density regions is the same as the number of partitions of $\{1,\dots,d+1\}$ into pairwise disjoint subsets, which is the *Bell number* B_{d+1}. This shows the combinatorial explosion of different possible outcomes: $B_2 = 2, B_3 = 5, B_4 = 15, B_5 = 52, B_6 = 203$, $B_7 = 877, B_8 = 4140$.

Bifurcation dynamics in the evolution of the bound of confidence

For each value of ε there is a certain *characteristic stabilized profile* under the assumption of a uniformly distributed initial profile. The number, the

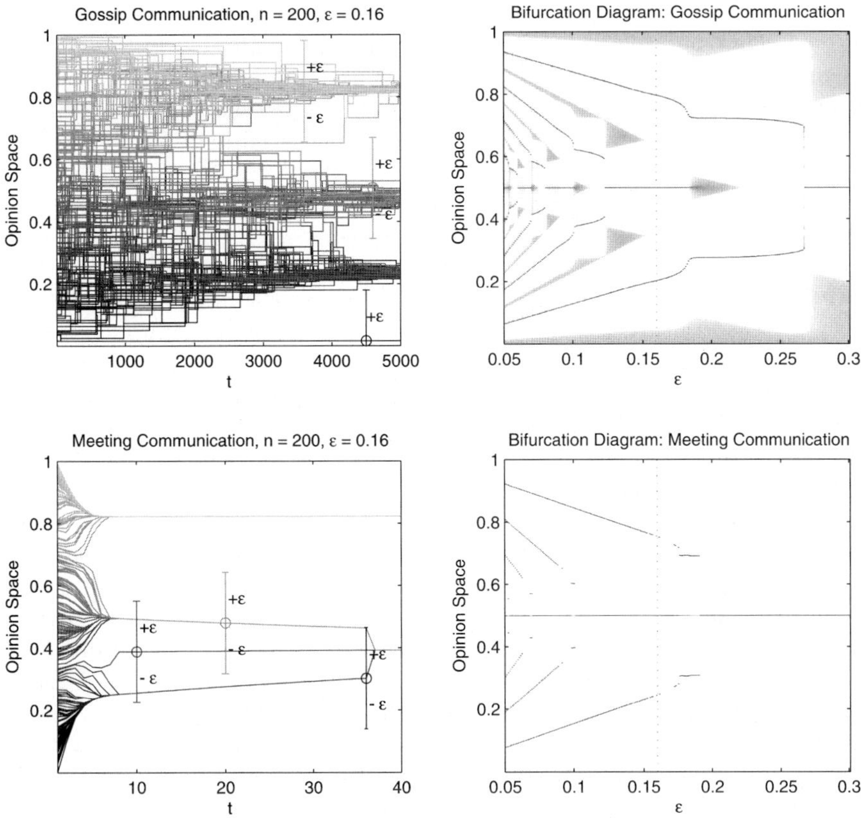

(a) Example processes for gossip and meeting communication in the interval $[0, 1]$ demonstrating the time evolution to a stabilized profile. Notice one outlier for gossip and the meta-stable state in meetings.

(b) Reverse bifurcation diagrams of characteristic states of the stabilized profile in the ε-evolution. Diagrams derived by interactive Markov chains. Black is a high number of agents, gray a low number of agents.

Fig. 3. Demonstration of general dynamical properties

size and the location of opinion clusters in this stabilized profile are of interest. In Figure 3(b) we see the reverse bifurcation diagrams for the attractive states of the meeting and the gossip process in $\square^1 = [0, 1]$ as relevant opinion space. These diagrams have been computed with interactive Markov chain that govern the evolution of the distribution of an idealized infinite population to a huge number of opinion classes in the opinion space (for details see [8, 9] and [1] for the inspiring differential equation approach). Such bifurcation diagrams should exist for moredimensional opinion spaces, too. A stabilized profile with 200 agents can be significantly blurred by low fluctuations in the

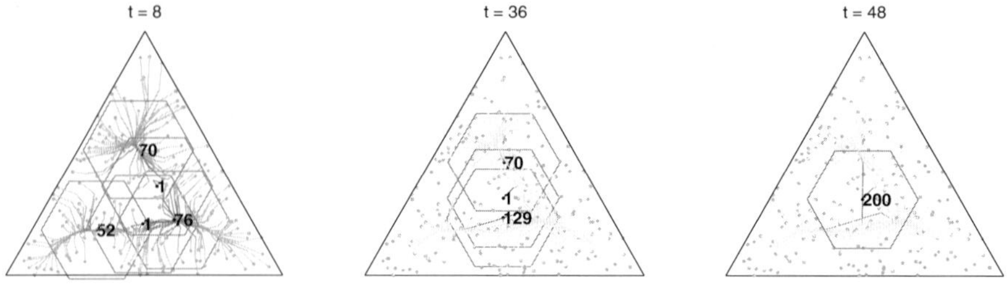

Fig. 4. Example for meeting communication in $\triangle^2$ for interesting time steps. Notice the successive joining of intermediate clusters

initial profile and thus does not behave as the bifurcation diagram predicts. But as simulation shows, a bifurcation diagram with attractive states and certain discontinuous changes when manipulating ε seems to underlie opinion dynamics under bounded confidence.

It is easy to accept that $\varepsilon = 1$ leads to a central consensus, while $\varepsilon \to 0$ leads to full plurality where no opinion dynamics happens. The behavior in between can be understood as *bifurcations* of the consensual central cluster into other configurations of clusters. In the gossip and the meeting process the main effect when going down with ε is that the central cluster bifurcates at certain values of ε into two equally sized major clusters left and right which drift outwards when lowering ε further. The central cluster vanishes (nearly) completely to get reborn and grow again until it bifurcates again. We call the interval between two bifurcation points an *ε-phase* for a characteristic stabilized profile. The length of the ε-phases scales with ε, so for lower ε the phases get shorter. This fact is the basis of the $1/2\varepsilon$-rule (see [11]) which determines the number of major clusters under gossip communication.

Besides the common behavior the gossip and the meeting process differ. Under gossip communication there are minor clusters at the extremes, a nucleation of minor clusters between the central and the first off-central clusters and minor clusters between two major off-central clusters. These minor clusters occur as a few outliers in agent based example processes, too. Meeting communication shows no minor clusters but the surprising phenomena of consensus striking back after bifurcation. Convergence in this phase takes very long (see [9]). The long convergence times to central consensus occurs also in front of each bifurcation of the central cluster. E.g. for $\varepsilon = 0.2$ we reach a meta-stable state of two off-central clusters and a small central cluster which attracts them very slowly to a consensus. The slow convergence due to meta-stable states close to bifurcation points occurs also in example processes.

In this study we focus on fostering consensus. So the most interesting point for us is the value of ε where the big central cluster bifurcates into two major clusters. This is the phase transition from polarization to consensus. We call this the *majority consensus transition.* Only 'majority' not total because of the extremal minor clusters in gossip communication. We call this point (in the style of [5]) the *majority consensus brink*.

4 Simulation Results

Simulation setup

Our simulation setup deals with initial profiles of random and equally distributed opinions with 200 agents. We run processes for the 24 settings of the opinion spaces $\Box, \triangle$ with dimensions $d = 1, 2, 3$, the areas of confidence for $p = 1, \infty$ and the communication regimes meeting and gossip. For each of these settings we took a big enough range of ε-values in steps of 0.01 so that we are sure that the majority consensus transition happens within this range. For each of this 24 settings and each value of the respective ε-range we run 250 simulation runs and collect the stabilized profiles for our final statistical analysis. We checked 50 and 500 agents with lower numbers of runs and verified that the results hold analog qualitatively and to a large extend quantitatively.

For each collection of stabilized profiles for a given point in the $\{\Box, \triangle\}$-$\{d = 1, 2, 3\}$-$\{p = 1, \infty\}$-{meeting/gossip}-ε-parameter space, we have to measure the degree of consensus. In earlier studies the most used measure was the average number of clusters. This is inappropriate because of the minor clusters at the extremes under gossip communication. We use the *average size of the biggest cluster*. If it is 200 we are for sure above the majority consensus brink. If it is slightly below this can have two reasons according to what we know from Section 3. First, some runs reach consensus, while some others polarize, or second, there is a big central cluster but also an amount of agents in minority clusters at the extremes. The second happens mostly for gossip communication.

Figure 5 shows the average size of the biggest cluster with respect to ε for all 24 parameter setting. We derive qualitative statements about *fostering consensus* from that. With fostering consensus we mean that the transition to a majority consensus appears for lower values of ε.

The impact of the communication regime (meetings vs. gossip)

Communication in repeated meetings is fostering consensus in comparison to gossip. But surprisingly Figure 6(a) gives strong evidence about the universal scale that a group of agents in meeting communication needs only 0.8ε to reach the same average size of the biggest cluster as the same group under gossip communication with ε. This holds for all our parameter settings. Only for very high sizes of the biggest cluster meeting communication gets even better, probably due to more minor clusters in gossip communication.

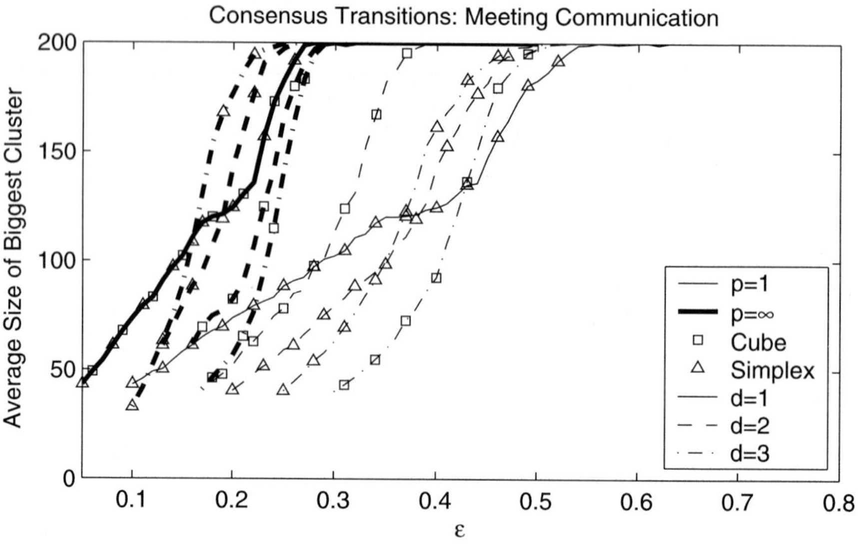

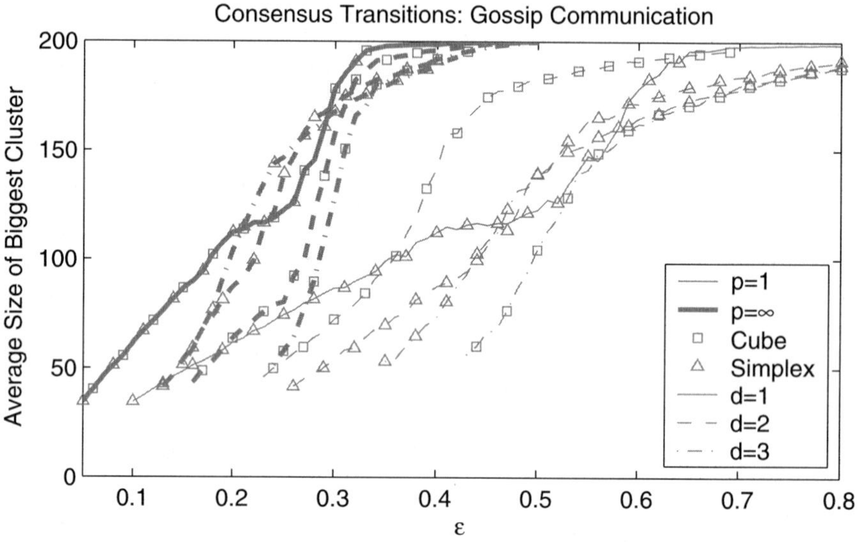

Fig. 5. The average size of the biggest cluster for □, △ (marker), $d = 1, 2, 3$ (line style), $p = 1, \infty$ (line width) and communication regime (*black, gray*)

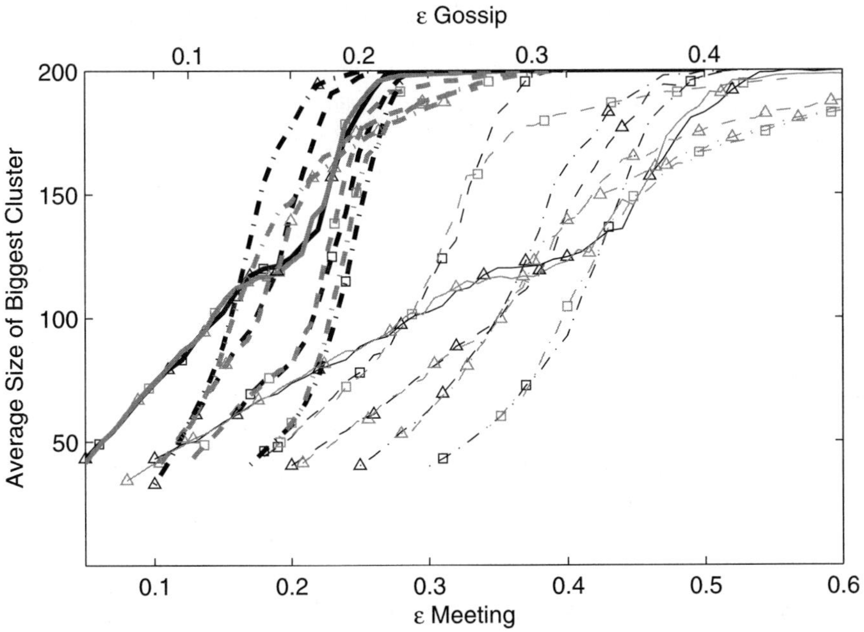

(a) The lines of figure 5 with ε-axis for lines of meeting communication scaled to 80%.

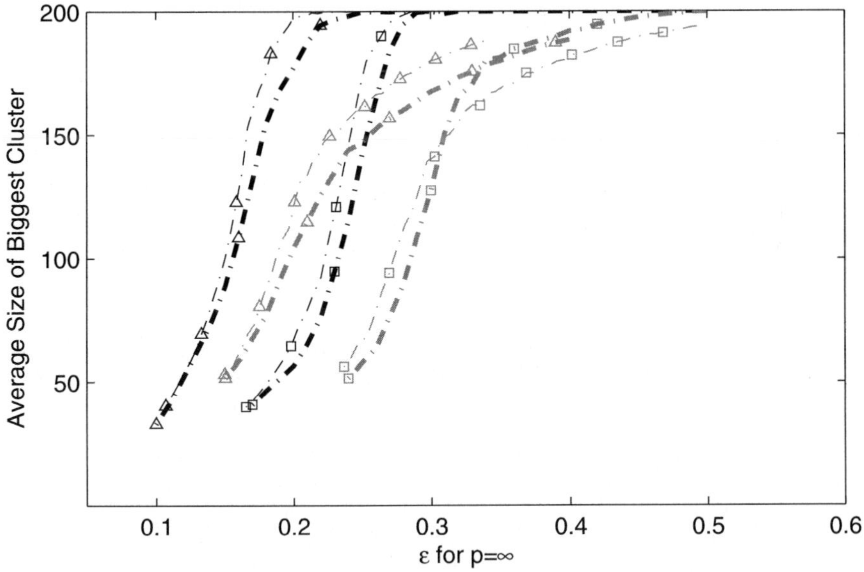

(b) The lines of figure 5 with ε-axis for lines of $p = 1$ scaled to volume equality with $p = \infty$.

Fig. 6. Further simulation results for the average size of the biggest cluster

The impact of the number of opinion issues d

What happens if we raise the number of issues? The answer is: It depends on the shape of the initial relevant opinion space. In a simplex, raising the number of issues fosters consensus. In a cube raising the number of issues dilutes consensus. Numerical values for fostering with meeting communication in a simplex and $p = \infty$: the biggest cluster contains at least 80% of the agents in 80% of the runs for $\varepsilon > 0.25$ with $d = 1$, $\varepsilon > 0.23$ with $d = 2$ and, $\varepsilon > 0.20$ with $d = 3$. One drawback is that under gossip communication we produce more and bigger extremal minor clusters in a simplex when raising d, one in each vertex. Thus, for fostering a complete consensus without outliers raising dimensionality under gossip dynamics is not good.

The impact of the shape of the relevant opinion space ($\triangle$ vs. $\square$)

What fosters consensus better: an opinion space of three independent issues ($\square^3$) or four issues under fixed budget constraints ($\triangle^3$)? Colloquial: Is it good to add a budget dimension. The simplex is better for all p and all communication regimes. But this does not hold for $d = 2$, where the square is better under $p = 1$ but the simplex is better under $p = \infty$. Both shapes are trivially equal for $d = 1$. We conjecture that the simplex is getting better in higher dimensions. Another question of similar type is: Does it foster consensus to break a problem of three independent issues ($\square^3$) down to a problem of three issues under budget constraints ($\triangle^2$)? The answer is yes. It holds also for breaking down from $\square^2$ to $\triangle^1$ under $p = \infty$, but it is the other way round for $p = 1$.

The impact of compensating vs. noncompensating ($p = 1, \infty$)

Imagine you appeal to your noncompensating ($p = \infty$) agents 'compensate: switch to $p = 1$'. This would imply that they should not tolerate distances of ε in each issue but only in the sum of all distances. Of course this will not foster consensus because their area of confidence is then only a smaller subset of their former. Perhaps you can appeal, that they should compensate in the way such that they should allow longer distances then ε in one issue in the magnitude as the other distances are short. This would lead to maximal distances of $d\varepsilon$ in one issue and perhaps the agent find this two much to tolerate. The 'mathematically correct' switching from noncompensating to compensating is to scale ε to that magnitude that the d-dimensional volumes of the areas of confidence would be equal. We did this for $d = 3$ in Figure 6(b). The scale for $\square^3$ is $\sqrt[3]{6} \approx 1.82$ and for $\triangle^3$ it is $\sqrt[3]{64/5} \approx 2.34$. This 'normalization' leads to the result that switching to compensating fosters consensus a little bit. Probably this result holds only in this configuration of the relevant opinion space and the area of confidence, there might be negative configurations.

5 Summary and Outlook

A colloquial summary: If we want to foster consensus and believe that agents adjust there opinions by building averages of other's opinions but have bounded confidence, then we should manipulate the opinion formation process in the following way (if possible):

- Install meetings (or publications) where everyone hears all opinions and do not rely only on gossip.
- Bring more issues in but put them under budget constraints.
- Release guidelines about compensation in the judgements of different issues.

Of course, our simple model neglects several properties of real opinion dynamics, e.g. rules about voting decisions, underlying social networks, heterogeneity of agent's confidence, long run ideologies or strategies and inflow of new information. All this are tasks for further analytical and experimental work. An unanswered question is also the reason for the universal 80% scale for meeting communication compared to gossip.

But we believe that under more realistic extensions there will be influence of the underlying bifurcation diagram and that critical consensus transitions will exist. Thus, it is worth to observe and design the structural properties of opinion dynamic processes, if one aims to foster consensus.

References

1. Ben-Naim, E., Redner, S. and Krapivsky, P.L.: Bifurcation and patterns in compromise processes. Physica **183** (2003) 190–204
2. Deffuant, G., Nadal, J.–P., Amblard, F. and Weisbuch, G.: Mixing beliefs among interacting agents. Advances in Complex Systems **3** (2000) 87–98
3. DeGroot, M.–H.: Reaching a consensus. Journal of the American Statistical Association **69** (1974) 118–121
4. Hegselmann, R. and Krause, U.: Opinion dynamics and bounded confidence, Models, Analysis and Simulation. Journal of Artificial Societies and Social Simulation **5** (2002) `http://jasss.soc.surrey.ac.uk/5/3/2.html`.
5. Hegselmann, R. and Krause, U.: Opinion dynamics driven by various ways of averaging. Computational Economics **25** (2004) 381–405
6. Hill, R.A. and Dunbar, R.I.M.: Social network size in humans. Human Nature **14** (2003) 53–72
7. Lehrer, K. and Wagner, C.: Rational Consensus in Science and Society D. Reidel Publishing Company, Dordrecht, Holland (1981)
8. Lorenz, J.: A stabilization theorem for dynamics of continuous opinions. Physica A **355** (2005) 217–223
9. Lorenz, J.: Consensus strikes back in the Hegselmann-Krause model of continuous opinion dynamics under bounded confidence. Journal of Artificial Societies and Social Simulation **9** (2006) `http://jasss.soc.surrey.ac.uk/9/1/8.html`.

10. Moreau, L.: Stability of multiagent systems with time-dependent communication links. IEEE Transactions on Automatic Control **50** (2005)
11. Weisbuch, G., Deffuant, G., Amblard, F. and Nadal, J.–P.: Meet, Discuss and Segregate! Complexity **7** (2002) 55–63

Multi-Stakeholder Governance - Emergence and Transformational Potential of a New Political Paradigm

Bertrand de La Chapelle

French Ministry of Foreign Affairs, 23, rue La Pérouse, 75116 Paris, France
bdelachapelle@gmail.com

1 Introduction

The Internet has changed the way we work, learn and entertain ourselves. It is now changing the way humans organize themselves in societies and forces a political paradigm change in their governance. Although it barely made headlines, the World Summit on the Information Society (WSIS) may well stay in history as having sanctioned, with the creation of an "Internet Governance Forum", a promising alternative political paradigm: the "multi-stakeholder approach".

In "The structure of scientific revolutions", Thomas Kuhn introduced the concept of scientific paradigm shift. He described how dominant scientific theories, such as the Ptolemaic astronomical system, become contested when they prove unable to explain new observations produced by more precise instruments, in that case, Galileo's telescope. A crisis period ensues: the underlying paradigm, e.g. the earth as center of the universe, loses credibility and alternative paradigms begin to compete, until one - the Copernican model - finally imposes itself because of its superior explicative and predictive power.

There is a natural analogy in the political sphere. Whereas scientific paradigms are the basis of theories that help humans understand the world and influence it, political paradigms form the basis of governance systems that help humans organize their societies and the relationships between them. Therefore, just as scientific paradigms are contested when the theories built on them cannot explain the world any more, established political paradigms naturally become contested when the governance structures they provide the foundations and legitimacy for prove unable to address the new challenges a society faces or do not allow human polities to organize themselves peacefully.

* The ideas expressed here are exclusively the views of the author and in no way represent an official position of the national government on behalf of which he participates in international negotiations.

The Internet has transformed social and economic activities in ways that were hard to predict only ten years ago. But it also created a global community of a billion people. The present paper exposes the potential of a new political paradigm for the governance of global polities: 1) why Internet-related issues are difficult to address within a United Nations system based on the paradigm of a community of nation-states; 2) how the notion of multi-stakeholder Internet governance emerged from the World Summit on the Information Society (WSIS), a four year United Nations process; 3) why this represents a potential paradigm shift that can ultimately simplify and transform the international system.

2 Why the Internet Challenges the Existing Intergovernmental System

The development of the Internet has changed the way human societies structure themselves, accelerating a trend towards the formation of complex networks. It also raises public policy issues that the present hierarchical and geographic country-based system of intergovernmental agencies has difficulties addressing.

2.1 The Internet as Complex Network

What we casually call "The Internet" is a unified network resulting from the aggregation of hundreds of thousands heterogeneous networks respecting common interoperability protocols. A few principles, such as the end-to-end principle or the technically layered structure, have allowed this unique artefact of human inventiveness to evolve into probably the most elaborate man-made structure, exhibiting emergent properties that were not initially planned.

The Internet of today is the emergent result of the cumulative efforts of millions of actors, individuals, companies and governments, guided by a common vision, and who self-organized. Studying such a multi-agent, dynamic, adaptive, non-linear and distributed system naturally interests complexity science, a discipline that has developed considerably in the recent decades.

Indeed, scientists like Laszlo Barabasi have explored the clustered, somehow fractal organization of this highly complex physical network as well as the main application it supports (the World Wide Web). Their studies revealed a scale-free and power law structure: a very limited number of major nodes aggregating most connections and an immense majority of nodes with very few links.

2.2 Global Social Networks

But beyond connecting machines or databases, the Internet connects people and transforms their social interactions. In particular, with the development of blogs and social networking tools, the academic network cum commercial marketplace has evolved during the recent years into a complete social, economic and political space. A global Internet Community of a billion people is today distributed - albeit still too unevenly - on the surface of the whole planet. Their social links are highly multi-dimensional, combining traditional family ties, tribal connections or national citizenships but also linguistic, cultural and religious attachments. This multiplies the number of communities, thematic groups and organizations that people are members of, and of course the issues they have a stake in.

Most importantly, the very existence of the Internet encourages the development of "weak links" between physically distant people based on thematic affinities, through a multitude of interest groups. Such relations, maintained on an episodic basis in spite of distances, sometimes nurtured by professional meetings or leisure travel around the world, dramatically reduce the number of "degrees of separation" between the members of this global community and make it densely interconnected.

Such trends existed or were latent before the Internet. But the global communications network has considerably accelerated them. As a result, individuals around the world now belong to multiple, sometimes overlapping, social networks. Each of these networks increasingly emerge and organize itself online like the Internet and the Web did: in particular, in large online community sites, some individuals represent a disproportionate part of the sales (the Power Sellers on eBay), have a disproportionate number of "friends" (on MySpace), or post a disproportionate number of contributions (on Wikipedia or Amazon Reviews).

Future studies of this global community will most likely confirm that the organization of global social networks exhibits properties comparable to those detected in the Internet and the World Wide Web: power law distributions and scale-free structures, with strong, emergent clustering. Indeed, these social connections form the embryo of a network of global "polities".

2.3 Complex Policy Issues

Members of these global communities nonetheless interact with one another in a global common space, and the growing usage and ubiquity of the Internet raise new public policy issues. Most of them, such as the fight against spam or cybercrime, protection of privacy and personal data or freedom of expression are highly transnational and involve potential conflicts of jurisdiction.

Such issues also usually involve a great diversity of actors: governments of course, but also private companies and non-governmental organizations (NGO). They also potentially articulate multiple levels: some local action somewhere on the planet can have a global impact and vice versa. Finally, the problems to be addressed are often non-linear as there is no proportionality between the cause and effects: for instance a few lines of code can replicate virally around the world and infect millions of users in a few hours.

Such transnational, multi- actor, multi- level and non-linear public policy issues raised by the usage of the Internet can legitimately be qualified as complex policy issues. They concern each global human polity but also the interactions between them.

2.4 Heterogeneous Value Systems

Unfortunately, at the very time when the growing usage of the Internet makes those issues more pressing, the increasingly heterogeneous value systems within and among each community make it even harder to address them in a uniform manner. As the Internet develops and engages more and more people, participating individuals have increasingly different backgrounds and cultural, political or religious reference systems. Some are uncomfortable being exposed online to value sets they do not encounter in real life. Also put in contact are very different national governance frameworks, or incompatible legal systems. Some governments feel threatened by the freedom of expression prevailing on the Internet because it was born and initially developed in western democratic societies. As a consequence, the common tool and space that so facilitates relations and that is rightly praised as a great unifier for humanity, runs the risk on the contrary, of provoking hatred and confrontations.

The critical question therefore is: does the present international architecture provide appropriate tools for these global communities to address the complex issues they are facing? In particular, is the present political paradigm able to offer a unifying framework for very heterogeneous value and governance systems?

2.5 The Present International System

The present international architecture is based on a community of nation-states. Their governments send diplomatic representatives for ad hoc conferences or on a permanent basis to participate in the work of intergovernmental agencies, the best known of which compose the United Nations system. It is a geography-based system, built on the principle of the equal and absolute sovereign right of each country and the a priori legitimacy of its government (whatever its mode of election or designation). Acceptance of this paradigm represented a clear improvement in the face of the devastations of the early XXth century and it allowed the development of a broad range of international organizations, agencies and conferences.

Still, all in all, this system bears the mark of an epoch when nations were relatively few (with colonial empires still existing in 1945) and neatly separated by frontiers; interactions between them were scarce or mainly at the level of governments - apart from wars and commerce designated as import-export, travel was difficult and telephone communications were extremely expensive, justifying the permanent posting of diplomats for each intergovernmental organization.

This system naturally reaches its limits in an era where any major city on the planet can be reached from another in less than a day, where people routinely travel and work in foreign countries, where instantaneous communication costs plummet and where traditional import-export is replaced by complex chains of outsourcing. Out of phase with the evolution of transportation means and communications, an international system based on the pure intergovernmental paradigm has more and more difficulties addressing the challenges of a global and interconnected world.

2.6 The Need for a Political Paradigm Shift

Nowhere is this clearer than on Internet-related issues. Their fundamentally trans-national nature run contrary to absolute sovereignty; their trans-disciplinary nature crosses mandate boundaries of existing United Nations agencies, leading either to turf battles among them or to disregard for those issues falling into the cracks. Representatives of national governments, even when they really represent their citizens, have difficulties handling causal chains spanning multiple levels, from the local to the global and are not always properly trained to handle the complex technical, social, economical and political dimensions of these subjects. Furthermore, the hierarchical and exclusively geographically-based architecture of the present system does not take into account the fact that many actors have multiple affiliations and interests, beyond their national belonging.

Finally, and maybe most importantly, existing rules of procedures make it very difficult to associate other actors than government representatives, such as business and civil society entities, although they have been and still are absolutely essential in the development and functioning of the Internet and the applications it supports. In a way, the complicated rules of accreditation elaborated to associate civil society and business actors in the work of intergovernmental organizations look somewhat like the constantly increasing complications that were added to the Ptolemaic system to make it artificially compatible with Galileo's observations.

The need for a new political paradigm to base the governance of Internet-related issues upon becomes obvious as the Web develops. It was the unanticipated result of a recent United Nations process, the World Summit on the Information Society (WSIS), to reveal the limits of the present system and to allow the emergence of a promising alternative political paradigm.

3 The Quiet Revolution of the "Internet Governance Forum" (IGF)

Between 2002 and 2005, a United Nations process called the World Summit on the Information Society (or WSIS), specifically addressed the various issues related to the Internet. On the surface, WSIS looked like a UN summit like all others: this four-year process was structured around two major events (Geneva in 2003 and Tunis in 2005) at the level of heads of states or governments. They gathered more than 15,000 people each, and produced four documents signed by more than 180 governments after lengthy diplomatic negotiations.

But upon closer look, the WSIS was a dynamic, emergent process that forced hundreds if not thousands of diplomats, business people, civil society actors and technical specialists to interact during four years. As a result they were forced to progressively recognize each other as legitimate actors in the debate. Although it barely made headlines, WSIS may well stay in history as having sanctioned, with the creation of an "Internet Governance Forum", an alternative political paradigm: the "multi-stakeholder approach".

3.1 Mutual Recognition of the Different Categories of Stakeholders

Initially, by fear of creating a dangerous precedent, most governments tried to preserve the strict intergovernmental nature of negotiations taking place under the leadership of the International Telecommunications Union (ITU), a specialized UN agency. This included physically throwing out of the negotiating rooms duly accredited representatives of business and civil society. But this attitude was not sustainable.

Progressively, governments were forced to recognize the undisputable competence, and therefore legitimacy and utility, of business and civil society actors. After all, they had not only invented but built and managed the now ubiquitous global Internet during a time when few governments were paying attention. And the knowledge of those actors was more than necessary for many diplomats initially lacking technical understanding of the issues.

But symmetrically, civil society and business actors were forced to recognize that the new complex issues raised by the growing use of the Internet, such as spam, cybercrime, protection of privacy and personal data or freedom of expression, could not be addressed without some involvement of governments. This ran contrary to early claims that the Internet had made governments obsolete and that Internet-related issues should be the sole province of private sector (via self-regulation) or the technical community alone.

This progressive mutual recognition during the two first years of the Summit allowed the emergence of the expression "stakeholders" as a generic term to designate the different categories of actors. It appeared repeatedly through the documents adopted in Geneva in 2003.

3.2 A Definition of Internet Governance (IG)

The WSIS also witnessed the progressive recognition within the diplomatic community of the expression "Internet Governance" . The term had its origin in the technical community to designate the management of the Internet's core resources under the responsibility of the Internet Corporation for Assigned Names and Numbers (ICANN), particularly IP addresses and domain names. But during the WSIS, Internet Governance emerged as an agreed "meme" among all actors to cover in a single term a broad variety of issues. IG progressively became understood as encompassing not only the technical management of the network but also the whole range of public policy issues related to its use. In other terms, "Internet governance" progressively meant both the governance "of" the Internet and governance "on" the Internet.

During the year 2004, a specific Working Group (the appropriately named Working Group on Internet Governance or WGIG) created by the Geneva Summit, was even tasked, among other things, to establish a working definition of Internet Governance. The Tunis Agenda for the Information Society (TAIS) formally included the resulting working definition: *"Internet Governance is the development and application by governments, civil society, business and international organizations, in their respective roles, of shared principles, norms, rules, decision-making procedures and programmes that shape the evolution and use of the Internet".*

The last part of this sentence clearly reaffirms the dual dimension of Internet governance as governance "of" the Internet ("the evolution") and "on" the Internet ("the use"). But more than anything, the very mention in a document signed by more than 180 countries of the expression "Internet Governance" and of the necessary involvement in it of the different categories of stakeholders was a considerable milestone.

3.3 The Internet Governance Forum (IGF)

The first phase of the WSIS in Geneva had recognized the importance of the different stakeholders, but they implicitly were supposed to remain separate. The second phase of the Summit, on the contrary, saw the emergence of the expression "multi-stakeholder", ultimately appearing more than 15 times in the Tunis Agenda for the Information Society (TAIS). A major additional step forward was the creation of an open "forum for multi-stakeholder policy dialogue".

This Internet Governance Forum (or IGF) held its successful inaugural meeting in Athens in October 2006. More than 1,300 participants with no other accreditation requirement than registering online took part in a four-day event. Official sessions were devoted to issues of Security, Openness, Diversity and Access and more than 30 workshops organized at the initiative of participants dealt with a very broad range of issues.

Although the Internet Governance Forum is formally attached to the United Nations Secretary General, the IGF is neither a new organization (it has no legal status so far), nor a traditional conference taking place during a fixed and short period of time. Established for at least five years and more than likely to be continued afterwards, it is an ongoing process, punctuated by annual meetings but with intersessional work likely to develop.

The creation of the IGF is a quiet but significant institutional innovation, very different from traditional UN processes.

3.4 Thematic Networks of Stakeholders

First of all, the whole approach is not based on a community of nations-states any more but on thematic networks of stakeholders.

In the present international system, based on the hierarchic, geographic paradigm and the preeminence of the national level, citizens are represented at the international level exclusively through their government, whatever its nature or the way it was designated. The present international system is an assembly of the "peoples" of the world and not of individual citizens. In that framework, thematic issues (health, culture, trade, security) are handled by ad hoc intergovernmental organizations or conferences by representatives of national governments. They may or may not associate other actors.

The Internet Governance Forum adopts an inverse approach, network-based and thematic rather than hierarchical and geographic. It gathers around specific issues (in this case Internet governance and its sub-themes) all concerned "stakeholders". This implies a voluntary move for each actor who can decide to participate or not. It is in many respects much more adapted to the structuring of the gobal community.

The spontaneous emergence during the first meeting of the IGF in Athens of "Dynamic Coalitions", gathering actors interested in specific sub-issues (such as spam, privacy, open standards...) illustrates the scalability and replicability of the concept at different levels.

3.5 Equal Footing of Stakeholders

Another essential difference between the IGF and existing intergovernmental organizations is the absence of separation between the different stakeholders and their participation on an equal footing. No organization in any way attached to the United Nations has ever attempted such a radical experiment. Even the International Labour Organization (ILO) keeps its tripartite representation of governments, employers and trade unions as separate constituencies.

This equal footing among all actors - at least in deliberative phases - may ultimately appear as the main advantage of the multi-stakeholder approach: the participation of actors coming from very different backgrounds is the best guarantee that the technical, social, economic and political dimensions of any issue will be taken into account in the early stages of discussion.

3.6 Participation of Individuals

Finally, participation in the work of the Internet Governance Forum is open to any actor interested, even individuals, in stark contrast not only with the traditional exclusive competence of governmental representatives but also with the heavy accreditation procedures for even the most open UN conferences or summits.

This new multi-stakeholder governance framework therefore implicitly establishes a revolutionary right for any individual to participate, in an appropriate manner, in the deliberative processes dealing with the issues he/she is concerned with or impacted by.

This should not mean that any individual will have the right to intervene at any stage in the elaboration of the principles, rules, norms, decision-making procedures and programmes mentioned in the definition of Internet Governance. The respective roles of the different categories of stakeholders may even vary depending on the issue, the venue where they are discussed or the stage of the discussion. But the fundamental principle is nonetheless implicit in the modalities adopted in the initial IGF meeting in Athens and it is a major innovation.

3.7 Towards a Notion of Stakeholdership?

A notion of "stakeholdership" could well represent for thematic governance networks what citizenship is for traditional intergovernmental processes: the basic unit of belonging and the foundation for a process legitimacy, the medium through which individuals with a common interest or concern gather to participate in its governance. But there is a major difference. Citizenship is geographical, usually exclusive (few dual nationalities) and received (via rules relating to parenthood or birth location) rather than chosen. By contrast, individuals can claim several stakeholderships, even on a single issue, according to the different interests or concerns they have in the subject or the different angles they choose to adopt in examining it.

So, although it constitutes today only a fragile experiment, the multi-stakeholder approach pioneered by the Internet Governance Forum is based on a radically different paradigm for dealing with public policy issues and structuring Policy Development Processes.

4 A Promising New Political Paradigm

Crisis periods are uncomfortable. Human groups strive for some coherence in their world views to avoid endlessly reopening debates upon each interaction. The new paradigm that finally imposes itself in science is the one that not only explains troubling observations better that alternative candidates but also provides an agreed basis for a broad range of useful new theories. Similarly, a new political paradigm will impose itself against other proposed solutions if it provides simpler ways to address complex issues and offers a good legitimacy basis for a broad diversity of governance regimes and frameworks.

Can multi-stakeholderism become a dominant paradigm? How can it spread? And what ultimate architecture would it produce?

4.1 A Simpler and more Flexible Framework

The "multi-stakeholder approach" solves the difficult question of the accreditation rules for the different categories of stakeholders that all organizations are presently struggling with. Non-decision-making fora for preliminary discussions allowing all stakeholders to participate on an equal footing are simpler to design and implement to address the complex Internet issues than elaborating complicated rules on an issue-by-issue basis.

Thematic Governance Networks could then emerge from those preliminary discussions. Their rules of procedure would be established progressively, with possible variations on an issue-by-issue basis, by the participants themselves, allowing a bootstrapping of the processes.

The endorsement by more than 180 governments of the multi-stakeholder and thematic approach in Internet Governance is largely motivated by these factors and the absence of a viable alternative. But the likelihood for a new world vision to rapidly be accepted also depends on the amount of efforts it requires from actors in their existing activities.

4.2 An Extension, Rather than a Replacement

Coming back briefly to Thomas Kuhn's analysis, it is important to note that some scientific paradigm shifts mean the pure replacement of one initial paradigm by a different one. For instance: either the sun rotates around the earth, or the earth around the sun; but both cannot be true at the same time. These are mutually exclusive principles. Transition from one to the other has important consequences for existing activities - including in Galileo's case, a dramatic change in the Church's teachings. This triggers natural reluctance to change.

But in other cases, the shift in scientific paradigms allows the newer paradigm not to replace but to extend the former, which remains valid under certain conditions, often as first order approximation. In such cases, the new paradigm is an enabling one, opening up new activities; transformations of existing activities are much more progressive and indirect, therefore more easily accepted. For instance, relativity or quantum physics only apply in important but non-trivial conditions: near the speed of light or at atomic scales. In most day-to-day applications, traditional Newtonian physics remain dominant.

The multi-stakeholder governance approach is such an evolutionary shift. It extends and mutates rather than replaces the present paradigm of the international architecture. It is indispensable for complex global issues, whose multi-dimensional nature strongly calls for an early involvement of all the different actors concerned on a world-wide basis. Nevertheless, existing governance frameworks (including national governments) remain fully legitimate in their respective domains of exclusive competence or sovereignty and are involved as such.

In particular, governments are explicitly designated in WSIS documents as one category of stakeholders for Internet Governance. Citizenship can even be considered as one sort of stakeholdership, and maybe a critical one. This should facilitate the acceptance of this new approach by existing actors (including governments) on a voluntary basis. It only requires a critical mass of actors to be useful and viable, and it only changes some of their modes of interaction, without requesting upfront any transformation of their internal working procedures. This makes it more easily acceptable by actors with very heterogeneous political, economical, cultural or religious value systems.

4.3 Growth and Percolation

The Internet and the World Wide Web have emerged in a bottom-up manner, from the relatively simple interactions of a myriad of agents. They simply endorsed a technical paradigm shift (packet switching and hypertext links respectively) and implemented open and shared rules and protocols (TCP/IP and HTML/HTTP). As a result, a first set of connections (or a first Web server) became like an initial germ progressively replicating to give birth to increasingly complex networks of machines or sites.

Similarly, we believe the paradigm shift towards multi-stakeholderism can allow an Internet Governance Architecture to progressively emerge at the global level, as the result of the convergent efforts of numerous autonomous actors using a set of common protocols. Generic and adaptive process rules for multi-stakeholder interaction and the functioning of thematic governance networks (a Multi-Stakeholder Protocol?) could provide the germ for a virally replicating revolution transforming the landscape of global governance. Dynamic Coalitions within the Internet Governance Forum (IGF) could be the first test beds for implementing such Protocols.

It is expected that actors in all domains of the Information Society, including various intergovernmental or international organizations, would then progressively adopt this approach for their Policy Development Processes and their interactions with other governance structures. It is even likely that other types of complex issues (such as health, the management of natural resources or the environment) will ultimately adopt a similar method. The Internet and the Web have progressively and peacefully transformed our societies. Global Governance will be transformed in the same manner, progressively and peacefully, like a genetic mutation expresses itself through morphogenesis, or as an adaptive trait reproduces in a gene pool.

Only time will tell how Global Governance will ultimately evolve. But there is a great likelihood it will self-structure as a complex network, in the image of the Internet itself: a scale-free and power-law distributed topology built from the aggregation of heterogeneous but interoperable multi-stakeholder processes. At least, that is what it probably should be and how the international community should try to build it.

4.4 The Special Responsibility of a "Small World" of Negotiators

Numerous meetings taking place around the world during four years put participants in the WSIS process in regular contact with one another. As the frequency of their interactions grew, some individuals, although coming from very different cultural and professional backgrounds, developed a better understanding of each other, probably creating a "small world" effect as the density of their connections passed a certain threshold.

Some have argued that even well-intentioned civil society actors got contaminated by a "Net-set" mentality (by analogy with the "jet-set"), hopping from one conference to the next held in posh hotels, and losing contact with their constituents. Accusations even appeared of being co-opted by an international elite class, disconnected with reality.

Although there could be a danger here, reality is more interesting. Bringing all actors into the same environment allowed for the rewiring of existing social networks and in particular, the development of "weak links". Individual "connectors" bridging gaps between constituencies began putting in contact groups that formerly had few relations - if any. This higher level of interaction and communication generated a better understanding of the different facets of each problem within this group of actors, and a growing recognition of the existence of issues of common interest or concern. It ultimately produced a sense of joint responsibility in finding the ways and means to address them.

In the follow-up phase of WSIS since 2005, this "small world" group of actors continues to interact in the numerous international fora dealing with Internet Governance, and particularly the ITU, ICANN and the IGF. They provide a unifying thread for distributed negotiations and in particular, they guarantee that the agreements reached in Tunis progressively percolate in all organizations they participate in. They avoid the reopening of debates that have already been addressed at length during WSIS and also explore together, particularly in the IGF, the ways to move forward on implementation.

Provided members of this small world group avoid the trap of an "exclusive club" mentality and, on the contrary, contribute to bridging gaps between constituencies and actors in all processes they participate in, they could form the core of an exemplary group of practitioners of multi-stakeholder governance and play a critical role in shaping the protocols that will allow multi-stakeholderism to establish itself as the dominant political paradigm of the XXIst century.

5 Conclusion

The multi-stakeholder governance approach is a major conceptual innovation. But it only became practicable at the global level because of the existence of online tools facilitating access to information (Web sites and real-time transcriptions), remote participation (webcasts, blogs), iterative consultation processes (mailing lists and forums) and soon, collaborative drafting (wikis). Indeed, multi-stakeholder governance requires a combination of physical interactions and "intersessional" online collaboration that only the Internet itself makes possible.

Internet Governance is therefore not only the governance "of" the Internet and "on" the Internet. It is also, in a certain way, governance "enabled by" the Internet, or in other terms, the embryo of "Governance for the Internet Age", which is also a complexity age. The global network demands a new type of governance; but it is also the tool that makes this new governance possible and shapes it in its own image: real-time, scalable, participatory, distributed and emergent.

Furthermore, the multi-stakeholder principles and methods the IGF experimented with, may very well be applicable in the future to other domains such as environment, health or natural ressources, by merely changing the concerned stakeholders. As a consequence, a pragmatic evolution of the whole international system could emerge through the progressive implementation of successive thematic governance regimes.

The Internet has transformed social and economic activities in ways that were hard to predict only ten years ago. It is now poised to also transform the way human communities organize themselves, that is: policy-making at the different levels. This recognition of the "multi-stakeholder principle" for Internet Governance is therefore a limited but essential step towards the global governance system our interdependent world needs, and the Internet Governance Forum constitutes a laboratory for new modalities to organize the international community. It should be allowed to evolve as a complex system, from the bottom-up and through the adoption of simple protocols, to develop its full peaceful but transformational potential.

This fragile experiment is just a beginning. Its future is not written and the IGF will certainly be confronted to major challenges. But it offers a glimmer of hope in the debate on global governance and offers the needed paradigm shift to progressively modify the way the international community addresses the global issues it is confronted with.

References

1. On the World Summit on the Information Society (WSIS) and its outcomes, in particular the Tunis Agenda for the Information Society (TAIS). URL: http://www.itu.int/wsis
2. About the International Telecommunications Union (ITU) in general. URL: http://www.itu.int
3. On the Internet Governance Forum (IGF) and its Dynamic Coalitions. URL: http://www.intgovforum.org
4. Kuhn, T.S.: The Structure of Scientific Revolutions, 3rd. ed. Univ. of Chicago Pr., Chicago (1999)

5. Barabasi, A.–L.: His key articles related to the World Wide Web scale-free topology. URL: http://www.nd.edu/ alb/
6. Granovetter, W.: On the concept of Weak Ties in social networks. URL: http://www.stanford.edu/dept/soc/people/faculty/granovetter/granovet.html
7. Dawkins's, R.: On the concept of "memes". URL: http://richarddawkins.net
8. Watts, Duncan J.: On the notion of "Small Worlds". URL: http://www.sociology.columbia.edu/fac-bios/watts/faculty.html

Part VII

Network Design

Evolutionary Engineering of Complex Functional Networks

Pablo Kaluza[1], Hiroshi Kori[2], Alexander S. Mikhailov[3]

[1] Abteilung Physikalische Chemie, Fritz-Haber-Institut der Max-Planck-Gesellaschaft kaluza@fhi-berlin.mpg.de
[2] Department of Mathematics, Hokkaido University, Kita 10, Nishi 8, Kita-Ku, Sapporo, Hokkaido, 060-0810, Japan kori@nsc.es.hokudai.ac.jp
[3] Abteilung Physikalische Chemie, Fritz-Haber-Institut der Max-Planck-Gesellschaft mikhailov@fhi-berlin.mpg.de

1 Introduction

The study of complex networks has applications in various fields and attracts growing attention. In the last decade, much progress has been reached in understanding complexity of network architectures. However, not only the architecture but also the dynamics taking place in a network is important. This dynamics determines functions of networks, related, e.g., to information processing in the brain, which is a network of neurons, or to the production process in a factory, which is a network of machines. Most of the real-world networks are constructed in such a way that they implement certain desired dynamical behaviors. Not only the individual components of a network such as single neurons, but also the network architecture, e.g., connections between neurons, should be appropriately designed. Thus, it should be expected that real-world networks have architectures reflecting their desired dynamical behaviors. If one can extract topological properties of the networks relevant for their desired dynamics, this may provide insights into the role of each topological property from the viewpoint of dynamics and help to understand the design principles of functional networks.

Engineering of functional networks is a major scientific challenge. Even the best modern rationally designed networks are still inferior as compared to the natural networks which result from an evolution. The fastest computers are well behind the information processing capacity of the brain, despite the fact it is based on the elements which are intrinsically very slow. Inside a single biological cell, thousands of different molecular manufacturing and delivery processes proceed in parallel, in a crowded environment in the presence of strong thermal fluctuations, and, nonetheless, characterized by the level of predictability and robust adaptability which is unprecedented for human-designed manufacturing and transportation systems. Therefore, it may be highly beneficial to use evolutionary optimization approaches for the design of artificial functional networks. Among many potential applications of such networks, manufacturing and delivery nanosystems should be mentioned.

This article provides a review of our recent studies exploring possibilities of evolutionary design of artificial dynamical networks with prescribed functional properties. Three examples of evolutionary engineering of such networks are presented. First, logistic delivery networks similar to signal transduction networks of a biological cells are considered. We show how networks with prescribed arbitrary delivery patterns can be constructed by evolutionary optimization. Furthermore, we demonstrate how such functional transportation networks robust against local damage (random removal of links or nodes) can be designed. Our second example deals with self-tuning dynamical networks which adjust their architecture via an evolution-like process, learning to imitate another dynamical system. Finally, a network of coupled phase oscillators in the presence of an external pacemaker is considered. By changing their patterns of connections, the oscillators evolve to a network architecture which makes their entrainment by the pacemaker possible.

2 Design of Functional Networks Robust against Local Damage

The problems of network robustness have been often discussed in an abstract context, aimed at developing optimal defense strategies for the Internet and the WWW [1, 2, 3]. In these studies, performed for large random networks, the emphasis was on the percolation phenomena, with robustness of a network viewed as its capacity to avoid disintegration under accidental or directed damage. In contrast to this, functional networks should maintain their *prescribed*, specific functions while possibly exposed to random damage or parameter variations. Is it possible, by adjustment of the network structure, to develop systems with prescribed functions that are furthermore robust against damage? How strongly would requirements of robustness against a particular kind of damage affect their architecture?

To analyze these questions, a study of functional flow distribution networks has been undertaken [4]. A cell should activate a fixed group of genes in response to each arriving stimulus, while a factory is designed to manufacture particular products from given resources and a logistic network must transport goods to particular destinations. As a simple abstraction of such systems, a toy model of a flow distribution network can be considered. The network transports material flows, applied to it input nodes, through a number of middle redistribution nodes to a set of output nodes. If a flow is applied to one of the input nodes, it reaches, in varying amounts, different output nodes. The set of such delivery patterns, corresponding to activation of various individual input nodes, defines the flow distribution function of a particular network. We show how, by an evolutionary optimization method, networks with prescribed arbitrary distribution functions can be designed and how, furthermore, they can be made robust against random local damage while fully retaining their functionality.

A flow distribution network consists of N_{in} input nodes, M middle nodes, and N_{out} output nodes (examples of such networks are shown in Fig. 1). Its architecture is specified by a directed graph of connections between the nodes with an adjacency matrix A_{ij} (we have $A_{ij} = 1$, if there is a link from node j to node i, and $A_{ij} = 0$ otherwise). Each link bears some flux u_{ij}. The sum of all incoming fluxes for any node is equal to the sum of all outgoing fluxes. For any node, all outgoing fluxes are equal in intensity and are obtained by splitting the total incoming flux in equal parts between the outgoing connections. Thus, we have

$$u_{ik} = \frac{1}{\sum_l A_{lk}} \sum_j A_{kj} u_{kj} \tag{1}$$

for any node k. Introducing total fluxes $x_i = \sum_j A_{ji} u_{ji}$ passing though nodes i, this distribution law can also be written as

$$x_i = \sum_j A_{ij} \frac{x_j}{\sum_k A_{kj}}. \tag{2}$$

External fluxes can be applied to the input nodes and external sinks are attached to the output nodes. A unit external flux $x_\alpha = 1$, applied to an input node α, becomes distributed after passing through the network and fractions $x_\beta = Q_{\alpha\beta}$ of the applied flux reach different output nodes β.

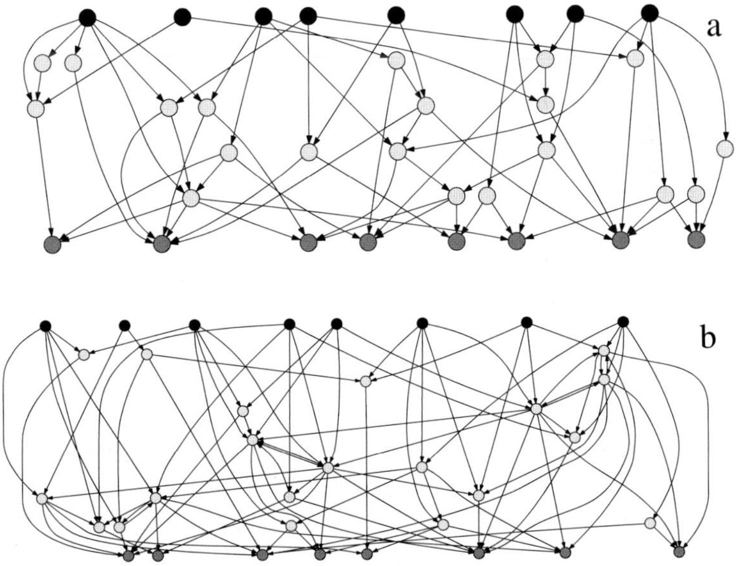

Fig. 1. Flow distribution networks robust against removal of nodes (**a**) and links (**b**)

The matrix $\mathbf{Q}$ with the elements $Q_{\alpha\beta}$ defines the distribution function $\mathbf{F}(G) = \mathbf{Q}$ of any given network G. The ideal performance of a network corresponds to some fixed output pattern $\mathbf{Q}_0$, specifying (for logistic applications) to which final destinations β and in what amounts $Q_{\alpha\beta}$ a particular kind α of goods must be supplied. A network G_0 is optimal, if $\mathbf{F}(G_0) = \mathbf{Q}_0$. The distance ϵ of any network from its ideal performance can be defined as $\epsilon = |\mathbf{F}(G) - \mathbf{Q}_0|$.

Networks with arbitrary distribution functions can be constructed by using an evolutionary optimization algorithm. Suppose that we want to design a network with the output pattern $\mathbf{Q}_0$. Then we define the flow error

$$\epsilon = \frac{1}{2N_{in}} \sum_{\alpha=1}^{N_{in}} \sum_{\beta=1}^{N_{out}} \left(Q_{\alpha\beta} - Q_{\alpha\beta}^0\right)^2 \tag{3}$$

as the distance between the output pattern $\mathbf{Q}$ of a network G and the ideal output pattern $\mathbf{Q}_0$ and try to minimize it through an evolution-like process.

A structural mutation applied to network G, thus obtaining a new network G' with the flow error ϵ'. If the error of the new network is smaller than that of the old network ($\epsilon' < \epsilon$), such a mutation is always accepted. If the error has increased ($\epsilon' \geq \epsilon$), the mutation is accepted with the probability

$$p(\epsilon) = \exp\left[-\frac{\epsilon' - \epsilon}{\epsilon\sigma}\right] \tag{4}$$

where σ is a fixed parameter. Thus, a variant of the stimulated annealing method is used here, with the effective temperature $\theta = \epsilon\sigma$ that decreases with the flow error. If a mutation has been accepted, this iteration step is repeated with the new network. If the mutation is rejected, the next iteration step is performed for the old network G. Details of the optimization algorithm, including the description of applied structural mutations, can be found in [4]. Note that the total number of middle nodes was not allowed to increase during the evolution.

Our simulations, performed for the networks with 8 input, 8 output and 20 middle nodes, show that flow networks whose output patterns are close to an arbitrary, randomly generated output pattern can be readily constructed by using this optimization method [4].

Next, networks whose performance is robust against local damage are constructed. To do this, the concept of *functional robustness* should be formulated. Only those networks whose performance does not deviate too much from the required ideal performance can be viewed as retaining their functionality. Thus, the fraction of all networks, which are obtained by applying a single structural mutation to a given (functional) network and whose performance lies within the tolerance window, can be taken as the quantitative measure of the robustness.

Let us consider a set S of all networks, obtained by applying local damage (removal of randomly chosen single links or nodes) to the network G. This set can be viewed as the *damage shell* of the network G. We introduce some tolerance threshold h and say that all networks with errors $\epsilon > h$ with respect to the ideal performance $\mathbf{Q}_0$ are *abortive.* Then, the functional robustness ρ of the network G is defined as the fraction of all networks in its damage shell which are *not* abortive, i.e.

$$\rho = \frac{1}{N_S} \sum_{G'' \in S} \Theta\left(h - |\mathbf{F}(G') - \mathbf{Q}_0|\right) \tag{5}$$

where N_S is the number of networks in the damage shell S and $\Theta(z)$ is the step function, such that $\Theta(z) = 1$ for $z > 0$ and $\Theta(z) = 0$ otherwise.

If a large number of middle nodes is available, a robust network can be easily constructed. Indeed, in this case different subsets of middle nodes can be used for generation of different responses. Moreover, each subnetwork, responsible for generation of a particular response, can be doubled, thus ensuring response safety if a local network damage has taken place. Increasing the number of middle nodes and employing parallelization would have been the standard engineering approach to this problem.

The situation becomes much more complicated if the number of middle nodes is restricted and is so small that most of the nodes should be involved in generation of several different responses. Such "crowded" functional networks are characteristic for biological systems (e.g., for the signal transduction system of a living cell). To design these networks in a rational way and to make them furthermore robust is extremely difficult. However, the solution can easily be found by using evolutionary optimization methods.

The evolutionary optimization method, described above, can be modified in such a way that optimization is switched to network robustness once functional networks with sufficiently small flow error are obtained (for details, see [4]). Figure 2 shows a typical evolution. When the flow error ϵ becomes smaller than the threshold h at time t_1, the optimization criterion is changed to robustness P. During the subsequent evolution, robustness increases from
les

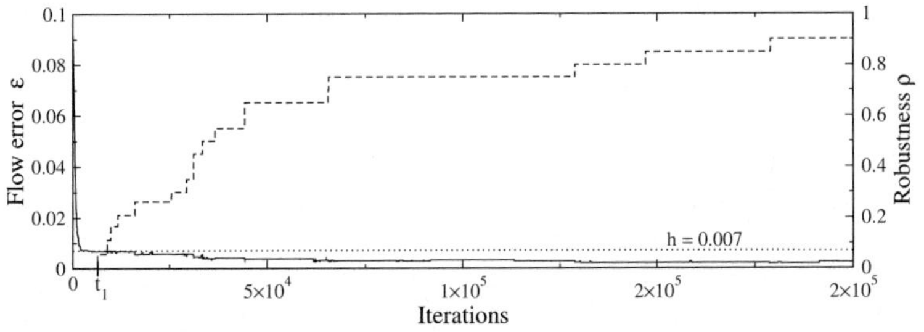

Fig. 2. Evolution of flow error (*solid line*) and robustness (*dash line*) in the optimization process. From [4]

the threshold h. Examples of functional networks, robust against deletion of links or nodes, are given in Fig. 1.

Evolution, based on the criterion of functional robustness, proceeds at the level of the entire damage shells of evolving networks. Figure 3 shows distributions of flow errors in statistical ensembles of 100 networks, optimized only with respect to the flow error (Fig. 3a) and for robustness against deletion of a node (Fig. 3b). Additionally, as gray filled histograms, we display here distributions of flow errors within the damage shells of these networks (also averaged over 100 networks each). While the distributions of errors for the networks themselves do not much differ for the two ensembles, clear differences are seen in the properties of their damage shells. For the general ensemble of functional networks (Fig. 3a), errors in the networks forming their damage shells are much larger and there is a long tail in their statistical distribution. In contrast to this, most of the networks in the damage shells in the ensemble of robust functional networks (Fig. 3b) are small and lie below the tolerance threshold h. Thus, not only a designed network itself, but also the networks obtained after a local damage, would typically be functional in the latter ensemble.

Requirements of robustness against a particular kind of damage have a pronounced influence on the network architecture. In Fig. 4, we compare, for different tolerance thresholds, statistical properties of the networks robust against deletion of links or nodes. The mean degree (i.e., the mean number of connections per node) of the networks robust against node deletions is not significantly different from that of the control ensemble of the networks, selected only on the basis of their flow error (Fig. 4a). On the other hand, networks robust against link deletions have a larger number of connections per node. The clustering coefficient is increased with respect to the purely functional networks for the networks robust against link removals and decreased

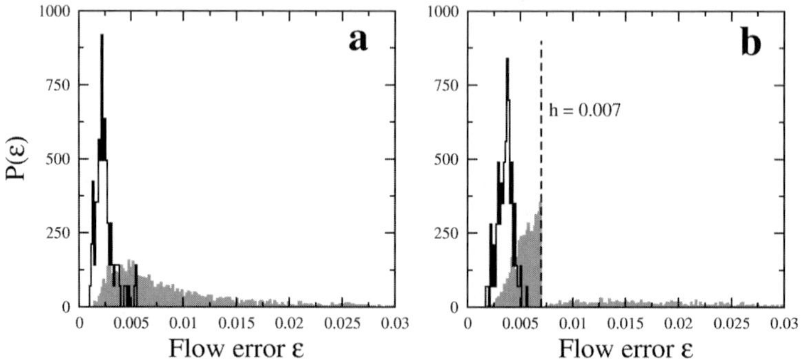

Fig. 3. Distributions of flow errors in the ensembles of 100 networks (**a**) optimized only by flow error and (**b**) optimized for robustness against deletion of a node. Gray histograms show distributions of flow errors after application of local damage. From [4]

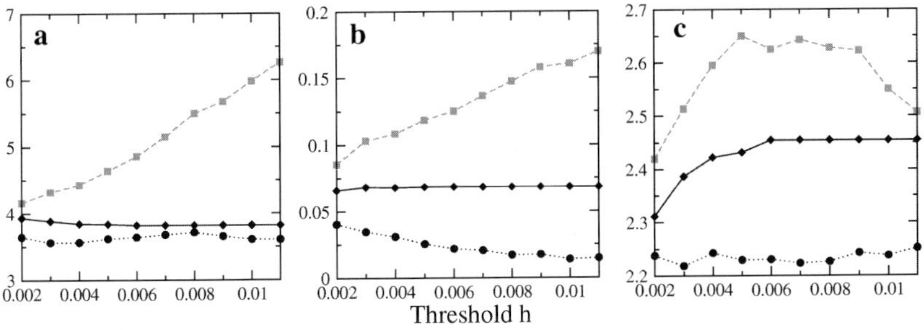

Fig. 4. Statistical properties of networks optimized only by flow error (*diamonds*), and robust against deletion of links (*squares*) or nodes (*circles*): (**a**) mean degree, (**b**) clustering coefficient, and (**c**) average shortest path length from input to output nodes

for the networks robust against the removal of nodes (Fig. 4b).The average path length, defined as the mean shortest path connecting input and output nodes, shows a similar behavior (Fig. 4c). To determine each point in Fig. 4, we have used an ensemble of 100 networks, each obtained by running evolution for 3×10^5 iterations with optimization only for the flow error or for two kinds of robustness with a given threshold h. In order to compute averages, only the subsets of networks with the error below h were taken from such ensembles (this explains why a dependence on h is also seen in Fig. 4 for the networks which were not optimized for robustness).

Distributions of structural motifs were determined [4] for the designed networks robust against link or node removals (Fig. 5). To do this, numbers of different three-node subgraphs in a given network were first found. These were further compared with the frequencies at which various three-node motifs are present in the respective randomized graphs and the scores of different motifs for the considered network were thus computed (the method is described in [5]). A positive (negative) score means that a particular motif is found more (less) frequently in the considered graph than in its randomized version. In this way, any graph can be characterized by a certain structural motif diagram. For each network type, an ensemble average over a subset of functional networks with the error less than $h = 0.007$ in the ensemble of 100 independently evolved networks with different, randomly generated, output patterns was performed. Bars show statistical dispersion of data.

Previously, motif distributions have been analyzed for various real-world networks and four distinctive network superfamilies have been identified [5]. For comparison, Fig. 5 shows the data [5] for two different network superfamilies. We find that the designed networks robust against link deletions have a remarkably good agreement with the second superfamily, including biological information processing networks (signal transduction networks and genetic developmental networks in multicellular organisms, the neural network of a primitive animal). The motif distributions of the networks robust against node

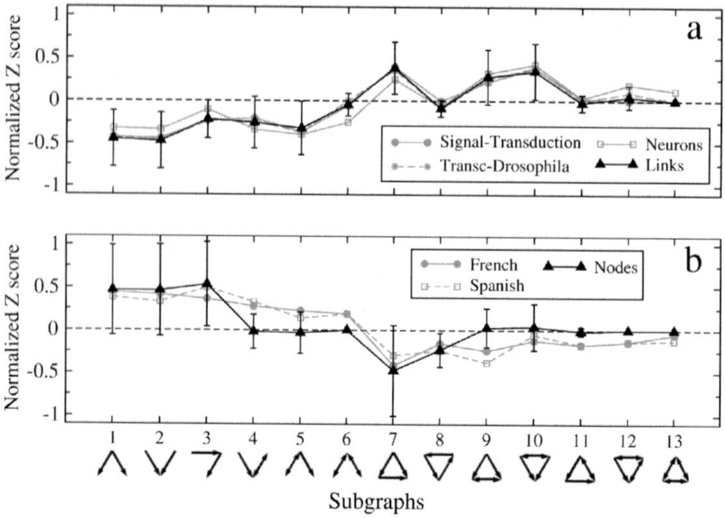

Fig. 5. Motif distributions of networks robust against removal of (**a**) links and (**b**) nodes. For comparison, data for (a) signal transduction networks, developmental genetic transcription networks in drosophila, neural networks of *C. elegans*, and (b) two linguistic networks is included. From [4]

deletions are qualitatively different; they are closer to the motif diagrams in the fourth superfamily including different language networks.

These findings suggest that biological information processing networks may have evolved to enhance their robustness with respect to connection breakups, whereas languages are robust against node deletions (so that the meaning can be conveyed even when some word is forgotten). Further discussion of such aspects can be found in [4]. In the present paper, our attention is focused on engineering of functional networks. The above examples show that evolutionary optimization methods are a powerful tool in the design of functional networks robust against local damage, where traditional rational design methods are hardly applicable.

Responses, generated by the networks considered in this section, were static. In the next section, evolutionary design of dynamical networks is discussed.

3 Design of Networks with Prescribed Dynamics

Design of dynamical systems with prescribed dynamics is a problem of fundamental importance [6, 7]. From an abstract perspective, the brain can be viewed as a large dynamical system which is able, by adjusting the pattern of its connections, to *emulate* the dynamic behavior of various other systems in the surrounding world, thus building internal models of them [8]. If principles of such operation are understood, similar approaches can be used in many applications. For instance, this would allow to develop powerful "predictor" devices for complex dynamical processes that would operate without knowing the actual evolution equations. This would also open a way for a new generation of robots able to learn complex movements and to produce complex motoric responses.

Networks are ideally suited for the design of dynamical systems. The number of possible networks with N nodes is about $2^{N(N-1)}$ and becomes astronomically large even for relatively small network sizes. For a dynamical system based on a network, each connection pattern would correspond to a different dynamics. It is highly plausible that, among the huge number of various network architectures, a connection pattern approaching any desired dynamics can be found. The problem is how to identify a particular needed architecture. Rational construction methods would certainly fail here. However, a solution based on some kind of evolutionary optimization and a learning process may be possible.

Usually, one would need to design nonlinear dynamical systems with prescribed attractors. Such systems should be able to produce required stable motions, starting from various initial conditions. While such a design still remains an open problem, we show below, following [9, 10, 11], how simple linear dynamical systems with prescribed properties can be designed by evolutionary optimization methods.

Let us consider a mechanical system that consists of N identical particles connected by identical elastic strings. The network of connections is a graph G defined by a symmetric adjacency matrix $\mathbf{A}$. An elastic bond between particles i and j is present, if $A_{ij} = 1$, and absent if $A_{ij} = 0$. This dynamical system is described by a set of linear differential equations

$$\ddot{x}_i + \sum_{j=1}^{N} A_{ij}(x_i - x_j) = 0, \tag{6}$$

where x_i is the coordinate of the particle i. Vibrational frequencies ω_α of such elastic molecule are given by eigenvalues $\lambda_\alpha = -\omega_\alpha^2$ of the associated matrix $\mathbf{T}$ with elements

$$T_{ij} = A_{ij} - \left(\sum_{j=1}^{N} A_{ij} \right) \delta_{ij}. \tag{7}$$

In mathematical literature[12], the set of eigenvalues λ_α is known as the Laplacian spectrum of the graph G. The spectrum always includes the eigenvalue $\lambda = 0$, which corresponds to rigid translation of the entire network.

The system (6) is conservative and any combination of normal vibrational modes with arbitrary amplitudes yields a solution. The dynamical property of this system, invariant of initial conditions, is its spectrum of vibrational frequencies ω_α or, equivalently, its Laplacian spectrum $\{\lambda_\alpha\}$. Hence, the design problem can be formulated in this case as the construction of a graph with a prescribed Laplacian spectrum.

It is convenient to introduce the spectral density $\rho(\omega)$ as

$$\rho(\omega) = K \sum_{\alpha=1}^{N-1} \frac{\gamma}{(\omega - \omega_\alpha)^2 + \gamma^2} \tag{8}$$

with some small γ and the normalization constant K, chosen in such a way that $\int_0^\infty \rho(\omega) d\omega = 1$. Thus, each graph becomes characterized by a certain function, having maxima at the locations of its vibrational frequencies ω_α.

Using spectral densities, distance ϵ between any two graphs G_1 and G_2 can be defined as

$$\epsilon = \sqrt{\int_0^\infty [\rho_1(\omega) - \rho_2(\omega)]^2 \, d\omega}. \tag{9}$$

Suppose that we want to construct a graph G_0 with a set of $N-1$ vibrational frequencies $\omega_\alpha^{(0)}$ and the corresponding spectral density $\rho_0(\omega)$. In order to do this, we generate an arbitrary initial graph of size N and introduce a stochastic process of mutations and selection. Mutations will represent random modifications of the connection pattern, whereas selection will be based on the spectral distance between two graphs. This evolutionary optimization process is intended to minimize the distance ϵ between the spectral density $\rho(\omega)$ of the evolving graph G and the target spectral density $\rho_0(\omega)$.

To perform a mutation, we choose at random some node i and delete it, together with all its connections. Then, a new node is introduced into the graph. To construct its connections, we first decide what total number m of connections should it have, by choosing m at random between 1 and $N-1$. After that, we decide, also at random, which m nodes of the new graph G' should be connected to the introduced node.

To decide whether a mutation is accepted, we compute the distance ϵ' between the spectral density of the new graph G' and the density $\rho_0(\omega)$. This is compared with the distance ϵ between the spectral density of the graph G before the mutation and the density $\rho_0(\omega)$. If $\Delta\epsilon = \epsilon' - \epsilon < 0$, the mutation is always accepted. If $\Delta\epsilon > 0$, the mutation is accepted with probability $p = \exp(-\Delta\epsilon/\sigma\epsilon)$. When the mutation has been accepted, the graph G is replaced by G'. Note that, thus, a variant of simulated annealing with the effective temperature $\theta = \epsilon\sigma$, depending on the distance ϵ to the ideal solution, is employed.

These two steps are applied iteratively and the evolution is continued until the spectra become identical ($\epsilon = 0$) or the spectral distance ϵ gets sufficiently small.

Applications of this evolutionary design method are given in [10, 1]. In these studies, the target Laplacian spectrum $\{\lambda_\alpha^{(0)}\}$ was chosen corresponding to some known graph G_0, to ensure that an exact solution of the optimization problem is present. In this formulation, the problem can also been as as the reconstruction of a graph from its Laplacian spectrum. Then, however, one should keep in mind that cospectral graphs (that is, different graphs having the same Laplacian spectrum) are possible. Therefore, the designed graph with a given spectrum may generally differ from the original graph G_0. The fraction of cospectral graphs is however small and, hence, this situation would only rarely be found. Below we give two examples of the network design, taken from [10].

In the first example, small graphs of size $N = 10$ are considered. The target graph G_0 and the initial graph G_{init} are shown in Fig. 6a. The spectral densities of these two graphs (with $\gamma = 0.08$) are displayed in Fig. 6b, as solid and dashed lines, respectively. By running evolutionary optimization with $\sigma = 0.044$, we find that at time $t = 3500$ the spectral densities of the evolving graph and of the target become identical (Fig. 7). The designed graph G_{final} coincides with the target graph G_0.

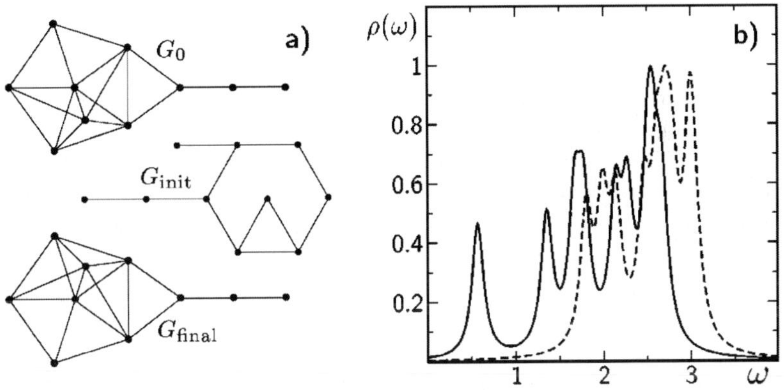

Fig. 6. Initial, target, and final graphs (**a**) and the spectral densities (**b**) of the target (*solid line*) and initial (*dashed line*) graphs. From [10]

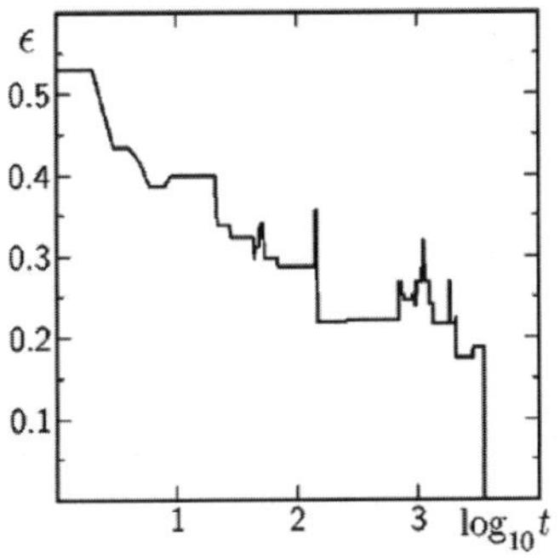

Fig. 7. Evolution of the spectral distance $\epsilon(t)$. From [10]

In the second example, larger graphs with $N = 40$ nodes are considered. As the target, a small-world graph G_0 is chosen. It consists of a ring of 40 nodes, each connected to its two neighbors. In addition, each node can be connected to a randomly chosen non-neighbor node with the probability 0.1. The adjacency matrix $\mathbf{A}_0$ of the target graph G_0 is visually displayed in the left frame in Fig. 8. Spectral densities ($\gamma = 0.08$) of the target graph and of the initial graph G_{init} are shown by solid and dashed lines in Fig. 9.

By running evolutionary optimization with $\sigma = 0.021$, we cannot find, within a reasonable computation time, the exact solution of the optimization problem for such larger graphs. However, a solution which is very close to the exact solution can be constructed. The spectral density of the final graph G_{final}, obtained after 10^5 iterations, is shown by the dotted line in Fig. 9. We see that it is almost indistinguishable from the spectral density of the target (solid line). The analysis reveals [10] that the statistical properties of the two graphs, such as the diameter, the clustering coefficient and the mean degree, are also very close.

Similarity between the graphs can furthermore be discussed in terms of their architectures, specified by the adjacency matrices. For any two graphs G_1

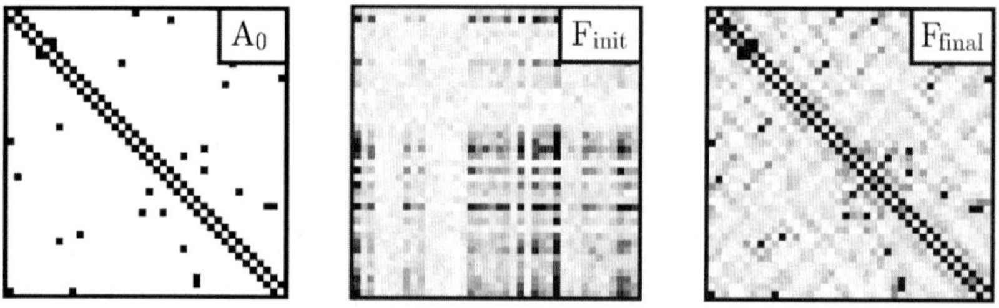

Fig. 8. Density plots of the adjacency matrix $\mathbf{A}_0$ of the target small-world graphs and of the matrices $\mathbf{F}_{init}$ and $\mathbf{F}_{final}$. From [10]

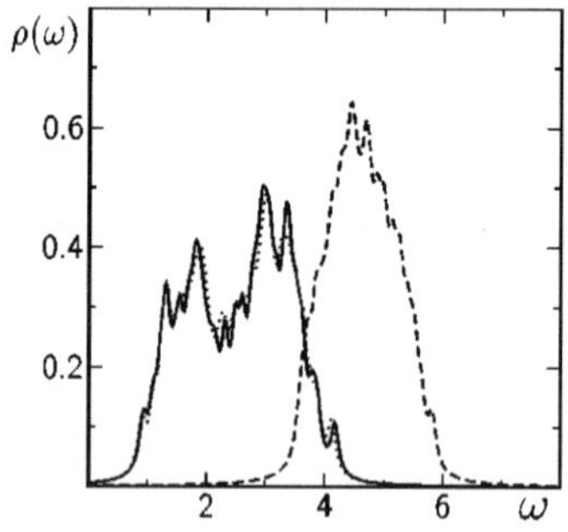

Fig. 9. Spectral densities of the target, initial, and final graphs (*solid, dashed and dotted lines*, respectively). From [10]

and G_2 with adjacency matrices $\mathbf{A}_1$ and $\mathbf{A}_2$, a transformation $\mathbf{F} = \mathbf{F}(\mathbf{A}_1, \mathbf{A}_2)$ can be introduced as

$$\mathbf{F} = \mathbf{U}_1^T \mathbf{U}_2 \mathbf{A}_2 \mathbf{V}_2^T \mathbf{V}_1. \tag{10}$$

Here, real matrices $\mathbf{U}_{1,2}$ and $\mathbf{V}_{1,2}$ are defined by the singular-value decomposition [13] of $\mathbf{A}_1$ and $\mathbf{A}_2$. If two graphs are identical and their adjacency matrices only differ because of a different enumeration of nodes, the identity $\mathbf{A}_1 = \mathbf{F}(\mathbf{A}_1, \mathbf{A}_2)$ holds and the difference $\boldsymbol{\Delta} = \mathbf{A}_1 - \mathbf{F}$ is zero. On the other hand, if two graphs do not coincide, the norm

$$\delta = \frac{1}{N} \left(\sum_{i,j=1}^{N} \Delta_{ij}^2 \right)^{1/2} \tag{11}$$

of this difference can be used as the measure of the distance between the graphs.

In Fig. 8, we have visually displayed the matrices $\mathbf{A}_0$, $\mathbf{F}_{init} = \mathbf{F}(\mathbf{A}_0, \mathbf{A}_{init})$ and $\mathbf{F}_{final} = \mathbf{F}(\mathbf{A}_0, \mathbf{A}_{final})$, where $\mathbf{A}_0$, $\mathbf{A}_{init}$ and $\mathbf{A}_{final}$ are the adjacency matrices of graphs G_0, G_{init} and G_{final}. The elements of a matrix are represented by a square array of pixels using gray-scale color maps whose limits are determined by minimum and the maximum values of matrix elements.

The initial graph G_{init} was strongly different from the target G_0, as seen in the large differences between matrices $\mathbf{A}_0$ and $\mathbf{F}_{init}$. The final graph G_{final} does not coincide with G_0, but is still very close to it, as seen by comparing $\mathbf{F}_{final}$ with $\mathbf{A}_0$. The distances δ for the initial and the final graphs are $\delta_{init} = 0.92$ and $\delta_{final} = 0.01$.

These results are remarkable. The total number of different symmetric graphs of size N can be roughly estimated as $M_N = 2^{N(N-1)/2}/N!$. Therefore, there are $M_{10} = 2^{50}/10! \approx 10^{20}$ possible graphs of size 10 and 10^{971} graphs of size 40 (the latter number is by many orders of magnitude larger than the total number of atoms in the Universe!). Nonetheless, an optimization trajectory, converging to a network with a prescribed dynamics or to a network with a very similar dynamics, can easily be found.

4 Design of Nonlinear Oscillator Networks

It is well known that coupled nonlinear oscillators can undergo spontaneous synchronization [14, 15]. However, such systems can be also entrained by external periodic signals applied to a subset of the elements. The most important example of this effect is provided by the circadian clock in mammals, responsible for the maintenance of their daily rhythm [16, 17]. This clock is formed by a population of oscillatory neurons in the special region of the brain, called suprachiasmatic nucleus (SCN). In the SCN, neurons mutually synchronize their physiological rhythms even in absence of any environmental assistance.

Additionally, a different kind of synchronization—i.e., entrainment to environmental rhythms—takes place here. The entrainment is essential for the proper functioning of this biological clock. Generally, the intrinsic period of the circadian rhythm is different from 24 h (in humans, it would typically be about 25 hours). Therefore, it must be tuned to the 24-hour period through some external influences. Moreover, the phase of circadian oscillations needs to be locked appropriately to the local time. The external periodic signal comes to such neurons from the eyes recording daily light variations. However, only a subset of neurons receives such signals, so that the rest of the neural network should be indirectly entrained via the interactions between neurons.

Does the ability to become entrained by local periodic external forcing (that is, by a pacemaker) depend on the network architecture? How networks, which can be easily entrained, may be designed? Recently, we have discussed [18, 19] these questions by using a simple model of a dynamical network formed by coupled phase oscillators. By running an evolutionary optimization process, networks capable of entrainment have been designed and their statistical properties have been investigated.

We consider a system of $N+1$ phase oscillators, including a special element representing a pacemaker. The basic model is given by a set of evolution equations for the oscillator phases ϕ_i ($1 \leq i \leq N$) and the pacemaker phase ϕ_0,

$$\dot{\phi}_i = \omega + \frac{\kappa}{m} \sum_{j=1}^{N} A_{ij} \sin(\phi_j - \phi_i) + \mu B_i \sin(\phi_j - \phi_i), \tag{12}$$

$$\dot{\phi}_0 = \omega + \Delta\omega. \tag{13}$$

The network architecture is determined by the adjacency matrix with elements A_{ij}. Since directed networks are considered, this matrix is asymmetric. The mean degree m is the average number of incoming connections per node in the considered network. The pacemaker (i.e., the element with $i = 0$) has the frequency $\omega + \Delta\omega$ which is different by $\Delta\omega$ from the rest of the elements. The pacemaker is connected to a subset of N_1 elements (for which we have $B_i = 1$). Coupling inside the network has strength κ; coupling between the pacemaker and the network elements has strength μ. The latter coupling is so strong that all N_1 directly connected network elements are always entrained and their frequency is $\omega + \Delta\omega$.

The presence of a pacemaker imposes hierarchical organization in the network architecture (see Fig. 10), which plays a crucial role in determining the entrainment ability. For any node i, its distance l_i with respect to the pacemaker is defined by the length of the *minimum forward path* separating this node from the pacemaker. Thus, the whole network is divided into a set of *shells*, each of which is composed of oscillators with the same distance h from the pacemaker. An important property of a network is its *depth* L, defined as the mean distance from the pacemaker,

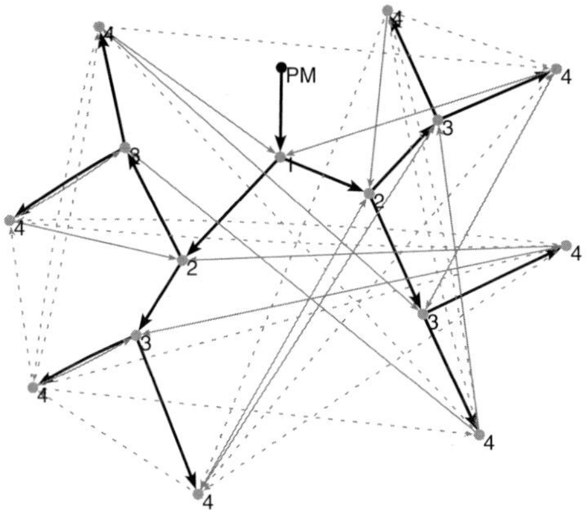

Fig. 10. An example of a hierarchically structured network. Thick lines, thin lines and dashed lines show, respectively, forward, backward and intrashell connections. Numbers indicate the hierarchical level of each node

$$L = \frac{1}{N} \sum_{i=1}^{N} l_i = \frac{1}{N} \sum_h h N_h \tag{14}$$

where N_h is the number of oscillators in the shell h.

All network connections can be classified into forward, backward and intrashell connections. They are, respectively, connections from the nodes in a certain shell h to the nodes in the next shell $h+1$, from the nodes in the shell h to the nodes in the upper shell $k < h$, and between the nodes within the same shell. Backward connectivity parameter ξ can be defined as the fraction of backward connections in the considered network; the forward connectivity χ is the fraction of forward connections.

Analytical and numerical investigations of random Erdös-Renyi (ER) networks have shown [18, 19] that the ability of such networks to undergo entrainment is strongly sensitive to the depth of a network and its backward connectivity. The entrainment window $\Delta\omega_c$ of these networks is given [19] by

$$\Delta\omega_c = \frac{\kappa}{m} (1 + \xi m)^{2-L} . \tag{15}$$

Only pacemakers with frequency differences $\Delta\omega$ within the window $-\Delta\omega_c < \Delta\omega < \Delta\omega_c$ can entrain the network, so that all its elements perform oscillations with the pacemaker frequency $\omega + \Delta\omega$, instead of their natural frequency ω.

Thus, for random ER networks the entrainment window should decrease exponentially with the network depth L and only shallow networks may be in practice entrained. Moreover, the ability to undergo entrainment in such networks improves as the backward connectivity ξ is decreased, approaching a feedforward network.

The above conclusions have been derived for a special class of random networks and a question remains whether a similar behavior can generally be expected. To answer this question, an evolutionary optimization study has been performed [19].

Two types of optimization have been undertaken. In the first method, the network structure evolves in such a way that the mean frequency of the whole network becomes closer to that of the pacemaker. In the second algorithm, each oscillator selects its incoming connections in such a way that its *own oscillation* frequency becomes increased. In both cases, coupling strength κ is chosen to be sufficiently low, so that the initial random networks are far from entrainment. The coupling strength is maintained fixed during the evolution. To avoid trivial results, the connection pattern from the pacemaker is maintained throughout the evolution process, so that the population N_1 of the first shell does not change.

In the first case, the long-time frequency

$$\langle \omega \rangle = \frac{1}{NT} \sum_{i=1}^{N} \left[\phi_i(2T) - \phi_i(T) \right], \tag{16}$$

averaged over the whole population and for a sufficiently long time T, is determined at each iteration step starting from the initial conditions with random phases. A structural mutation representing a single network rewiring is performed. We choose at random an existing directed link and eliminate it. Next, we choose at random a missing link in the network and add a new connection there (thus, the total number of connections is preserved in any mutation). After the mutation, a numerical simulation is again run starting from random initial phases and the new average frequency $\langle \omega \rangle'$ is determined. If $\langle \omega \rangle'$ is closer to $\omega + \Delta\omega$ than $\langle \omega \rangle$, the mutation is accepted. Otherwise, the network is reset to its structure before the mutation. The iterations are repeated until until the long-time frequency $\langle \omega \rangle$ becomes equal to $\omega + \Delta\omega$.

A typical dependence of the average frequency $\langle \omega \rangle$ and of the two topological properties of a network during its evolution is shown in Fig. 11a. Note that the entrainment frequency $\omega + \Delta\omega$ is equal to unity here. It can be observed that both the network depth L and the backward connectivity ξ are decreasing as the entrainment is approached.

In the second case, at each iteration step one oscillator i is randomly chosen. Numerical simulation is run and the long-time frequency $\omega_i = \left[\phi_i(2T) - \phi_i(T) \right] / T$ of this oscillator is determined. The same structural (rewiring) mutation, as in the first case, is applied. After the mutation, numerical simulation is repeated and the new frequency ω_i' of the oscillator i is measured. The mutation is accepted, if $\omega_i' > \omega_i$, and rejected otherwise. At the next step, another oscillator is randomly chosen and the same procedure is repeated.

In this evolutionary process, selection is based on individual activities of the oscillators. Nonetheless, it also leads to global network architectures that

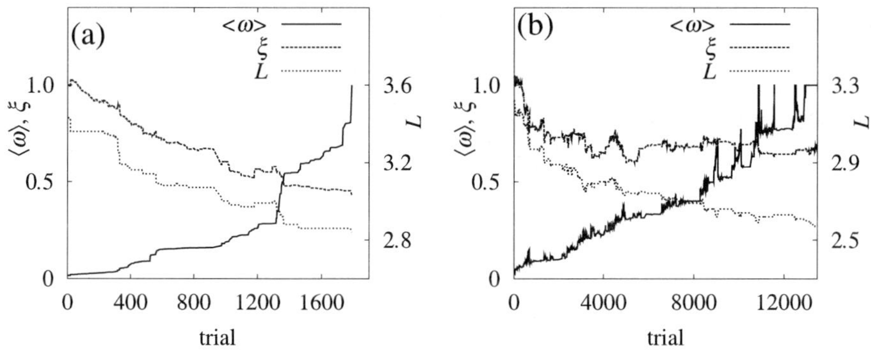

Fig. 11. Evolutions under the global (**a**) and local (**b**) optimization algorithms. Parameter values are $N = 100, m = 10, N_1 = 1$, and $\kappa = 20$. From [19]

make entrainment possible. Figure 11b displays the average frequency $\langle \omega \rangle$ and topological properties L and ξ of the network during the evolution. Global network entrainment is eventually achieved. The network architecture changes during the evolution similar to what is found in the previous case.

During the evolution, the network depth becomes smaller via an increase of the outgoing degrees of nodes in the upper shells. Through this process, hubs of outgoing connections develop. On the other hand, the development of small backward connectivity is mainly due to a decrease of the outgoing degrees in the deep shells. Consequently, heterogeneity of outgoing degrees tends to become stronger as the optimization process goes on.

5 Conclusions

Three different examples, presented in the article, demonstrate that evolutionary optimization methods provide a powerful tool in network engineering. Using such methods, functional networks which are robust to a particular kind of random damage, can be constructed. Running artificial evolution processes, networks with prescribed dynamical properties can be designed. Finally, dynamical networks with an enhanced ability to become enslaved by external controls can be obtained by using such methods.

Evolutionary optimization methods can be employed to systematically generate large ensembles of networks with special functional properties. Statistical investigations of such artificially constructed functional networks can yield insights on the topological network properties imposed by particular functions. Comparing results of such statistical investigations with the data for real biological or social networks which have emerged through a natural evolution, one can better see what is accidental in their properties and what is implied by general requirements of their operation. On the other hand, it would be also interesting to use similar evolutionary optimization methods for engineering of practical networks, which may find application in industrial manufacturing or in logistics.

The authors are grateful to U. Alon, Y. Kuramoto and M. Vingron for stimulating discussions.

References

1. Albert, R., Jeong, H. and Barabasi, A.–L.: Nature **406** (2000) 378
2. Callaway, D.S., Newman, M.E.J., Strogarz, S.H. and Watts, D.J.: Phys. Rev. Lett **85** (2000) 5468
3. Tanizawa, T., Paul, G., Cohen, R., Havlin, S. and Stanley, H.E.: Phys. Rev **E 71** (2005) 047101
4. Kaluza, P., Ipsen, M., Vingron, M. and Mikhailov, A.S.: Phys. Rev **E 75** (2007) 015101
5. Milo, R., Itzikovitz, S., Kashtan, N., Levitt, R., Shen-Orr, S., Ayzenstat, I., Sheffer, M. and Alon, U.: Science **303** (2004) 1538
6. Mikhailov, A.S.: Engineering of dynamical systems for pattern recognition and information processing. In: Engelbrecht, J. and Rabinovich, M.I.:(eds) Nonlinear Waves II Springer, Berlin (1989) 38–51
7. A. S. Mikhailov, Artifcial life: An engineering perspective In: Friedrich, R. and Wunderlin, A.: (eds) Evolution of Dynamical Structures in Complex Systems. Springer, Berlin (1992) 301–312
8. Mikhailov, A.S.: Foundations of Synergetics I.Distributed Active Systems. Springer, Berin (1990) 2nd revised edition (1994)
9. Mikhailov, A.S.: J. Phys. **A 21** (1988) L487
10. Ipsen, M. and Mikhailov, A.S.: Phys. Rev. **E66** (2002) 046109
11. Calenbuhr, V. and Mikhailov, A.S.: From Cells to Societies: Models of Complex Coherent Action. Springer, Berlin (2002), 2nd updated printing (2006)
12. Chung, F.: Spectral Graph Theory American Mathematical Society, Providence (1997)
13. Gollub, G. and Loan, C.: Matrix Computations John Hopkins University press, Baltimore (1996)
14. Kuramoto, Y.: Chemical Oscillations, Waves, and Turbulence Springer, Berlin (1984)
15. Manrubia, S.C., Mikhailov, A.S.and Zanette, D.H.: Emergence of Dynamical Order: Synchronization Phenomena in Complex Systems World Scientific, Singapore (2004)
16. Reppert,S.M. and Weaver, D.R.: Nature **418** (2002) 935
17. Aton, S.J. and Herzog, E.D.: Neuron **48** (2005) 531
18. Kori, H. and Mikhailov, A.S.: Phys. Rev. Lett. **93** (2004) 254101
19. Kori, H. and Mikhailov, A.S.: Phys. Rev. E **74** (2006) 066115

Path Length Scaling and Discrete Effects in Complex Networks

Julian Sienkiewicz, Agata Fronczak, Piotr Fronczak, Krzysztof Suchecki, Janusz A. Hołyst

Faculty of Physics and Center of Excellence for Complex Systems Research, Warsaw University of Technology, Koszykowa 75, PL-00-662 Warsaw, Poland
julas|agatka|jholyst|suchecki@if.pw.edu.pl

1 Introduction

Complexity has thousands of faces. This mere fact is commonly known to wide group of scientists working in such different areas as physics, chemistry, meteorology or logistics. It is rather impossible to see at first glance if a problem we are tackling with can be counted as a complex one. In the 18th century, when the problem of seven bridges in Königsberg was solved by Leonhard Euler [1], people gained an extremely useful and simple tool for difficult problems - the *graph theory* - which has now been dynamically developing since 1950's works of Erdös [2]. Late 1990's brought us another "revolution" in understanding of sophisticated problems - *complex networks*. The discoveries of Watts and Strogatz [3] as well as those of Barabási and Albert [4] set up a new direction in complexity. The networks have shown us directly how a structure which initially seems to be complicated, might be modelled by simple rules if we map the ingredients of this system onto nodes and links in a complex network. The power of the complex networks lays in the fact that such distant relations as acquaintances between people [5], collaboration of scientists [3], protein interactions [6], Internet [7], stock assets [8] or city public transport [9] may all be treated as one topological object.

In this work we would like to show two different studies: the first devoted to explanation of certain scaling laws observed in real-world systems and the second one that was set up to give exact analytical expression for average path length in some network models. Those two studies combined together enabled us to discover and to explain effects of oscillations on average path length - a phenomenon emerging from the discrete structure of complex networks . At the end we present a direct application of the oscillation effect to a simple optimization problem [10] that may be used in real-life situations.

2 Basic Network Characteristics

An object called *network* consists of *nodes* (vertices) and *edges* (links) that connect nodes. The most natural definition is that of *degree* - the number of edges connected to a vertex. To describes statistical properties of network one uses *degree distribution* $p(k)$ - the probability that a randomly chosen vertex is characterized by degree k. As any other science, also network science has its own fundamental observables which can be used to distinguish between different types of networks [11].

The *internode distance* l_{ij} between nodes i and j is depicted at Fig. 1 and can be defined as the shortest number of edges one has to pass getting from one node to another. The sum of all internode distances in the network divided by the total number of pairs is called *average path length*.

The *Clustering coefficient* is a parameter that describes the local transitivity of the network. For a single node such coefficient is defined as the ratio of the number of connections e_i among the first neighbors of node i to the maximal possible number of such links (see Fig. 2). It can be written as

$$c_i = \frac{2e_i}{k_i\,(k_i - 1)}. \tag{1}$$

An average value taken over all nodes in the network gives the global clustering coefficient .

To investigate correlations between nodes' degrees one uses *assortativity coefficient* [12] that takes a following form:

$$r = \frac{\sum_i j_i k_i - \frac{1}{M}\sum_i j_i \sum_i k_i}{\sqrt{\sum_i j_i^2 - \frac{1}{M}(\sum_i j_i)^2}\sqrt{\sum_i k_i^2 - \frac{1}{M}(\sum_i k_i)^2}}, \tag{2}$$

where M - number of pairs of nodes (twice the number of edges), j_i, k_i - degrees of vertices at both ends of i-th pair and index i goes over all pairs of nodes in the network. A positive value of r means that nodes with high degree tend to link among each other (see Fig. 3A) and similarly low degree

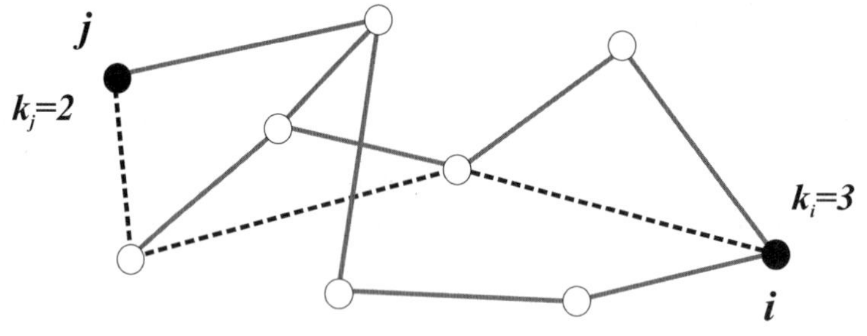

Fig. 1. Internode distance $l_{ij} = 3$ between nodes i and j - marked as a dashed line

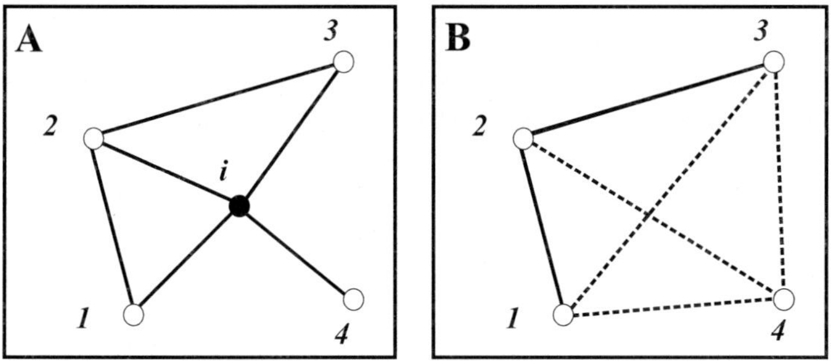

Fig. 2. Construction used to calculate clustering coefficient for node i: (**A**) node i with its nearest neighbors (**B**) connections among nearest neighbors of node i - existing are solid lines, possible ones are dashed lines. One can calculate that in this example $c_i = 2/6 = 1/3$

nodes are connected to other low degree nodes, while negative values - that high degree nodes are mostly connected to low degree ones (see Fig. 3B).

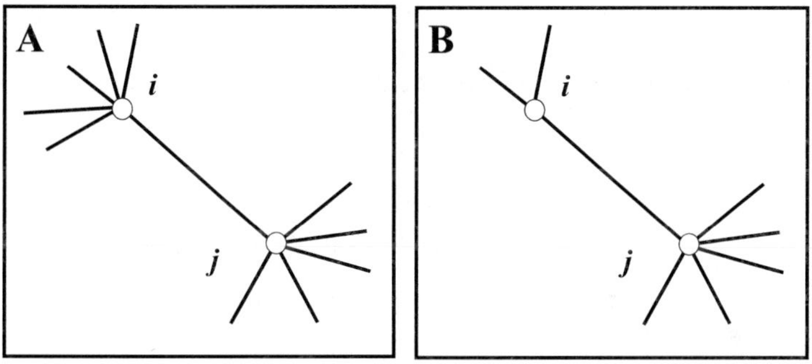

Fig. 3. Visualization of typical connections in (**A**) assortative (**B**) disassortative network

3 Internode Distances

In 2005 our group observed [13] a universal scaling for distances $\langle l_{ij} \rangle$ between nodes possessing degrees k_i and k_j. The distances behave as

$$\langle l_{ij} \rangle = a - b \log(k_i k_j), \tag{3}$$

where the mean value is taken over all pairs of nodes having a *fixed* product $k_i k_j$. Figure 4 presents mean distance $\langle l_{ij} \rangle$ between pairs of nodes i and

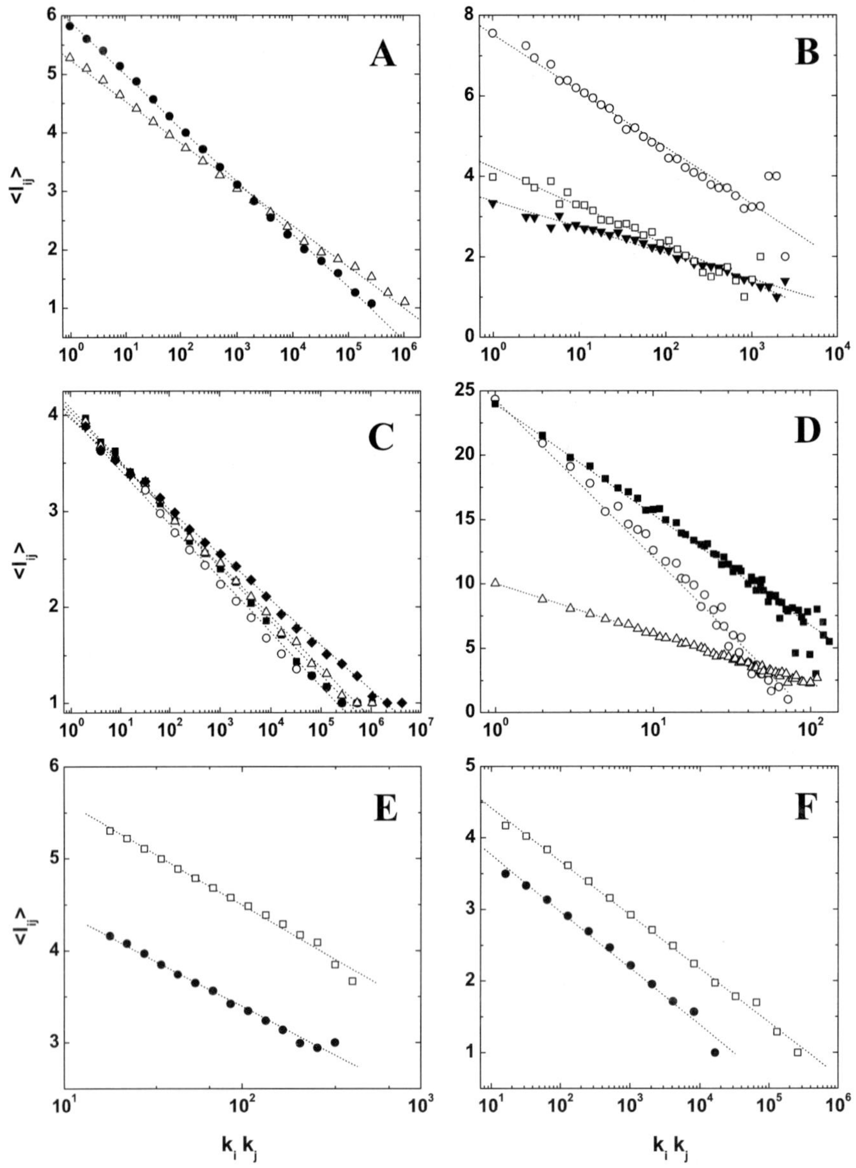

Fig. 4. Mean distance $\langle l_{ij} \rangle$ between pairs of nodes i and j as a function of a product of their degrees $k_i k_j$. (**A**) Co-authorship networks: *Astro* (triangles), *Cond-mat* (circles). (**B**) Biological networks: *Silwood* (squares), *Yeast* (circles), *Ythan* (triangles). (**C**) Internet Autonomous Systems, years: *1997* (circles), *1998* (squares), *1999* (triangles), *2001* (diamonds). (**D**) Public transport networks in Polish cities: *Gorzów Wlkp.* (circles), *Łódź* (squares), *Zielona Góra* (triangles). (**E**) *Erdős-Rényi* random graphs: $\langle k \rangle = 8$ and $N = 1000$ (circles) $N = 10000$ (squares), (**F**) *Barabási-Albert* networks: $\langle k \rangle = 8$ and $N = 1000$ (circles) $N = 10000$ (squares). In (A), (C), (E) and (F) data are logarithmically binned with the power of 2, in case of (B) with the power of 1.25 and in case of (D) data are not binned

j versus the product of their degrees $k_i k_j$ in various complex networks belonging to very different types: Erdős-Rényi random graphs, Barabási-Albert evolving networks, biological networks [14, 15, 16] (*Silwood*, *Ythan*, *Yeast*), social networks [17, 18] (co-authorship groups *Astro* and *Cond-mat*), Internet Autonomous Systems [19] as well as selected networks for public transport in Polish cities [20, 9](Gorzów Wlkp., Łódź, Zielona Góra). The relation (3) is very well fulfilled over several decades for all our data. We would like to underline that the networks mentioned above display a wide variety of basic characteristics such as degree probability, clustering coefficient or degree-degree correlations and their origins are quite different in each single case. The only apparent common feature of all above systems is the small-world effect.

3.1 Model

To justify the relation (3) we use a simple approach basing on the concept of branching trees exploring the space of a random network. The goal is to find the mean shortest path between a node i of degree k_i and a node j of degree k_j. One can immediately notice that following a random direction of a randomly chosen edge one arrives at node j with a probability $p_j = k_j/(2E)$, where $2E = N\langle k\rangle$ is a double number of links. In consequence one needs (in average) $M_j = 1/p_j = 2E/k_j$ of random trials to arrive at the node j.

Now let us consider a branching process represented by the tree T_i (Fig. 5) that starts at the node i where an average branching factor is κ (all loops are neglected). If a distance between the node i and the surface of the tree equals to x then in average there are $N_i = k_i\kappa^{x-1}$ nodes at such a surface and there is the same number of links ending at these nodes. It follows that in average the tree T_i touches the node j when $N_i = M_j$ i.e. when

$$k_i k_j \kappa^{x-1} = N\langle k\rangle. \qquad (4)$$

Since the mean distance from the node i to the node j is $\langle l_{ij}\rangle = x$ thus we get the scaling relation (3) with

$$a = 1 + \frac{\log(N\langle k\rangle)}{\log\kappa} \quad \text{and} \quad b = \frac{1}{\log\kappa}. \qquad (5)$$

The result (5) is in agreement with the paper [21] where the concept of generating functions for random graphs has been used. One should be aware of the fact that in the above presented estimations we have omitted degree-degree correlations or loops, and we have treated the branching level x as a continuum variable to fulfill the relation (4).

The mean branching factor κ is a mean value over all local branching factors and over all trees in the network. In the first approximation it could be estimated as the mean arithmetic value of a nearest neighbor degree less one (incoming edge): $\kappa = \langle k\rangle_{nn} - 1$. Such a mean value is however not exact because local branching factors in every tree are *multiplied* one by another in

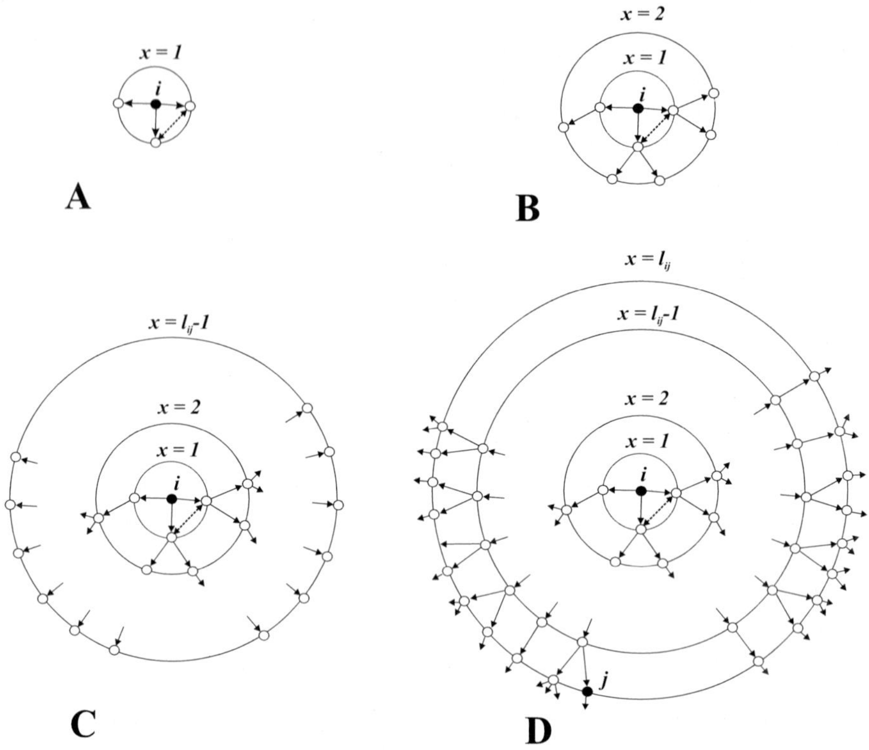

Fig. 5. Tree formed by a random process, starting from the node i and approaching the node j. Parts **A, B, C** and **D** depict consecutive steps of the random tree formation process

(4). The corrected mean value of κ should be taken as an arithmetic mean value over all geometric values from different trees, what is very difficult to perform numerically. We calculate arithmetic mean branching factor over nearest neighborhood of every node m, i.e. $\kappa^{(m)} = \langle k \rangle_{nn}^{(m)} - 1$, and then average it geometrically over all nodes m, i.e. $\kappa = \langle \kappa^{(m)} \rangle_m$. Although our approach is not exact, it leads to a good agreement between coefficients a_e, b_e taken from real networks (see Table 1) and a, b calculated from our model.

The good agreement between theory based on random networks and empirical data suggests that the considered real networks exhibit a large level of randomness.

3.2 Clustering Coefficient

The influence of loops of the length three on the relation (3) can be estimated as follows. Let us assume that in the branching process forming the tree T_i two nodes from the nearest neighborhood of the node i are *directly* linked (the dashed line at Fig. 5). In fact such a situation can occur at any point of the branching tree T_i. Since such links are useless for further network exploration by the tree T_i thus an *effective* contribution from both connected nodes to the mean branching factor of the tree T_i is decreased. Assuming that clustering coefficients of every node are the same, the corrected factor for the branching process equals to $\kappa_c = \kappa - c\kappa$ where c is the network clustering coefficient. This equation is not valid for the branching process around the node i where $\kappa'_i = \kappa - c(k_i - 1)$. A similar situation arises around the node j. Replacing k_i and k_j with $\langle k \rangle$ in κ'_i and κ'_j one gets

$$k_i k_j [\kappa(1 - c')]^2 [\kappa(1 - c)]^{x-3} = N\langle k \rangle, \tag{6}$$

where $c' = c(\langle k \rangle - 1)/\kappa$. It follows that instead of (5) we have

$$a' = 3 + \frac{\log(N\langle k \rangle) - 2\log[\kappa(1 - c')]}{\log[\kappa(1 - c)]} \quad \text{and} \quad b' = \frac{1}{\log[\kappa(1 - c)]}. \tag{7}$$

Corrections due to clustering effects give a better fit for the coefficient a', while for some networks the coefficient b is closer to experimental value b_e than b' (see Table 1).

3.3 Degree-degree Correlations

One can see from Table 1 some of the examined systems share the property of high degree-degree correlations, either negative or positive. We would like to take that fact into account in our considerations [22]. Degree-degree correlations mean that average degrees $k_i^{(nn)}$ of nodes in the neighborhood of a node i depend on the degree k_i. Let us assume that this relation can be written as

$$\kappa_i \equiv k_i^{(nn)} - 1 = D k_i^{\phi - 1}. \tag{8}$$

The value of ϕ being over 1 means the network is assortative, while $\phi < 1$ is a sign of disassortative nature of the system. If we neglect higher order correlations then Eq. (4) should be replaced by

$$k_i k_j \kappa_i \kappa_j \kappa^{x-3} = N\langle k \rangle. \tag{9}$$

Taking into account Eq. (8) we can replace parameters a and b given by the Eq. (5) with

$$a_\phi = a + 2 - 2b \log D \quad \text{and} \quad b_\phi = \phi b. \tag{10}$$

Figure 6 illustrates the estimation of ϕ coefficient for two different networks: assortative *Cond-mat* collaboration network (Fig. 6A) and disassortative *Internet Autonomous System* network taken in year 1998 (Fig. 6B). After obtaining the histogram of $\langle k_{nn} - 1 \rangle$ for each node i and plotting the dependence of those values on the node degree we perform the linear approximation

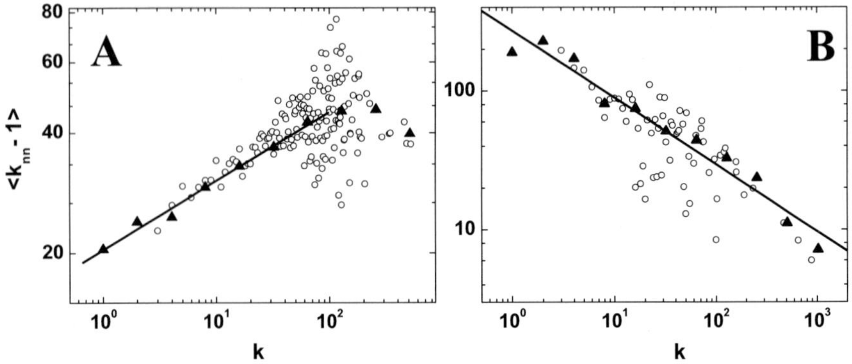

Fig. 6. Estimation of ϕ coefficient for (**A**) *Cond-mat* co-authorship network, (**B**) *Internet Autonomous System 1998*

in the log-log scale. A slope calculated in this way corresponds to the exponent $\phi - 1$ (with accordance to the formula $\langle k_{nn} - 1 \rangle \sim k^{\phi-1}$). One can see, that the scaling (8) is not so obvious as for the relation (3).

Table 1 shows the comparison between experimental data collected from the examined networks and the results obtained from Eqs (5) and (10). One can notice that the values of a_ϕ and b_ϕ are more accurate for the networks characterized by a ϕ coefficient above unity (assortative).

3.4 Rescaling

We are also able to present all the data points at one plot only - this can be done using simple re-scaling methods which as a reference points take *effective degree value* k_{eff} and *maximal intervertex distance* $l_{\max}$. Defining rescaled values $\widetilde{\langle l_{ij} \rangle}$ and $\widetilde{k_i k_j}$ as:

$$\widetilde{\langle l_{ij} \rangle} = 2 \frac{\langle l_{ij} \rangle}{l_{\max}} \log k_{\text{eff}} \tag{11}$$

$$\widetilde{k_i k_j} = \frac{k_i k_j}{k_{\text{eff}}^2} \tag{12}$$

we get (see Appendix A) a simplified formula for the scaling relation:

$$\widetilde{\langle l_{ij} \rangle} = - \log \widetilde{k_i k_j}, \tag{13}$$

Table 1. Comparison between experimental and theoretical data. *Astro* and *Cond-mat* are co-authorship networks, *Silwood*, *Yeast* and *Ythan* are biological networks and *AS* stands for the Internet Autonomous Systems with number meaning the year data were gathered, *Gorzów Wlkp.*, *Łódź* and *Zielona Góra* are public transport networks in corresponding Polish cities. c is the clustering coefficient, r - assortativity value, and ϕ coefficient is obtained using procedure described in the text. a_e and b_e mean experimental values (Fig. 4), a and b are given by (5), a' and b' by (7) while a_ϕ and b_ϕ by (10). In case of models due to the lack of correlations ϕ parameters should be equal to 1, so we omitted values of a_ϕ and b_ϕ

network	c	r	ϕ	a_e	a	a'	a_ϕ	b_e	b	b'	b_ϕ
ER $N = 10^3$	0.007	0	-	5.43	5.46	5.48	-	1.017	1.143	1.147	-
ER $N = 10^4$	0.001	0	-	6.77	6.60	6.61	-	1.136	1.143	1.143	-
BA $N = 10^3$	0.038	0	-	4.54	4.24	4.27	-	0.813	0.830	0.842	-
BA $N = 10^4$	0.007	0	-	5.17	4.81	4.81	-	0.778	0.777	0.779	-
Astro	0.609	0.055	1.23	5.24	4.30	4.98	4.41	0.707	0.595	0.786	0.732
Cond-mat	0.604	0.053	1.19	5.90	5.09	6.38	5.05	0.908	0.786	1.150	0.935
Silwood	0.142	-0.316	0.71	4.22	3.69	3.78	3.19	0.955	0.941	1.004	0.668
Yeast	0.068	-0.158	0.59	7.53	6.66	6.87	5.71	1.406	1.552	1.629	0.916
Ythan	0.216	-0.254	0.61	3.39	3.35	3.45	2.81	0.649	0.765	0.832	0.466
AS 1997	0.182	-0.229	0.46	3.99	3.39	3.42	2.58	0.562	0.596	0.629	0.274
AS 1998	0.250	-0.200	0.48	4.08	3.41	3.45	2.65	0.555	0.575	0.620	0.276
AS 1999	0.250	-0.183	0.49	4.03	3.35	3.38	2.55	0.532	0.540	0.579	0.265
AS 2001	0.289	-0.185	0.45	3.96	3.23	3.25	2.50	0.471	0.481	0.518	0.217
Gorzów Wlkp.	0.082	0.385	1.44	24.36	16.06	19.76	16.67	12.270	5.333	6.651	7.679
Łódź	0.065	0.070	1.19	24.01	11.67	12.70	11.89	8.621	3.084	3.389	3.670
Zielona Góra	0.067	0.238	1.41	10.03	8.96	9.63	9.62	3.908	2.682	2.917	3.781

which enables us to present several different datasets at one plot (see Fig. 7) grouping along one common line.

4 Log-periodic Oscillations on Path Lengths

In the previous sections we have shown that our results nicely cover the relation (3), both in case of network models as well as for real-world examples. However, our recent results (see [23]) leave no doubt that for some specific situations the relation between $\langle l_{ij} \rangle$ and $k_i k_j$ is not so straightforward. As it is presented at Fig. 8, in several examples we have encountered signs of a log-periodic behavior of internode distances, that is:

$$\langle l_{ij} \rangle = a - b \log k_i k_j + c \sin(\log k_i k_j). \tag{14}$$

We will try to explain that phenomenon using hidden variable formalism and its applications we have used in our previous work [24].

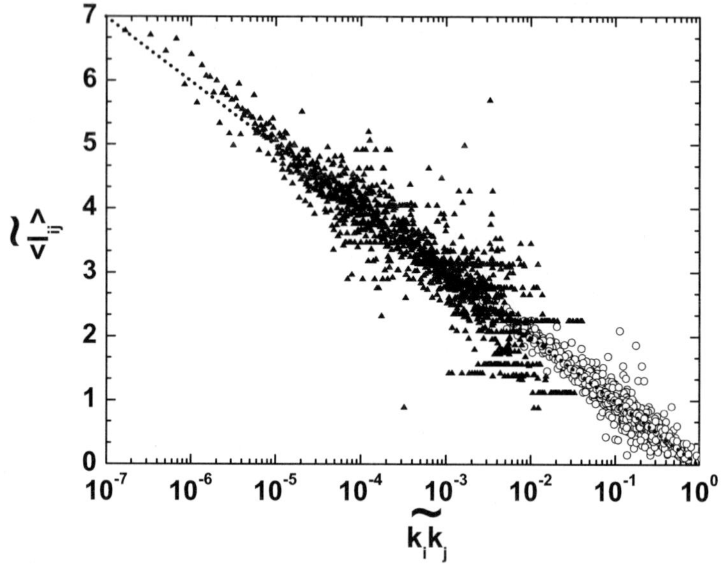

Fig. 7. Rescaled data from biological networks (*filled triangles*) and Public Transport Networks (*open circles*)

4.1 Hidden Variables

In [24] Fronczak et *al.* have derived exact expressions for average path lengths in case of uncorrelated random networks using a hidden variable approach. In this method every node i is assigned with a *hidden variable* h_i randomly drawn from a distribution $\rho(h)$ and the connection probability between any pair of nodes is proportional to $h_i h_j$. The resulting node degree distribution $P(k)$ is given by [25]:

$$P(k) = \sum_h \frac{\mathrm{e}^{-h} h^k}{k!} \rho(h). \tag{15}$$

It has been proved [24] that probability $p^*_{ij}(x)$ of nodes i and j being exactly x-th neighbors can be written as $p^*_{ij}(x) = F(x-1) - F(x)$, where:

$$F(x) = \exp\left[-\frac{h_i h_j}{\langle h^2 \rangle N}\left(\frac{\langle h^2 \rangle}{\langle h \rangle}\right)^x\right] \tag{16}$$

In the above equation N is network size and:

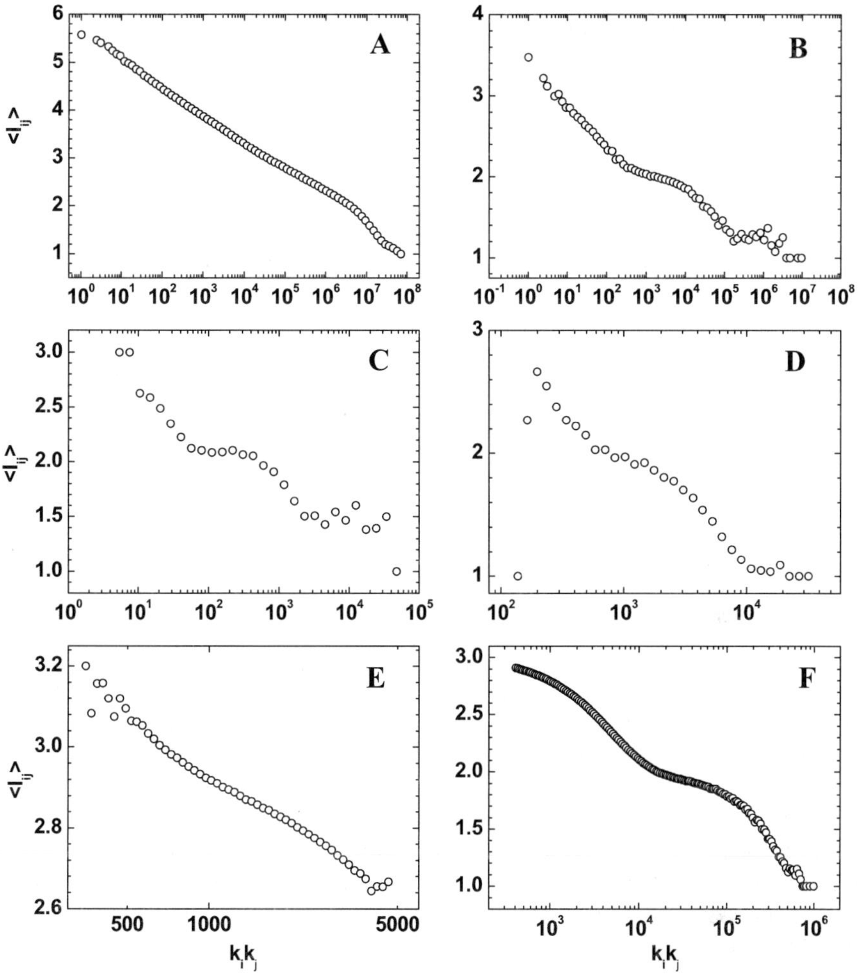

Fig. 8. Average distance $\langle l_{ij} \rangle$ between nodes i and j for real-world networks (A - D) and network models (E - F). (**A**) Movie actor network $N = 382219$, $\langle k \rangle = 86.64$ (**B**) Spanish language word cooccurrence network $N = 11857$, $\langle k \rangle = 7.43$ (**C**) Caribbean food web network $N = 249$, $\langle k \rangle = 25.73$ (**D**) Opole public transport network $N = 205$, $\langle k \rangle = 50.19$ (**E**) Erdös-Rényi random graph $N = 10000$, $\langle k \rangle = 40$. (**F**) Barabási-Albert evolving network $N = 10000$, $m = 20$. All data are logarithmically binned. Sources for data are following: data for movie actor have been taken from A.-L. Barabási web page http://www.nd.edu/ alb, Spanish cooccurrence dataset has been downloaded from V. Batagelj WWW page http://vlado.fmf.uni-lj.si/pub/networks/pajek/ and Caribbean food web data have been taken from The Integrative Ecology Group http://ieg.ebd.csic.es. Opole data have been collected for so called "space P" [9] which is defined as follows: nodes are bus and tram stops and an edge means that there is a direct route linking them

$$\langle h \rangle = \int_{h_{min}}^{h_{\max}} h\rho(h)\mathrm{d}h\,, \tag{17}$$

$$\langle h^2 \rangle = \int_{h_{min}}^{h_{\max}} h^2\rho(h)\mathrm{d}h\,. \tag{18}$$

As consequence of these assumptions, Fronczak et *al.* have been able to express the average distance $\langle l_{ij} \rangle$ between nodes i and j as:

$$\langle l_{ij} \rangle = \frac{-\ln h_i h_j + \ln N\langle h \rangle - \gamma}{\ln \frac{\langle h^2 \rangle}{\langle h \rangle}} + \frac{3}{2} + R, \tag{19}$$

where $\gamma \simeq 0.5772$ is the Euler's constant and R is a sum of cosine Fourier transforms of $F(x)$ function:

$$R = \sum_{n=1}^{\infty} R_n \equiv 2\sum_{n=1}^{\infty}\left(\int_0^{\infty} F(x)\cos(2n\pi x)\mathrm{d}x\right), \tag{20}$$

After integrating Eq. (19) over all pairs $h_i h_j$ one can obtain an equation for average path length $\langle l \rangle$ which is in fact dependent only on N and on specific distribution $\rho(h)$:

$$\langle l \rangle = \frac{-2\langle \ln h \rangle + \ln N\langle h \rangle - \gamma}{\ln \frac{\langle h^2 \rangle}{\langle h \rangle}} + \frac{3}{2} + S, \tag{21}$$

here S is the term R integrated over all hidden variables $h_i h_j$.

4.2 Average Internode Distance

Using appropriate hidden variable distributions $\rho(h)$ for different types of networks, one can obtain exact expressions for average internode distance from Eq. (3):

- *Scale-free (SF) networks.* A hidden variable distribution $\rho_{sf}(h)$ for asymptotically SF networks ($k \gg 1$) is given by the following expression [26]:

$$\rho_{SF}(h) = \frac{(\alpha-1)m^{\alpha-1}}{h^{\alpha}}. \tag{22}$$

As in this work we pay attention only to the case, where $\alpha = 3$, thus after using Eqs (17) and (18) and substituting $h_{min} = m$ and $h_{\max} = m\sqrt{N}$ one can easily show that:

$$\langle l_{ij}^{SF} \rangle = \frac{-\ln h_i h_j + \ln 2mN - \gamma}{\ln(m\ln\sqrt{N})} + \frac{3}{2} + R, \tag{23}$$

- *Exponential (EXP) networks.* In the case of the exponential networks the distribution of hidden variables takes a form of

$$\rho_{EXP}(h) = \frac{1}{\langle k \rangle} e^{-\frac{h}{\langle k \rangle}}. \tag{24}$$

Using Eq. (3) one gets:

$$\langle l_{ij}^{EXP} \rangle = \frac{-\ln h_i h_j + \ln N \langle k \rangle - \gamma}{\ln(2\langle k \rangle)} + \frac{3}{2} + R, \tag{25}$$

- *Erdös-Rényi (ER) random graphs.* As the only way to obtain Poisson distribution from Eq. (15) is to define $\rho_{ER}(h) = \delta_{\langle k \rangle, h}$ (i.e. every node i is characterized by a hidden variable $h_i = \langle k \rangle$), thus examining Eq. (3) is useless.

A comparison between theoretical predictions of Eqs (23) and (25) and numerical simulations performed using algorithm [26] for scale free networks with the hidden variable distributions (22) and (24) is presented at Fig. 9. It is to notice that for small values of $\langle k \rangle$ (Fig. 9A and 9B) the scaling relation is well described with Eq. (3) while for larger average degree values oscillations appear.

4.3 Average Path Length

An average path length is a variable of much interest, because it is a main observable for many complex networks.

Once again, applying hidden variable distributions defined in the previous Section to Eq. (21) we can obtain exact expressions for average path length:

- *Scale-free networks.*

$$\langle l_{SF} \rangle = \frac{\ln \frac{2N}{m} - 1 - \gamma}{\ln(m \ln \sqrt{N})} + \frac{3}{2} + S, \tag{26}$$

- *Exponential networks.*

$$\langle l_{EXP} \rangle = \frac{\ln \frac{N}{\langle k \rangle} + \gamma}{\ln 2\langle k \rangle} + \frac{3}{2} + S, \tag{27}$$

- *Erdös-Rényi.*

$$\langle l_{ER} \rangle = \frac{\ln \frac{N}{\langle k \rangle} - \gamma}{\ln \langle k \rangle} + \frac{3}{2} + S. \tag{28}$$

Figure 10 gathers numerical simulations and theoretical predictions for all three types of networks, showing that for small values of average degree the dependence of $\langle l \rangle$ can be well described using Eqs (26-28) with neglected term S (Fig. 10A). However, as $\langle k \rangle$ increases (Fig. 10B), this relation is no longer valid and it necessary to include correction caused by term S. One should notice, that for exponential networks even $\langle k \rangle = 40$ does not imply the appearance of oscillations.

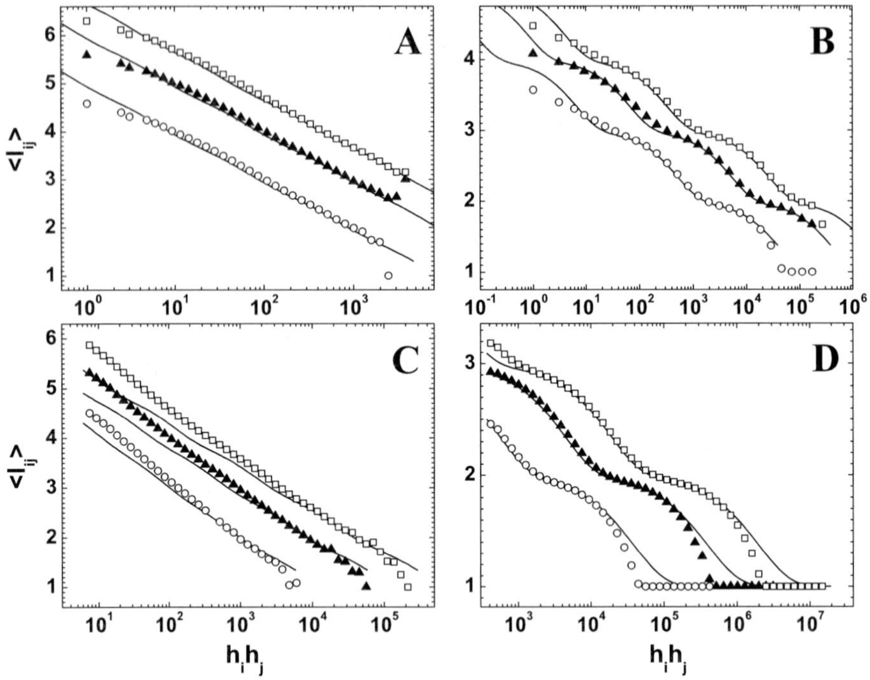

Fig. 9. Average internode distance $\langle l_{ij} \rangle$ between nodes i and j as a function of nodes' hidden variables product $h_i h_j$ for exponential networks (A - B) and scale-free networks with $\alpha = 3$. Solid lines are calculated using Eq. (3) (with numerical integration in order to obtain term R) while scatter data (circles - $N = 1000$, triangles - $N = 10000$ and squares - $N = 50000$) are numerical simulations using algorithm presented in [26]. Average degree $\langle k \rangle = 5$ for (**A**) and (**C**) and $\langle k \rangle = 40$ for (**B**) and (**D**).

4.4 Oscillations

From the previous sections one can spot that as long as the average number of links remains relatively small then, due to the generalized mean value theorem, the term R can be neglected. Otherwise one must take into account at least the first term from the infinite series in Eq. (20) what leads to log-periodic oscillation$\langle l_{ij} \rangle$ with the period $\Delta \ln(h_i h_j) = \ln B$ (see discussion below). In the next consideration we use symbols $A = (h_i h_j)/(\langle h^2 \rangle N)$ and $B = \langle h^2 \rangle / \langle h \rangle$ to avoid too complicated formulas.

Figure 11 shows a comparison of such oscillations in sparse ($m = 2$, upper row) and dense ($m = 40$, lower row) scale-free networks characterized by a hidden variable distribution $\rho(h) = (\alpha - 1) m^{\alpha - 1} h^{-\alpha}$ with $\alpha = 3$. The networks have been generated following the procedure C in [26] and represent the class of random networks with asymptotic scale-free connectivity distributions

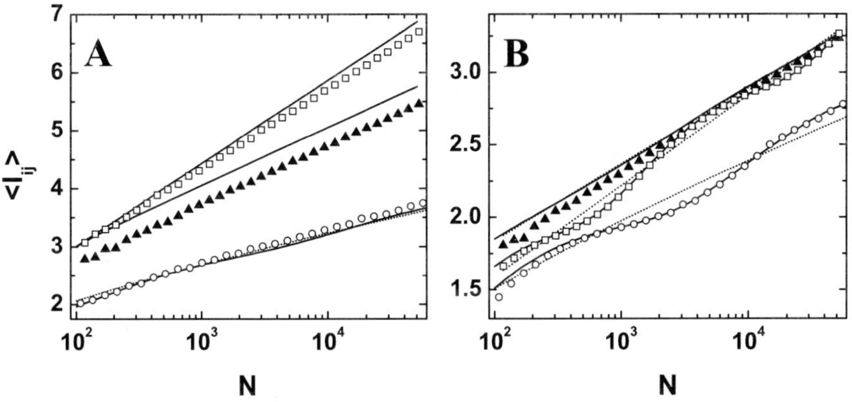

Fig. 10. Average path length $\langle l \rangle$ versus network size N. Scatter data are numerical simulations for scale-free networks with $\alpha = 3$ (*circles*), exponential networks (*triangles*) and Erdös-Rényi random graphs (*squares*), solid lines are calculated using Eq. (21) (with numerical integration in order to obtain term R), dotted lines are also obtained from Eq. (21), but term R is neglected. (**A**) $\langle k \rangle = 5$ for ER graphs and exponential networks and $\langle k \rangle = 8$ for SF networks. (**B**) $\langle k \rangle = 40$ for ER graphs and exponential networks and $\langle k \rangle = 60$ for SF networks. Solid and dotted lines corresponding to exponential and ER networks at (A) and to exponential networks at (B) have the same values, thus only the solid ones are visible

characterized by an arbitrary scaling exponent $\alpha > 2$. Figure 11A presents $F(x)$ (dotted line) and p^*_{ij} (solid line) are along with points corresponding to discrete values of those functions. One can notice that for $m = 40$ probability p^*_{ij} is much more narrow than for $m = 2$, so the slope of $F(x)$ decays more rapidly in the first case. Figure 11B shows the cosine transform of $F(x)$ given by Eq. (20). Depending on the shape of $F(x)$, the amplitude of this transform can take small/large values resulting in small/large values of R. As R is a sum of discrete values of a given transform taking only the first term in the sum (i.e. $n = 1$) is sufficient to obtain well approximated value of R (cf. points corresponding to discrete values of R_n at Fig. 11B). Figure 11C shows different behavior of resulting average distance $\langle l_{ij} \rangle$ between nodes i and j versus the product $h_i h_j$ for $m = 2$ (upper plot) and $m = 40$ (lower plot).

To obtain more quantitative results one should perform the integral in Eq. (20), however it is not analytical, so we made an approximation of $F(x)$ (see Appendix B) receiving a following estimation for R:

$$\widetilde{R}_n = -\frac{\ln B \sin\left(\frac{\pi n e}{\ln B}\right)}{\pi^2 n^2 \mathrm{e}} \sin\left[\frac{\pi n}{\ln B}\left(2\ln A - 2 + \mathrm{e}\right)\right] . \tag{29}$$

As one can see taking only the first term (i.e. $n = 1$) from Eq. (29) is justified because next terms decay as $1/n^2$. Equation (29) allows us to make an immediate observation that deviations from Eq. (3) take the form of regular

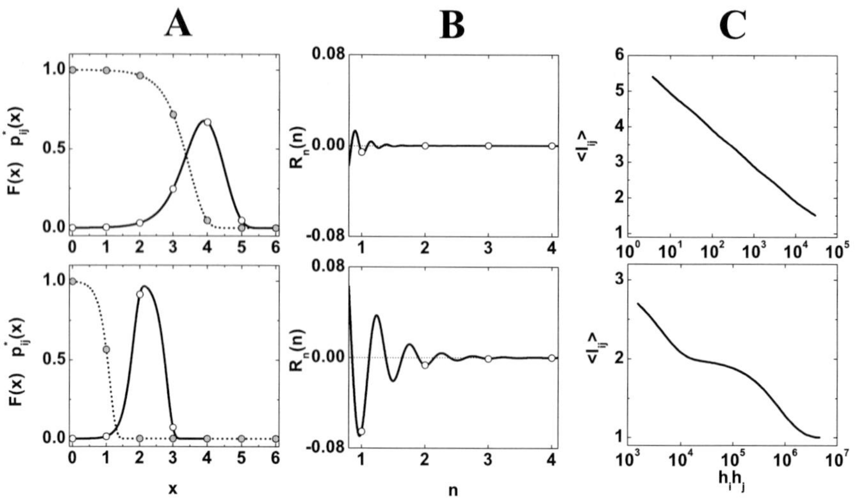

Fig. 11. Comparison of Two Networks Characterized by Hidden Variable Distribution $\rho(h) = (\alpha - 1)m^{\alpha-1}h^{-\alpha}$ for $\alpha = 3.0$ and $N = 10000$ - upper row $m = 2$, lower row $m = 40$. (**A, B, C**) - detailed description in text and. In case of plots (A) and (B) values of A have been chosen in such a way that the deviation is maximal

oscillations along $h_i h_j$ axis with period equal to $\ln B$ which increases with the heterogeneity of the networks. For dense networks the amplitude of oscillations grows monotonically with B - that is why the effect of oscillations is visible only in sufficiently dense networks. Similar oscillations effects are also observed for average path length $\langle l \rangle$ (see Fig. 10).

4.5 Cost Function

To show a practical way to make use of the phenomenon described in the previous Sections, we will concentrate on network optimization- a problem so far described in many works [10, 27, 28]. Optimization issues are well known and appear in different fields such as telecommunication and road construction. One of the most simplest model that assumes minimal costs in transportation includes two main aspects of network performance: the cost of constructing and maintaining the links between nodes and the cost of communication speed among them. The first part is proportional to the total number of links and the second one is proportional to the sum of shortest connections between each pair of nodes:

$$C = (1 - \lambda)\frac{N}{2}\langle k \rangle + \lambda \binom{N}{2} \langle l \rangle. \tag{30}$$

where λ controls the linear combination of two demands in Eq. (30): fully connected network with the shortest connections and a tree with the smallest

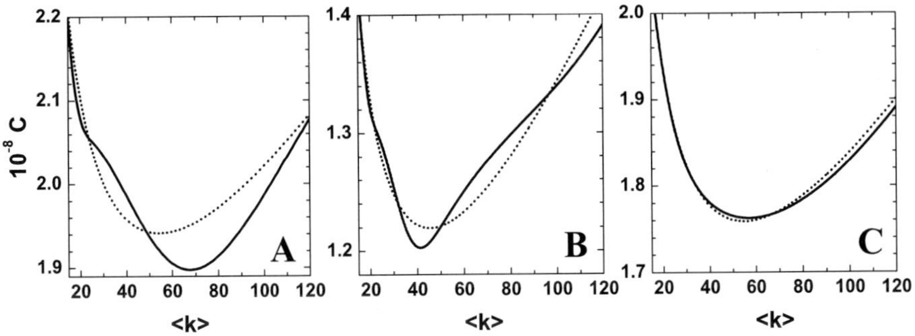

Fig. 12. Cost function C versus average degree $\langle k \rangle$ for three different networks of $N = 10^6$ nodes each. Solid lines represent cost function defined in Eq. (30) while dotted lines neglect the existence of S term. (**A**) Scale-free network with exponent $\alpha = 3$ $\lambda = 10^{-4}$ (**B**) Erdös-Rényi random graph $\lambda = 5 \cdot 10^{-5}$ (**C**) Exponential network $\lambda = 2 \cdot 10^{-5}$

number of links. A standard solution of this problem is a cost function with minimum at some specific value of $\langle k \rangle$ (see [10]).

Effects of discretization presented in the previous Sections change the shape of the cost function. Figure 12 depicts cost function for three different types of networks of $N = 10^6$ nodes each: scale-free with exponent $\alpha = 3$, exponential and Erdös-Rényi random graph. One can see that effects of oscillations induce a shift in the position and a decrease in value of the minimum of the cost function in the case of scale-free and Erdös-Rényi networks.

Figure 13 shows that different values of parameter λ can lead to even more interesting situation, where instead of one global minimum we obtain two well separated minima. The network adaptation to lower cost conditions can lead now to a temporal increase of costs since one has to pass over a cost barrier.

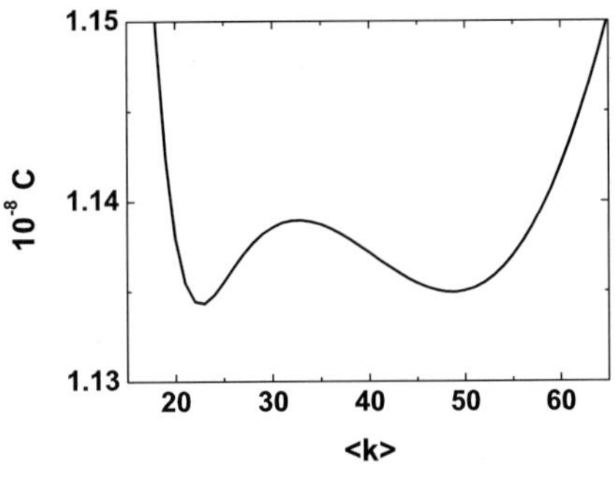

Fig. 13. Cost function C versus average degree $\langle k \rangle$ for scale-free network ($\alpha = 3$) of $N = 10^6$ nodes with $\lambda = 5.4 \cdot 10^{-4}$

5 Summary

In the last Section we presented results that possess applications for commonly known problems - optimization of such structures as telecommunication or public transport networks plays a crucial role in the present life. Without an extended analysis of these systems as complex networks we would not have been able to spot this phenomenon. In other words, the problem itself may not be seen as complex one at first glance, however a closer examination of its ingredients can lead to a reformulation and thus to a significant change of the results.

Acknowledgements

J.S., K.S. and J.A.H. acknowledge a support from the EU Grant MMCOM-NET No. FP6-2003-NEST-Path-012999 and from Polish Ministry of Science and Higher Education (Grant No. 13/6.PR UE/2005/7). P.F. acknowledges a support from the EU Grant CREEN FP6-2003-NEST-Path-012864 and from Polish Ministry of Science and Higher Education (Grant No. 134/E-365/6.PR UE/DIE 239/2005-2007).

A Rescaling of $\langle l_{ij} \rangle$ and $k_i k_j$.

For the majority of examined networks the average intervertex distance $\langle l_{ij} \rangle$ described with relation (3) takes its maximal value (i.e the maximal intervertex distance) $l_{\max}$ for the minimal possible values of k_i and k_j, that is $k_i = k_j = 1$. Then applying the above conditions to (3) we get

$$a = l_{\max}. \tag{31}$$

In order to obtain the second parameter of (3), we observe that the straight line described by (3) crosses X-axis for some effective value k_{eff}^2 i.e. $k_i = k_j = k_{\text{eff}}$. This allows as to express b as

$$b = \frac{l_{\max}}{2 \log k_{\text{eff}}}. \tag{32}$$

Applying a and b to Eq. (3) after some algebra we arrive at the following expression:

$$\frac{\langle l_{ij} \rangle}{l_{\max}} 2 \log k_{\text{eff}} = -\log \left(\frac{k_i k_j}{k_{\text{eff}}^2} \right). \tag{33}$$

Introducing characteristic values $\widetilde{\langle l_{ij} \rangle}$ and $\widetilde{k_i k_j}$ (see Eqs (11) and (12)) we get a universal formula for the scaling relation :

$$\widetilde{\langle l_{ij} \rangle} = -\log \widetilde{k_i k_j}. \tag{34}$$

B Approximation of function $F(x)$

In order to calculate the term R one can approximate $F(x)$ with the following piecewise linear function $\widetilde{F}(x)$:

$$\widetilde{F}(x) = \begin{cases} 1 & x < x_0, \\ \frac{1}{\mathrm{e}}\left(1 - \ln A - x \ln B\right) & x \in < x_0, x_1 >, \\ 0 & x > x_1, \end{cases} \tag{35}$$

where $x_0 = (1 - \ln A - \mathrm{e}) / \ln B$ and $x_1 = (1 - \ln A) / \ln B$. Since the function $F(x)$ is translationally invariant with respect to the argument x after rescaling the parameter A ($F(x; A) = F(x - x'; A')$) one can freely choose the point in which the slope coefficient is calculated as the tangent of $F(x)$. In our case in order to simplify the calculation we have chosen the inflexion point x_i of $F(x)$. Using Eq. (35) one can approximate terms R_n with

$$\widetilde{R}_n = -\frac{\ln B \sin\left(\frac{\pi n \mathrm{e}}{\ln B}\right)}{\pi^2 n^2 \mathrm{e}} \sin\left[\frac{\pi n}{\ln B}\left(2 \ln A - 2 + \mathrm{e}\right)\right]. \tag{36}$$

and thus one obtains an expression responsible for the oscillating term in $\langle l_{ij} \rangle$ scaling.

References

1. Euler, L.: Comment. acad. sc. Petrop. **8** (1741) 128
2. Erdös, P. and Rényi, A.: Publ. Math. Debrecen **6** (1959) 290
3. Watts, D. J. and Strogatz, S.H.: Nature **393** (1998) 440
4. Albert, R., Jeong, H., Barabási, A.–L.: Nature **401** (1999) 130
5. Milgram, S.: Psychology Today **2** (1967) 60
6. Jeong, H., Tombor, B., Albert, R., Oltvai, Z.H. and Barabási, A.–L.: Nature **407** (2000) 651
7. Pastor-Satorras, R. and Vespignani, A.: Evolution and Structure of the Internet: A Statistical Physics Approach. Cambridge University Press, Cambridge, 2004
8. J.-P. Onnela, A. Chakraborti, K. Kaski, J. Kertész, and A. Kanto: Phys. Rev E **68** (2003) 056110
9. Sienkiewicz, J., Hołyst, J.A.: Phys. Rev E **72** (2005) 046127
10. Schweitzer, F.: Brownian Agents and Active Particles. Springer, Berlin (2003)
11. Albert,R. and Barabási, A.–L.: Rev. Mod. Phys. **74** (2002) 47
12. Newman, M.E.J.: Phys. Rev. Lett. **89** (2002) 208701
13. Hołyst, J.A., Sienkiewicz, J., Fronczak, A., Fronczak, P. and Suchecki, K.: Phys. Rev E **72** (2005) 026108
14. Data for Foodwebs (*Silwood* and *Ythan*) [15] have been taken from www.cosin.org/extra/data/foodwebs/ whereas data for protein interaction network (*Yeast*) [16] has been taken from www.nd.edu/networks/database/index.html.
15. Garlaschelli, D., Caldarelli, G. and Pietronero, L.: Nature **423** (2003) 165
16. Jeong, H., Mason, S.P., Barabási, A.–L. and Oltvai, Z.N.: Nature **411** (2001) 41

17. Scientific collaboration network [18] data have been collected by P. Wójcicki from two publicly available databases of papers http://arxiv.org/archive/astro-ph (*Astro*) and arxiv.org/archive/cond-mat (*Cond-mat*) for the period 1995-2001.
18. Newman, M.E.J.: Phys. Rev. E **64** (2001) 016131
19. Data for the Internet [7] has been taken from www.cosin.org/extra/data/internet/.
20. In those networks vertices are bus- and tramstops while an edge exists if at least one public transport line crosses two stops. [9].
21. Motter, A.E., Nishikawa, T. and Lai, Y.C.: Phys. Rev. E **66** (2002) 065103(R)
22. Hołyst, J.A., Sienkiewicz, J, Fronczak, A., Fronczak, P. and Suchecki, K.: Physica A **351** (2005) 167
23. J. Sienkiewicz, P. Fronczak, J. A. Hołyst: e-print cond-mat/0608273 (2006)
24. Fronczak, A., Fronczak, P. and Hołyst, J.A.: Phys. Rev. E **70** (2004) 056110
25. Boguñá, M. and Pastor-Satorras, R.: Phys. Rev. E **68** (2003) 036112
26. Fronczak, A. and Fronczak, P.: Phys. Rev. E **74** (2006) 026121
27. Gastner, M.T. and Newman, M.E.J.: Eur. Phys. J. B **49** (2006) 247
28. Ferrer i Cancho, R. and Solé, R.: Statistical Physics of Complex Networks, Lecture Notes in Physics. Springer, Berlin (2003)

Index

Printing: Krips bv, Meppel, The Netherlands
Binding: Stürtz, Würzburg, Germany